London Mathematical Society Student Texts

Managing editor: Professor J.W. Bruce, Department of Mathematics,
University of Liverpool, Liverpool L69 3BX, United Kingdom

3 Local fields, J.W.S. CASSELS
4 An introduction to twistor theory, second edition, S.A. HUGGETT & K.P. TOD
5 Introduction to general relativity, L.P. HUGHSTON & K.P. TOD
7 The theory of evolution and dynamical systems, J. HOFBAUER & K. SIGMUND
8 Summing and nuclear norms in Banach space theory, G.J.O. JAMESON
9 Automorphisms of surfaces after Nielsen and Thurston, A. CASSON & S. BLEILER
11 Spacetime and singularities, G. NABER
12 Undergraduate algebraic geometry, MILES REID
13 An introduction to Hankel operators, J.R. PARTINGTON
15 Presentations of groups, second edition, D.L. JOHNSON
17 Aspects of quantum field theory in curved spacetime, S.A. FULLING
18 Braids and coverings: Selected topics, VAGN LUNDSGAARD HANSEN
19 Steps in commutative algebra, R.Y. SHARP
20 Communication theory, C.M. GOLDIE & R.G.E. PINCH
21 Representations of finite groups of Lie type, FRANÇOIS DIGNE & JEAN MICHEL
22 Designs, graphs, codes, and their links, P.J. CAMERON & J.H. VAN LINT
23 Complex algebraic curves, FRANCES KIRWAN
24 Lectures on elliptic curves, J.W.S. CASSELS
25 Hyperbolic geometry, BIRGER IVERSEN
26 An introduction to the theory of *L*-functions and Eisenstein series, H. HIDA
27 Hilbert space: Compact operators and the trace theorem, J.R. RETHERFORD
28 Potential theory in the complex plane, T. RANSFORD
29 Undergraduate commutative algebra, M. REID
31 The Laplacian on a Riemannian manifold, S. ROSENBERG
32 Lectures on Lie groups and Lie algebras, R. CARTER, G. SEGAL & I. MACDONALD
33 A primer of algebraic *D*-modules, S.C. COUTINHO
34 Complex algebraic surfaces, A. BEAUVILLE
35 Young tableaux, W. FULTON
37 A mathematical introduction to wavelets, P. WOJTASZCZYK
38 Harmonic maps, loop groups, and integrable systems, M. GUEST
39 Set theory for the working mathematician, K. CIESIELSKI
40 Ergodic theory and dynamical systems, M. POLLICOTT & M. YURI
41 The algorithmic resolution of diophantine equations, N.P. SMART
42 Equilibrium states in ergodic theory, G. KELLER
43 Fourier analysis on finite groups and applications, AUDREY TERRAS
44 Classical invariant theory, PETER J. OLVER
45 Permutation groups, P.J. CAMERON
46 Riemann surfaces: A primer, A BEARDON
47 Introductory lectures on rings and modules, J. BEACHY
48 Set theory, A. HAJNÁL & P. HAMBURGER
49 K-theory for C^*-algebras, M. RORDAM, F. LARSEN & N. LAUSTSEN
50 A brief guide to algebraic number theory, H.P.F. SWINNERTON-DYER
51 Steps in commutative algebra, second edition, R.Y. SHARP
52 Finite Markov chains and algorithmic applications, O. HAGGSTROM
53 The prime number theorem, G.J.O. JAMESON
54 Topics in graph automorphisms and reconstruction, J. LAURI & R. SCAPELLATO
55 Elementary number theory, group theory, and Ramanujan graphs, G. DAVIDOFF, P. SARNAK & A. VALETTE
56 Logic, induction and sets, T. FORSTER
57 Introduction to Banach algebra, operators and harmonic analysis, H. G. DALES et al.
58 Computational algebraic geometry, H. SCHENCK
59 Frobenius algebras and 2-D topological quantum field theories, J. KOCK
60 Linear operators and linear systems, J. R. PARTINGTON

London Mathematical Society Student Texts 62

Topics from One-Dimensional Dynamics

KAREN M. BRUCKS
University of Wisconsin, Milwaukee

HENK BRUIN
University of Surrey

Shaftesbury Road, Cambridge CB2 8EA, United Kingdom

One Liberty Plaza, 20th Floor, New York, NY 10006, USA

477 Williamstown Road, Port Melbourne, VIC 3207, Australia

314–321, 3rd Floor, Plot 3, Splendor Forum, Jasola District Centre, New Delhi – 110025, India

103 Penang Road, #05–06/07, Visioncrest Commercial, Singapore 238467

Cambridge University Press is part of Cambridge University Press & Assessment, a department of the University of Cambridge.

We share the University's mission to contribute to society through the pursuit of education, learning and research at the highest international levels of excellence.

www.cambridge.org
Information on this title: www.cambridge.org/9780521547666

First published 2004

A catalogue record for this publication is available from the British Library

ISBN 978-0-521-54766-6 Paperback

Contents

List of Figures

Preface

One-dimensional dynamics owns many deep results and avenues of active mathematical research. Numerous inroads to this research exist for the advanced undergraduate or beginning graduate student. It is precisely these students whom we target. Several glimpses into one-dimensional dynamics are provided with the hope that the results presented illuminate the beauty and excitement of the field. Many topics covered appear nowhere else in "textbook format," some are mini new research topics in themselves, and for nearly all topics we try to provide novel connections with other research areas both inside and outside the text. Among these topics are kneading theory and Hofbauer towers; detailed structure of ω-limit sets; topological entropy; lapnumbers and Markov extensions; the 2^∞ map (Feigenbaum-Coullet-Tresser), interplay amongst continued fractions, adding machines, circle maps, and unimodal maps; irrational rotations as factors of unimodal maps; connections between β-transformations and unimodal maps; Ledrappier's three-dot example; and itineraries for complex quadratic maps and Hubbard trees. The flavor is largely combinatoric, symbolic, and topological. The material presented is *not* meant to be approached in a linear fashion. Rather, we strongly encourage readers to pick and choose topics of interest. Trail routes (other than $n \mapsto n+1$) are indicated in Figure 1; more explicit information is provided at the beginning of each chapter. Suggested uses for the text include: dynamics courses, master theses, reading courses, research experiences for undergraduates (REUs), seminars, senior projects, and summer courses.

As mentioned, the topics covered are *not* the typical topics seen in undergraduate/graduate dynamics texts. Rather, the material is a filtering from the research literature of currently active topics that can be made accessible to the targeted student audience. Frequent references to the literature are made to enable the reader to further investigate the topic under study. Occasionally new or simplified proofs for existing results are given. The level of "mathematical sophistication" required by the reader varies between chapters. The text is also designed to provide the reader with a "critical mass" of results and tools, allowing for further investigation into the field.

Although this text can be used in a "lecture format" course, we have purposely employed a style such that a more active role can be played by students. Exercises are not reserved for the end of a section but rather are sprinkled throughout a section. The student is encouraged to complete exercises as s/he progresses through the material. More challenging exercises are tagged with a $\diamondsuit$. For some exercises, having referenced literature at hand is suggested and such exercises are tagged with a $\clubsuit$. The Java Applets, `www.uwm.edu/~kmbrucks/Dyntext.html`, assist with the exploration of many topics.

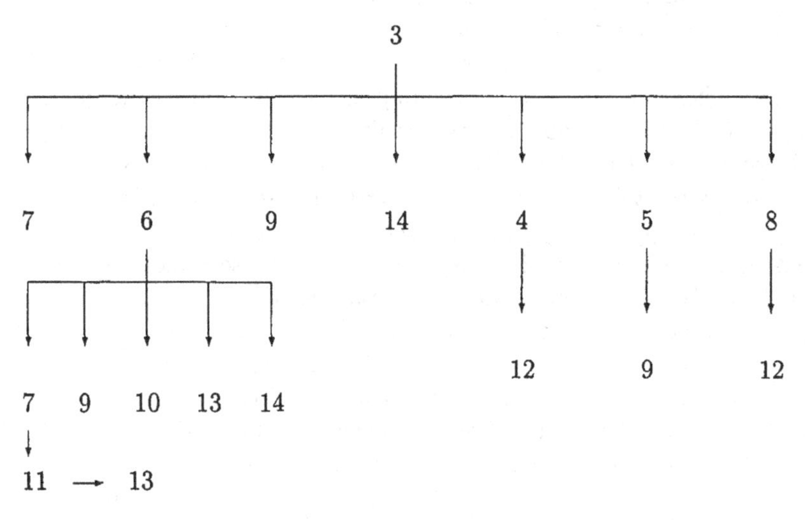

Figure 1: Trail guide

Historically, it was in the 1970s that dynamical systems had a "surge" of activity (which continues today). This surge was due largely to the accessibility of computers and the multitude of results and questions achieved with the aid of computers. However, a genealogical tree of dynamics would include G. D. Birkhoff (1884–1944), M. L. Cartwright (1900–1998), P. Fatou (1878–1929), G. Julia (1893–1978), and H. Poincaré (1854–1912). Historical discussions can be found in [4, 5, 108, 115].

Research in one-dimensional dynamics was ignited in the 1960s and 1970s by American, French, and Ukrainian mathematicians [58, 68, 69, 107, 116, 122, 155, 160]. Metric properties of the Coullet-Tresser (or Feigenbaum) map (see Chapter 5) were studied in the late 1970s [58, 68, 69] and somewhat earlier (and more narrowly) in [127]. The famous theorem of Sharkovsky (see Theorem 3.1.5) appeared in a Ukrainian journal in 1964 [155]. Work of Stefan [160] in 1977 helped bring Sharkovsky's theorem to the west. Around this same time, Li and Yorke published their famous paper "Period three implies chaos" [107]. Published works on the combinatorics of one-dimensional dynamics first appeared in the early 1970s [116, 122]. From these roots, one-dimensional dynamics has flourished into an active and exciting area of mathematical research.

Chapters 1 and 2 provide background material from topology and measure theory, respectively. Rather than begin with these chapters, we suggest the reader refer to these chapters as needed. These chapters are purposely succinct and provide multiple references for the reader looking for more detail/discussion. Lebesgue measure is the only measure used, and it appears infrequently. Hence the reader unfamiliar with measure theory can still access most of the text.

Chapter 3 provides an introduction to symbolic and topological dynamics. This material provides a dynamical foundation for the topics covered in later chapters. We suggest that the reader who is unfamiliar with this material work through the first four sections and move on, returning for the remaining material (recurrence and shift spaces) as needed. A beginning discussion on measurable dynamics is given in Chapter 4. Pieces of this material are referenced in Chapter 12; however, the material is not required for Chapter 12.

The dynamics of a 2^∞ map are investigated in Chapter 5. In particular, the dynamics of this map are shown to be that of the dyadic adding machine. The discussion (in Chapter 5) is accessible to an undergraduate student with minimal experience doing proofs. Generalized versions of the dyadic (triadic, etc.) adding machine are investigated in Chapters 11 and 13.

Chapters 6 and 7 provide combinatoric, number theoretic, and symbolic machinery to investigate one-dimensional dynamical systems. Chapters 8 to 14 target the student with a higher level of mathematical sophistication. Any one of these chapters (8–14) could serve as a component of a dynamics course or as material for a reading course, masters thesis, seminar, summer course, or term project. Of course, this text is not a comprehensive work on one-dimensional dynamics. Other texts in dynamics include [4, 5, 18, 53, 56, 64, 74, 96, 108, 115, 139, 156, 162, 168].

We thank Kurt Cogswell, Beverly Diamond, Jane Hawkins and her students, Chris Sears, and Lubo Snoha for their careful reading of the text; Jane Hawkins for contributing Chapters 2 and 4; Karsten Keller for the figure files, Peter Raith for e-mail discussions on Hofbauer towers; and Christopher Sears for the Java Applets.

This text grew out of courses taught (1997, 1999, 2001) at the Carleton Summer Mathematics Program (SMP) for undergraduate women in mathematics. Each summer, the SMP draws 18 talented undergraduate women interested in mathematics to Carleton College for four weeks of intensive mathematics study. We applaud Deanna Haunsperger and Steven Kennedy for running a program that so positively impacts all involved.

The first author thanks her family for their ongoing support, which occurs with full measure and densely on her life timeline. The second author was supported by a fellowship of the Royal Netherlands Academy of Sciences and Arts (KNAW). Both authors thank their colleagues and friends for many enjoyable and fruitful communications, all of which make this text possible. Comments are most welcome; send comments to kmbrucks@uwm.edu.

Chapter 1

Topological Roots

This chapter provides some basics from topology. Our style is purposely succinct as our objective is to collect, for the reader to reference as s/he works through the text, the results from topology needed for the text. The reader with minimal topological experience may want to keep an introductory topology text at hand, for example, [126, 170]. We suggest that others reference the chapter as needed.

Some notation: $\mathbb{N} = \{1, 2, 3, \dots\}$, $\mathbb{Z} = \{\dots, -2, -1, 0, 1, 2, \dots\}$, $\mathbb{R}$ denotes the real numbers, and $\mathbb{Q}$ denotes the rationals. For ease of notation, $\langle a; b\rangle$ denotes the closed (unless otherwise stated) interval $[a, b]$ when $a < b$ and denotes $[b, a]$ when $b < a$. Let $\#A$ denote the cardinality of the set A and A^c the complement of A. For $A \subset X$ we also use $X \setminus A$ to denote the complement of A.

1.1 Basics from Topology

Definition 1.1.1. A *metric space* is an ordered pair (E, ρ) consisting of a set E along with a function $\rho : E \times E \to \mathbb{R}$ satisfying the following for $x, y, z \in E$:

1. $\rho(x, y) \geq 0$;

2. $\rho(x, x) = 0$ and $\rho(x, y) = 0$ implies $x = y$;

3. $\rho(x, y) = \rho(y, x)$;

4. $\rho(x, y) + \rho(y, z) \geq \rho(x, z)$ (*triangle inequality*).

The function ρ is called a *metric* on E. If only conditions 1, 3, and 4 hold, we call ρ a *semimetric*.

Example 1.1.2. *The real line* $\mathbb{R}$ *with the distance function* $\rho(x, y) = |x-y|$ *is a metric space.*

Example 1.1.3. *The Euclidean space* $\mathbb{R}^n$ *with the distance function*

$$\rho(\langle x_1, \dots, x_n\rangle, \langle y_1, \dots, y_n\rangle) = \sqrt{\sum_{k=1}^{n} (x_k - y_k)^2}$$

is a metric space.

Example 1.1.4. *The Euclidean space* $\mathbb{R}^2$ *with the distance function*

$$\rho(x, y) = \max\{|x_1 - y_1|, |x_2 - y_2|\}$$

is a metric space.

Example 1.1.5. *Set*

$$E = \{0,1\}^{\mathbb{N}} = \{x_1, x_2, \dots \mid x_i \in \{0,1\} \text{ for all } i\}.$$

Thus, E consists of all one-sided infinite strings of 0's and 1's. Let $x = \{x_i\}_{i\geq 1}$ and $y = \{y_i\}_{i\geq 1}$ be elements from E and set:

$$\rho(x, y) = \sum_{i\geq 1} \frac{|x_i - y_i|}{2^i}.$$

Then (E, ρ) is a metric space.

Definition 1.1.6. Let (E, ρ) be a metric space and $x \in E$. For $\epsilon > 0$ we define the ϵ-*ball* about x as:

$$B(x, \epsilon) = \{y \in E \mid \rho(x, y) < \epsilon\}.$$

For subsets $S, T \subset E$, the distance between S and T is defined as:

$$\rho(S, T) = \inf\{\rho(x, y) \mid x \in S,\ y \in T\}.$$

The ϵ-ball about the set S is given as:

$$B(S, \epsilon) = \{y \in E \mid \rho(S, y) < \epsilon\}.$$

Definition 1.1.7. Let (E, ρ) be a metric space and $M \subset E$. We call M *open* provided that, for each $x \in M$, there is an ϵ-ball containing x and contained in M. A set is *closed* provided it is the complement of an open set.

Exercise 1.1.8. *Prove each of the following.*

1. *Any union of open sets is open.*
2. *Any finite intersection of open sets is open.*
3. *Construct an infinite collection of open sets whose intersection is NOT open.*

What happens if you replace "open" with "closed"? **HINT:** *For item (3), let* $U_n = (-\frac{1}{n}, \frac{1}{n})$ *for* $n \geq 1$. *Each* U_n *is open and yet* $\cap_n U_n = \{0\}$. *When considering closed sets, recall the following equalities from set theory:*

$$E \setminus (\cap_{\lambda \in \Lambda} U_\lambda) = \cup_{\lambda \in \Lambda} (E \setminus U_\lambda) \quad \text{and} \quad E \setminus (\cup_{\lambda \in \Lambda} U_\lambda) = \cap_{\lambda \in \Lambda} (E \setminus U_\lambda).$$

Definition 1.1.9. Let (E, ρ) be a metric space and $M \subset E$. The *interior* of M, denoted M°, is defined as $M^\circ = \cup\{U \subset E \mid U \text{ is open and } U \subset M\}$.

Definition 1.1.10. Let (E, ρ) be a metric space and $M \subset E$. A point $x \in E$ is said to be an *accumulation point of the set* M provided that, for every open set $U \ni x$, we have $M \cap (U \setminus \{x\}) \neq \emptyset$.

Definition 1.1.11. Let (E, ρ) be a metric space, $M \subset E$, and M' be the set of accumulation points of M. The *closure* of the set M, denoted $\overline{M}$, is defined to be $M \cup M'$.

Exercise 1.1.12. *Let* (E, ρ) *be a metric space,* $M \subset E$, *and* M' *be the set of accumulation points of* M. *Prove* M' *is closed.*

Definition 1.1.13. Let (E, ρ) be a metric space and $M \subset E$. The *boundary* of M, denoted ∂M, is defined as $\partial M = \overline{M} \cap \overline{E \setminus M}$.

Definition 1.1.14. Let (E, ρ) be a metric space. A sequence $\{x_n\}_{n \geq 1}$ in E *converges to* x in E provided $\lim_{n \to \infty} \rho(x_n, x) = 0$. If $\{x_n\}$ converges to x, we write $\lim_{n \to \infty} x_n = x$ or, more shortly, $x_n \to x$.

Exercise 1.1.15. *Let* (E, ρ) *be a metric space and suppose that* $\lim_{n \to \infty} x_n = x$ $(x_n \to x)$ *and* $\lim_{n \to \infty} x_n = y$ $(x_n \to y)$. *Show* $x = y$.

Exercise 1.1.16. *We say a sequence* $\{x_n\}_{n \geq 1}$ *of real numbers is* monotone increasing *(*monotone decreasing*) provided* $x_i \leq x_{i+1}$ *for all* $i \geq 1$ *(provided* $x_i \geq x_{i+1}$ *for all* $i \geq 1$*). We call* $\{x_n\}$ *monotone if it is either monotone increasing or decreasing. Prove that every monotone bounded sequence of real numbers converges.* **HINT:** *Assume the sequence is monotone increasing. Let* L *be the least upper bound of* $\{x_n\}_{n \geq 1}$. *Show* $\lim_{n \to \infty} x_n = L$.

Exercise 1.1.17. *Prove that every sequence of real numbers contains a monotone subsequence. Conclude (using Exercise 1.1.16) that every bounded sequence of real numbers has a convergent subsequence.* **HINT:** *Call x_m a* turn-point *provided $x_n \leq x_m$ for all $n > m$. If there exist $m_1 < m_2 < \dots$ such that each x_{m_i} is a turn-point, then $x_{m_1} \geq x_{m_2} \geq \dots$ is a monotone subsequence. Show that if there are only finitely many (or no) turn-points, then one may choose $x_{n_1} < x_{n_2} < \dots$.*

Definition 1.1.18. Let (E_1, ρ_1) and (E_2, ρ_2) be metric spaces. A function $f : E_1 \to E_2$ is *continuous* at $x \in E_1$ provided that, for each $\epsilon > 0$, there is some $\delta > 0$ such that $\rho_2(f(x), f(y)) < \epsilon$ whenever $\rho_1(x, y) < \delta$.

Exercise 1.1.19. *Let (E_1, ρ_1) and (E_2, ρ_2) be metric spaces and $f : E_1 \to E_2$. Prove that the map f is continuous at $x \in E_1$ iff, for each open set $V \subset E_2$ containing $f(x)$, there is an open set $U \subset E_1$ containing x such that $f(U) \subset V$. Conclude that if f is continuous for all x in E_1 and $W \subset E_2$ is open (in E_2), then $f^{-1}(W)$ is open (in E_1).*

Exercise 1.1.20. *Let $(E_1, \rho_1), (E_2, \rho_2)$ be metric spaces and $f : E_1 \to E_2$. Prove that the map f is continuous if, whenever $x_n \to x$ in E_1, then $f(x_n) \to f(x)$ in E_2.*

Definition 1.1.21. Let (E, ρ) be a metric space and $M \subset E$. We say M is *nowhere dense in E* provided that $\overline{M}$ contains no nonempty open set.

The integers $\mathbb{Z}$ in $\mathbb{R}$ are nowhere dense in $\mathbb{R}$. The rationals $\mathbb{Q} \in [0, 1]$ are *not* nowhere dense in $[0, 1]$. We will soon get nontrivial examples of nowhere dense sets, namely, Cantor sets. First, some definitions.

Definition 1.1.22. Let (E, ρ) be a metric space and $M \subset E$. We say M is *perfect* provided M is closed and each point of M is an accumulation point of M.

Definition 1.1.23. Let (E, ρ) be a metric space. We call E *disconnected* provided there are disjoint nonempty open sets H and K in E such that $E = H \cup K$. If no such disconnection exists, then we call E *connected.*

Exercise 1.1.24. *Let (E_1, ρ_1) and (E_2, ρ_2) be metric spaces and $f : E_1 \to E_2$ a continuous onto map. Prove that if E_1 is connected, then E_2 is also connected.* **HINT:** *Suppose to the contrary that E_2 is not connected. Then there are nonempty open sets H and K in E_2 such that $E_2 = H \cup K$. However, then E_1 would be disconnected by $f^{-1}(H)$ and $f^{-1}(K)$, a contradiction.*

Definition 1.1.25. Let (E, ρ) be a metric space. We call E *totally disconnected* provided the only nonempty connected subsets of E are the one-point sets.

Notice that $M \subset \mathbb{R}$ is totally disconnected iff M contains no nonempty open set. For example, the rationals $\mathbb{Q}$ are totally disconnected.

Definition 1.1.26. A metric space E is *compact* if every sequence in E has a convergent subsequence in E (the limit of the convergent subsequence is, of course, in E).

Exercise 1.1.27. *Let (E, ρ) be a compact metric space and $f : E \to E$ continuous. Prove that $f(E)$ is compact.* **HINT:** *Recall Exercise 1.1.20.*

Exercise 1.1.28. *Let (E, ρ) be a compact metric space. Prove that $M \subset E$ is compact if and only if M is closed.* **HINT:** *Assume M is compact. If M is not closed, then $E \setminus M$ is not open, and hence one can choose $x \in E \setminus M$ and $x_n \in M \cap B(x, \frac{1}{n})$ for each $n \in \mathbb{N}$. As M is compact, a subsequence of $\{x_n\}$ converges to some $y \in M$, contradicting $\lim_{n\to\infty} x_n = x \notin M$. Next assume M is closed and that $\{x_n\}$ is a sequence in M. We need to find a convergent (in M) subsequence. As E is compact, we obtain a convergent (in E) subsequence $\{x_{n_j}\}$; say $\lim_{j\to\infty} x_{n_j} = y \in E$. We want to show that $y \in M$. If not, then (since M is closed) there is $B(y, \epsilon) \subset E \setminus M$; this contradicts $\{x_{n_j}\}$ converging to y with each $x_{n_j} \in M$.*

The interested reader should see a standard topology text for proofs of Theorems 1.1.29 and 1.1.34 [126, 170]. We prove Theorem 1.1.29 for the case $n = 1$.

Theorem 1.1.29 (Bolzano-Weierstrass [170]). *A subset of $\mathbb{R}^n$ is compact iff it is closed and bounded.*

Proof. We do only the case $n = 1$.

Assume $E \subset \mathbb{R}$ is closed and bounded, and let $\{x_n\}_{n\geq 1}$ be a sequence from E. By Exercise 1.1.17, we may choose a monotone subsequence $\{x_{n_k}\}$. As E is bounded, so is the sequence $\{x_n\}$ and the subsequence $\{x_{n_k}\}$. By Exercise 1.1.16, the subsequence converges. The limit of the subsequence is in E as E is closed. Hence, E is compact.

Assume $E \subset \mathbb{R}$ is compact. If E were unbounded, then one could construct a sequence $\{x_n\}_{n\geq 1}$ such that $x_n \to \infty$ (or $x_n \to -\infty$), contradicting E being compact (as no subsequence converges). Thus, E is bounded. If E is not closed, then we may choose an accumulation point z of E in $\mathbb{R}\setminus E$. As z is an accumulation point of E, we may choose a sequence $\{x_n\}$ from E converging to z. This sequence has no subsequence converging to a point in E, contradicting E being compact. Thus, E is indeed closed and bounded. □

Exercise 1.1.30. *Let (E, ρ) be a compact metric space and $M, N \subset E$ with $M \cap N = \emptyset$ and M, N closed. Show that $\rho(M, N) > 0$. Is this true if E is not compact?* **HINT:** *For the case E not compact, let $E = \mathbb{R}^2$, $M = \mathbb{R}$, and N be the graph of $\frac{1}{x}$.*

Definition 1.1.31. Let (E, ρ) be a metric space and $f : E \to E$ continuous. We say that f is *uniformly continuous* provided that, for every $\epsilon > 0$, there exists $\delta > 0$ such that $x, y \in E$ and $|x - y| < \delta$ imply $|f(x) - f(y)| < \epsilon$.

Exercise 1.1.32. *Let (E, ρ) be a compact metric space and $f : E \to E$ continuous. Prove that f is uniformly continuous.* **HINT:** *Suppose to the contrary that f is not uniformly continuous. Choose $\epsilon > 0$ and for each $n \in \mathbb{N}$ choose p_n, q_n such that $\rho(p_n, q_n) < \frac{1}{n}$ and $\rho(f(p_n), f(q_n)) \geq \epsilon$. As E is compact, choose $\{n_k\}$ and z such that $\lim_{k\to\infty} p_{n_k} = z$. Then $\lim_{k\to\infty} q_{n_k} = z$ also. Hence, $\lim_{k\to\infty} f(p_{n_k}) = \lim_{k\to\infty} f(q_{n_k}) = f(z)$; this contradicts $\rho(f(p_{n_k}), f(q_{n_k})) \geq \epsilon$ for all k.*

Exercise 1.1.33. *Let $f : (0, 1] \to [1, \infty)$ be given by $f(x) = 1/x$. Is f uniformly continuous?*

Theorem 1.1.34 (Heine-Borel [170]). *A subset M of a metric space (E, ρ) is compact iff, whenever M is contained in the union of a collection of open sets of E, then M is also contained in the union of a finite subcollection of these sets.*

Lemma 1.1.35 (Lebesgue covering lemma [170]). *Let (E, ρ) be a compact metric space and $\{U_1, \ldots, U_n\}$ be a finite open cover of E, that is, $\cup_{i=1}^{i=n} U_i = E$ and each U_i is open. There exists $\delta > 0$ such that, if A is any subset of E of diameter less than δ, then $A \subset U_i$ for some i.*

Proof. Suppose not. For each $n \in \mathbb{N}$, let A_n be a set of diameter less than $\frac{1}{n}$ such that $A_n \not\subseteq U_i$ for any i. Let $x_n \in A_n$ for each n. As E is compact, there exists $\{n_k\}$ and $x \in E$ with $\lim_{k\to\infty} x_{n_k} = x$. Choose i such that $x \in U_i$ and $\delta > 0$ such that $B(x, \delta) \subset U_i$. Choose k such that $\frac{1}{n_k} < \frac{\delta}{2}$ and choose $l > k$ such that $x_{n_l} \in B(x, \frac{\delta}{2})$. Then $A_{n_l} \subset B(x, \delta) \subset U_i$, a contradiction. □

Exercise 1.1.36. *Let (E, ρ) be a compact metric space and $f : E \to E$ a continuous map. Prove, directly, that f is uniformly continuous.* **HINT:** *Fix $\epsilon > 0$. For each $x \in E$, choose $\delta_x > 0$ such that $\rho(x, y) < \delta_x$ implies $\rho(f(x), f(y)) < \frac{\epsilon}{2}$. Use Theorem 1.1.34 to choose $x_1, x_2, \ldots, x_n \in E$ such that $\cup_{i=1}^{i=n} B(x_i, \delta_{x_i}) \supset E$. Use Lemma 1.1.35 to choose $\delta > 0$ such that $\rho(x, y) < \delta$ implies $x, y \in B(x_i, \delta_{x_i})$ for some i. Conclude that $\rho(x, y) < \delta \Rightarrow \rho(f(x), f(y)) < \epsilon$.*

Definition 1.1.37. Let (E_1, ρ_1) and (E_2, ρ_2) be compact metric spaces and $f : E_1 \to E_2$. We call f a *homeomorphism* provided f is one-to-one and both f and f^{-1} are continuous (the domain of f^{-1} being $f(E_1)$). If, additionally, f is onto, we call f an *onto homeomorphism*.

Exercise 1.1.38. ♣ *[170, Theorem 17.14] Let $h : X \to X$ be continuous, onto, and one-to-one. Assume X is compact. Prove h^{-1} is continuous and hence is a homeomorphism.*

Definition 1.1.39. Let (E, ρ) be a metric space. We call E a *Cantor set* provided E is compact, totally disconnected, and perfect.

In Section 1.2 we construct the *Middle Third Cantor set.* This set is a standard first example of a Cantor set. Any two Cantor sets are homeomorphic [170]. However, we have the stronger result that Cantor sets are *homogeneous*; see Exercise 1.1.40. This property is used in the proof of Proposition 1.1.52.

Exercise 1.1.40. ♣ *[170, page 218] Let K_1 and K_2 be Cantor sets. Fix $x_1 \in K_1$ and $x_2 \in K_2$. Prove there exists a homeomorphism $h : K_1 \to K_2$ with $h(x_1) = x_2$. Due to this property, we say Cantor sets are* homogeneous.

Definition 1.1.41. We call a set M *countable* if one of the following hold.

- There is a bijection (one-to-one and onto map) between M and $\mathbb{N}$.
- The set M is finite or empty.

Exercise 1.1.42. *Prove*

1. *$\mathbb{Q}$ is countable.*
2. *$\mathbb{R}$ is not countable.*

Exercise 1.1.43. *Let M be a countable set. Prove that $M^2 = \{(m_1, m_2) \mid m_i \in M \text{ for each } i\}$ is countable. More generally, show that, for each $n \in \mathbb{N}$, the set M^n is countable. Is $M^{\mathbb{N}}$ countable?*

Remark 1.1.44. Let (E, ρ) be a compact metric space and suppose $M \subset E$ is closed (and therefore compact) and perfect. Then M is not a countable set and hence is uncountable [170, Exercise 30B-2]. For further discussion see [170, Chapter 30].

Definition 1.1.45. Let (E, ρ) be a metric space. We say $D \subset E$ is *dense in E* provided $\overline{D} = E$. We call E *separable* provided E contains a countable dense subset.

The rationals $\mathbb{Q}$ are dense in $\mathbb{R}$ and hence $(\mathbb{R}, \rho)$ (ρ given in Example 1.1.3) is a separable metric space.

Exercise 1.1.46. *Let (E, ρ) be a separable metric space. Show there exists a countable collection of open sets, $\mathcal{U}$, with the property that, given $x \in E$ and $W \ni x$ open in E, there exists $U \in \mathcal{U}$ with $x \in U \subset W$.* **HINT:** *Let $\{e_1, e_2, e_3, \dots\}$ be a countable dense subset from E. Set $\mathcal{U} = \{B(e_i, \frac{1}{m}) \mid i, m \in \mathbb{N}\}$.*

Exercise 1.1.47. ◊ *Prove that every compact metric space is separable.*

Definition 1.1.48. Let (E, ρ) be a metric space, $M \subset E$, and $x \in M$. We call x a *condensation point of M* provided each open $U \ni x$ contains uncountably many points of M.

Lemma 1.1.49. *Let (E, ρ) be a separable metric space and $M \subset E$. Then M can be expressed as $M = F \cup H$, where H is countable and every point of F is a condensation point of M.*

Proof. The result is clear if M is countable (simply take $H = M$ and $F = \emptyset$). Hence, assume M is uncountable.

Let $\mathcal{U}$ be as in Exercise 1.1.46, and set

$$H = \{x \in M \mid x \text{ is not a condensation point of } M\}.$$

We show H is countable. For each $x \in H$, there is $U_x \in \mathcal{U}$ such that $x \in U_x$ and $U_x \cap M$ is countable. As $H \subset M$, we have that $U_x \cap H$ is countable. Lastly, as $\mathcal{U}$ is countable, we have H countable. Set $F = M \setminus H$. □

Exercise 1.1.50. *Let (E, ρ) be a compact metric space and $M \subset E$. Write $M = F \cup H$ as in Lemma 1.1.49. If $F \neq \emptyset$, prove F is perfect.*

Exercise 1.1.51. *Let $M \subset [0, 1]$ be closed, uncountable, and contain no nonempty open set. Then one can express M as $M = F \cup H$, where F is a Cantor set and H is countable (it can be that $H = \emptyset$).*

Proposition 1.1.52. *Let (E, ρ) be a compact metric space, and let $M \subset E$ be closed, uncountable, and totally disconnected. Then M can be uniquely expressed as the disjoint union of a Cantor set and a countable set.*

Proof. Express $M = F \cup H$ as in Lemma 1.1.49. Thus, every point of F is a condensation point of M, F is a Cantor set (recall Exercise 1.1.50), and H consists of those points of M that are not condensation points of M. Note that $F \cap H = \emptyset$. Suppose $M = F_1 \cup H_1$ with F_1 a Cantor set, H_1 countable, and $F_1 \cap H_1 = \emptyset$.

We show $H_1 \subset H$. Suppose not, that is, suppose $z \in H_1 \cap F$. As F_1 is closed and $z \notin F_1$, we may choose open $U \ni z$ such that $U \subset H_1$. Thus U is a countable set. However, $z \in F$ and U open implies $U \cap F$ is uncountable. This contradiction shows that indeed $H_1 \subset H$.

Lastly, we show $H_1 = H$. If H_1 is a proper subset of H, then $F_1 \cap H \neq \emptyset$. Let $x \in F \cap F_1$ and $y \in F_1 \cap H$. By Exercise 1.1.40, we may choose a homeomorphism $h : F \to F_1$ with $h(x) = y$. Since $y \in H$, it is the case that y is a not a condensation point of M, and hence we may choose an open set $U \ni y$ such that U is countable. Then $U \cap F_1$ is countable and thus $h^{-1}(U) \cap F \ni x$ is open and countable, contradicting $x \in F$. □

Exercise 1.1.53. ◊ *Let (E, ρ) be a compact metric space, $M = F \cup H$ be as in Proposition 1.1.52, and $g : M \to M$ be a continuous map. Suppose there exists a positive integer L such that, for all $x \in M$, $g^{-1}(x)$ has no more than L elements. Show $g(F) \subset F$.*

The next theorem is a special case of the Baire Category Theorem, namely, when the space involved is a compact metric space; see [55, Theorem D.37] and [170, Problem 24B-4, Section 25].

Theorem 1.1.54. *Let E be a compact metric space. Then E cannot be written as the union of a sequence of nowhere dense sets. Moreover, if $\{A_n\}$ is a sequence of nowhere dense subsets of E, then $E \setminus (\cup_n A_n)$ is dense in E.*

1.2 The Middle Third Cantor Set

Cantor sets arise often in dynamics. As a first example of a Cantor set, we construct the *Middle Third Cantor set.* Some notation: For each $m \in N$ we let $\{0,1\}^m$ denote the 2^m finite strings or words of length m on the alphabet $\{0,1\}$. Similarly, $\{0,1\}^{\mathbb{N}}$ denotes collection of all one-sided infinite strings of 0's and 1's. For $\gamma = \langle \gamma_1, \gamma_2, \gamma_3, \ldots \rangle \in \{0,1\}^{\mathbb{N}}$, let $\gamma|n = \langle \gamma_1, \ldots, \gamma_n \rangle$.

Middle Third Cantor Set: We recursively define sets K_γ for each $\gamma \in \cup_{m \geq 1} \{0,1\}^m$. Set

$$K_0 = [0, \tfrac{1}{3}] \quad \text{and} \quad K_1 = [\tfrac{2}{3}, 1].$$

If $m \geq 1$ and $K_\gamma = [a,b]$ for $\gamma \in \{0,1\}^m$ are given, set

$$K_{\gamma 0} = [a, a + \frac{1}{3^{m+1}}] \quad \text{and} \quad K_{\gamma 1} = [b - \frac{1}{3^{m+1}}, b].$$

Lastly set

$$K = \cap_{m \geq 1} \cup_{\gamma \in \{0,1\}^m} K_\gamma.$$

The Middle Third Cantor set is K. See Figure 1.1.

Exercise 1.2.1. *Prove each of the following.*

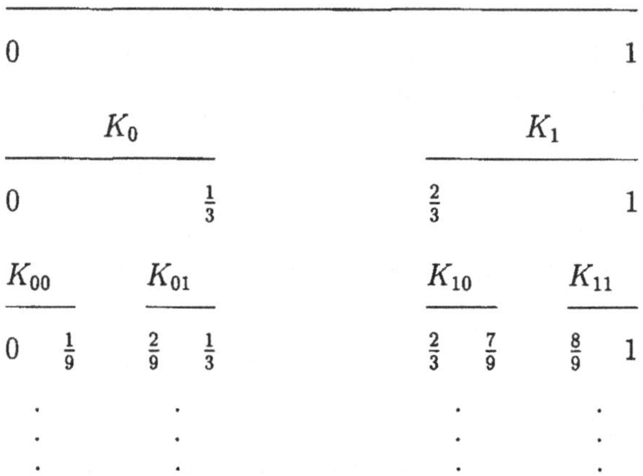

0 1

K_0 K_1

0 $\frac{1}{3}$ $\frac{2}{3}$ 1

K_{00} K_{01} K_{10} K_{11}

0 $\frac{1}{9}$ $\frac{2}{9}$ $\frac{1}{3}$ $\frac{2}{3}$ $\frac{7}{9}$ $\frac{8}{9}$ 1

Figure 1.1: Construction of the Middle Third Cantor set

1. *For $\gamma \in \{0,1\}^m$, we have $|K_\gamma| = \frac{1}{3^m}$ and hence*

$$\sum_{\gamma\in\{0,1\}^m} |K_\gamma| = (\frac{2}{3})^m.$$

2. $\cup_{\gamma\in\{0,1\}^{m+1}} K_\gamma \subset \cup_{\gamma\in\{0,1\}^m} K_\gamma$.

3. $K \neq \emptyset$.

4. *K is closed.*

5. *K contains no nondegenerate open intervals; hence it follows that K is nowhere dense.*

6. *K is perfect.*

7. *K is compact.*

8. *K is totally disconnected.*

9. *The Lebesgue measure of K (see Chapter 2 for a definition of Lebesgue measure) is zero.*

Exercise 1.2.2. *Each $x \in [0,1]$ has an expansion, $\langle x_1, x_2, x_3, x_4, \ldots \rangle$, in ternary form (i.e., $x_i \in \{0,1,2\}$ for all i) obtained by writing $x = \sum_{i\geq 1} \frac{x_i}{3^i}$. These expressions are unique, except that numbers (other than 1) expressible in a ternary expansion ending in a sequence of 2's can be reexpressed in an* *expansion ending in a sequence of 0's (e.g., $\frac{1}{3}$ can be written as $\langle 1, 0, 0, 0, \ldots \rangle$ or $\langle 0, 2, 2, 2, 2, \ldots \rangle$).*

Prove: The Middle Third Cantor set is precisely the set of points in $[0, 1]$ having a ternary expansion without 1's. **HINT:** *Try the following.*

1. *Show that, for each $x \in K$, there exists a unique $\gamma_x = \langle \gamma_1, \gamma_2, \gamma_3, \ldots \rangle \in \{0, 1\}^{\mathbb{N}}$ such that*
$$x = \textstyle\bigcap_{i \geq 1} K_{\gamma|i}.$$

2. *Let $x \in K$ and γ_x be as in (1). What is*
$$\sum_{i \geq 1} \frac{2\gamma_i}{3^i} ?$$

Exercise 1.2.3. *For $\gamma = \langle \gamma_1, \gamma_2, \cdots \rangle$ and $\tau = \langle \tau_1, \tau_2, \cdots \rangle$ in $\{0, 1\}^{\mathbb{N}}$, set $\rho(\gamma, \tau) = \sum_{i \geq 1} \frac{|\gamma_i - \tau_i|}{2^i}$. Prove that ρ is a metric and that $(\{0, 1\}^{\mathbb{N}}, \rho)$ is a compact metric space.*

Exercise 1.2.4. *Show that for each $\gamma = \langle \gamma_1, \gamma_2, \cdots \rangle \in \{0, 1\}^{\mathbb{N}}$ there exists a unique $x \in K$ such that $x = \bigcap_{i \geq 1} K_{\gamma|i}$. Hence we can define a map $h : \{0, 1\}^{\mathbb{N}} \to K$. Prove h is an onto homeomorphism.*

It follows from Exercise 1.2.4 that K is homeomorphic to $\{0, 1\}^{\mathbb{N}}$.

Exercise 1.2.5. $\diamondsuit$ *Fix $\alpha \in (0, 1)$. Construct a set $K_\alpha \subset [0, 1]$ such that properties (2)–(7) of Exercise 1.2.1 hold but such that the Lebesgue measure of K is precisely α. See also Section 2.3.2.*

The proofs of Theorems 1.2.6 and 1.2.7 are beyond this text. Neither theorem is used but, rather, are included for general interest [170].

Theorem 1.2.6. *The Middle Third Cantor set is the only totally disconnected, perfect compact metric space (up to homeomorphisms).*

Theorem 1.2.7. *Every compact metric space E is a continuous image of the Middle Third Cantor set.*

Chapter 2

Measure Theoretic Roots

If asked what is the measure of the interval $(a, b) \subset \mathbb{R}$, we reply $b-a$ without thinking. Indeed, the Lebesgue measure of any interval in $\mathbb{R}$ is its length; sets of the form $(-\infty, a]$ or (b, ∞) have infinite measure, while bounded intervals have finite measure. But what about a set that is much more irregular, such as the Middle Third Cantor set defined in the previous chapter? How do we compute its size? This is the question that Henri Lebesgue addressed in his Ph.D. thesis in 1902. His goal was to come up with a tool for integrating functions that were horribly discontinuous, unbounded, or both, functions not covered by the classical Riemann integration theory. In order to reach that goal, he designed a simple means for determining the "length" or measure of a set of real numbers that is not necessarily the union of intervals. We present the basic ideas here of Lebesgue measure on $\mathbb{R}$ and refer to measure theory texts for generalizing Lebesgue's ideas to other spaces. We will return to the ideas of integration in the chapter on measurable dynamics (Chapter 4). See [55, 70] for further discussion. Material from this chapter is used in Chapter 4.

2.1 Basics of Lebesgue Measure on $\mathbb{R}$

There are many ways one could generalize length, and perhaps this is why measure theory seems like a complicated subject to many. In addition, there is no reasonable way to preserve the desirable properties of length and have it work on *every* subset of $\mathbb{R}$, as the example in the next section reveals.

Following Lebesgue's simple approach to the subject, our starting point is that the measure of an interval should be its length. We denote Lebesgue measure on $\mathbb{R}$ by m; then

$$m([a, b]) = m((a, b)) = m([a, b)) = m((a, b]) = b - a.$$

Of course we want to extend m to more general sets, so next we cover an arbitrary set by a countable union of intervals, add up the lengths of the covering subintervals, and take the infimum of all possible such covers.

Definition 2.1.1. Let $S \subset \mathbb{R}$ be any subset, and let (a_j, b_j) denote any interval. We define the *outer Lebesgue measure* of S, denoted m^*, by

$$m^*(S) = \inf\{\sum_{j=1}^{\infty}(b_j - a_j) : S \subset \cup_{j=1}^{\infty}(a_j, b_j)\}.$$

Clearly, since $b - a$ gives the length of $(a, b), [a, b], (a, b]$, and $(a, b]$, any of these types of intervals can be used interchangeably in the definition.

Exercise 2.1.2. *Prove that outer Lebesgue measure satisfies the following properties.*

1. $m^*(\emptyset) = 0$; *and if* p *is any point, then* $m^*(\{p\}) = 0$.
2. *For any interval* (a, b), $m^*((a, b)) = b - a$; *that is,* m *and* m^* *agree on intervals.*
3. *If* U *is open, then* U *is a disjoint union of open intervals, say* $I_1 = (a_1, b_1), \ldots, I_k = (a_k, b_k), \ldots,$ *and* $m^*(U) = \sum_k m^*(I_k) = \sum_k b_k - a_k$.
4. *If* $A \subset S$, *then* $m^*(A) \leq m^*(S)$.

There is one additional property of m^* of great importance in measure theory, called *countable subadditivity,* which we prove here.

Proposition 2.1.3. *For any countable collection of subsets* $S_1, S_2, \ldots \subset \mathbb{R}$,

$$m^*(\bigcup_{j=1}^{\infty} S_j) \leq \sum_{j=1}^{\infty} m^*(S_j).$$

Proof. Consider $S_1, S_2, \ldots \subset \mathbb{R}$; define $S = \bigcup_{j=1}^{\infty} S_j$. Given any $\epsilon > 0$, for each j there exist intervals $I_1^j = (a_1^j, b_1^j), \ldots, I_k^j = (a_k^j, b_k^j), \ldots,$ such that $S_j \subset \cup_k I_k^j$ and $\sum_k(b_k^j - a_k^j) \leq m^*(S_j) + \epsilon 2^{-j}$. Then $S \subset \bigcup_{j,k} I_k^j$ and $\sum_{j,k}(b_k^j - a_k^j) \leq \sum_j m^*(S_j) + \epsilon$, so it follows that

$$m^*(S) \leq \sum_{j=1}^{\infty} m^*(S_j) + \epsilon.$$

Since ϵ can be chosen to be arbitrarily small, this proves the result. □

Remark 2.1.4. It follows from Exercise 2.1.2 that if the S_j's are disjoint intervals, we have

$$m^*(\bigcup_{j=1}^{\infty} S_j) = \sum_{j=1}^{\infty} m^*(S_j);$$

this is called *countable additivity.* Much as we might be tempted to stop here and call m^* a measure, there are consequences of measuring a set S only from the outside, which is what we do when we cover S with small intervals. We could be measuring far more than is in the set. Indeed, Lebesgue measure should be countably additive on *all* disjoint measurable sets, not just countably subadditive. So we consider the outer measure of the complement as well to see if the two measures add up to "the correct answer" in order to get a collection of measurable sets.

Definition 2.1.5. Consider the intervals $I_n = [n, n+1), n \in \mathbb{Z}$. For a set $S \subset I_n$, we define *the inner measure* of S by

$$m_*(S) = 1 - m^*(I_n \setminus S).$$

We say $S \subset I_n$ is *(Lebesgue) measurable* iff $m^*(S) = m_*(S)$. In this case, we define

$$m(S) = m^*(S) = m_*(S).$$

For an arbitrary $S \subset \mathbb{R}$, we say S is (Lebesgue) measurable iff $S \cap I_n$ is measurable for all n, and we define

$$m(S) = \sum_{n\in\mathbb{Z}} m(S \cap I_n).$$

In the definition above, we could just as easily consider any countable collection I_n of bounded intervals on $\mathbb{R}$ with pairwise intersection at most two endpoints and whose union is $\mathbb{R}$.

We collect various properties of Lebesgue measure here and leave most of the proofs as exercises. All can be proved from the definitions given or found in any measure theory text. Recall that A^c denotes the complement of the set A in $\mathbb{R}$

Theorem 2.1.6. *Let $\mathcal{L}$ denote the collection of sets in $\mathbb{R}$ that are Lebesgue measurable. Then:*

1. *All compact sets and all open sets are in $\mathcal{L}$.*

2. *If $A \in \mathcal{L}$, then $A^c \in \mathcal{L}$.*

3. *If* $A_1, A_2, \ldots, A_n, \ldots \in \mathcal{L}$, *then* $\cup_{j=1}^{\infty} A_j \in \mathcal{L}$.

4. $\emptyset \in \mathcal{L}$ *and* $\mathbb{R} \in \mathcal{L}$.

Since Lebesgue measure on an open or compact set in $\mathbb{R}$ is frequently easy to work with, we have the following useful regularity result about the sets in $\mathcal{L}$.

Proposition 2.1.7. *For a bounded interval* I, *a set* $S \subset I$ *is measurable iff, for each* $\epsilon > 0$, *there exists a compact set* K *and an open set* U *such that* $K \subset S \subset U$ *and* $m(U \setminus K) < \epsilon$.

We have constructed a collection of measurable sets defined precisely so that the next theorem holds. Notice that these properties are obvious when we apply them to intervals.

Theorem 2.1.8. *Consider Lebesgue measure on the measurable subsets of* $\mathbb{R}$, *which we denote by* $(\mathbb{R}, \mathcal{L}, m)$. *In other words, consider the map* $m : \mathcal{L} \to \mathbb{R}$ *defined by Lebesgue measure. Then*

1. $m(\emptyset) = 0$.

2. *If* $\{S_j\}_{j=1}^{\infty}$ *is a sequence of disjoint sets in* $\mathcal{L}$, *then*

$$m\Big(\bigcup_{j=1}^{\infty} S_j\Big) = \sum_{j=1}^{\infty} m(S_j).$$

3. *If* S *is a measurable set and* $r \in \mathbb{R}$, *then* $S + r = \{x + r : x \in S\}$ *is also measurable and* $m(S) = m(S + r)$.

2.2 A Nonmeasurable Set

This classical example shows that the collection of sets $\mathcal{L}$ we constructed in the previous section cannot contain **every** subset of the real numbers.

The set we define is a subset of $I = [0, 1)$; we begin by defining two points $x, y \in I$ to be equivalent if $x - y$ is rational. (For example, all the rational numbers are equivalent to 0, etc.) We write $x \sim y$ to show they are equivalent and define the *equivalence class of* x by

$$[x] = \{y \in I : x \sim y\}.$$

Let S be a subset of I that contains exactly one element from each equivalence class; finding such a set S requires the Axiom of Choice.

Proposition 2.2.1. *The set S is not Lebesgue measurable.*

Proof. For each rational $r \in \mathbb{Q} \cap I$, define

$$\begin{aligned} S_r &= S + r \bmod 1 \\ &= \{x + r : x \in S \cap [0, 1-r)\} \cup \{x + r - 1 : x \in S \cap [1-r, 1)\}. \end{aligned}$$

We claim that $[0,1) = \bigcup_{r \in \mathbb{Q}} S_r$, and that $S_r \cap S_t = \emptyset$ if $t \neq r$. (This is Exercise 2.2.2 below.)

Now suppose that S is Lebesgue measurable. Then

$$1 = m([0,1)) = \sum_{r \in \mathbb{Q}} m(S_r),$$

since the S_r's are disjoint and by applying Theorem 2.1.8(2). In addition, by Theorem 2.1.8(3), we have that $m(S) = m(S_r)$ for all r. However, if $m(S) > 0$, this means that $\sum_{r \in \mathbb{Q}} m(S_r) = \infty$, which is a contradiction; and if $m(S) = 0$, this means that $\sum_{r \in \mathbb{Q}} m(S_r) = 0$, which is also contradiction. Hence S cannot be in $\mathcal{L}$. □

Exercise 2.2.2. *Show that $[0,1) = \bigcup_{r \in \mathbb{Q}} S_r$, and that $S_r \cap S_t = \emptyset$ if $t \neq r$ for $t, r \in \mathbb{Q}$. That is, show that each $x \in [0,1)$ belongs to precisely one S_r.*

2.3 Lebesgue Measure of Cantor Sets

In Section 1.2, we defined the Middle Third Cantor set K as a subset of $[0,1] \subset \mathbb{R}$. We calculate its Lebesgue measure here and show that we can change the set K to have any measure we choose between 0 and 1.

2.3.1 The Middle Third Cantor Set

A close look at the construction in Section 1.2 shows that K is obtained from $[0,1]$ by removing the open middle third $(1/3, 2/3)$ (resulting in $K_0 \cup K_1$); and then removing the open middle thirds $(1/9, 2/9)$ and $(7/9, 8/9)$, leaving $K_{00} \cup K_{01} \cup K_{10} \cup K_{11}$. This is illustrated in Figure 1.1.

Using Theorem 2.1.6, 1–3, we see immediately that $K^c = [0,1]/K$ and hence K are Lebesgue measurable. We calculate that the Lebesgue measure of K^c is obtained by adding up the lengths of the disjoint open middle thirds removed at each level. At the mth step we remove 2^{m-1} intervals of length $\frac{1}{3}^m$. (See Exercise 2.2.1.1.) Therefore

$$m(K^c) = \sum_{m\geq 1} \frac{2^{m-1}}{3^m} = \frac{1}{3}\sum_{m\geq 0}(\frac{2}{3})^m = \frac{1}{3}\cdot 3 = 1.$$

Using Theorem 2.1.8 we conclude that $m(K) = 0$. This discussion provides the solution to Exercises 1.2.1–9.

2.3.2 Other Cantor Sets

The construction of the Middle Third Cantor set and the computation of its measure suggests an obvious generalization in which we successively remove open intervals of lengths γ^n in place of $\frac{1}{3}^n$. As an easy variation, suppose we construct a new Cantor set C as follows. The set C is obtained from $[0,1]$ by removing the open middle quarter $(3/8, 5/8)$ (resulting in $C_0 = [0, \frac{3}{8}]$ and $C_1 = [\frac{5}{8}, 1]$). Next remove from each of C_0 and C_1 the open middle interval of length $\frac{1}{16}$, leaving $C_{00} \cup C_{01} \cup C_{10} \cup C_{11}$ as before. Here $\gamma = \frac{1}{4}$.

At the mth step we remove 2^{m-1} intervals of length $\frac{1}{4}^m$. Therefore

$$m(C^c) = \sum_{m\geq 1} \frac{2^{m-1}}{4^m} = \frac{1}{4}\sum_{m\geq 0}(\frac{1}{2})^m = \frac{1}{4}\cdot 2 = \frac{1}{2}.$$

Using Theorem 2.1.8 we conclude that $m(C) = \frac{1}{2}$.

In this way we can find a set C whose Lebesgue measure is α for any $0 < \alpha < 1$ by removing the open middle intervals of appropriate lengths. C is called a *generalized Cantor set* and, like K, is compact, nowhere dense, and totally disconnected.

Exercise 2.3.1. *For any $n \in \mathbb{N}$, construct a generalized Cantor set C such that $m(C) > 1 - \frac{1}{n}$. Recall Exercise 1.2.5.*

This construction cannot be extended to produce a generalized Cantor set of measure 1. The reason for this is that there does not exist a nowhere dense set $S \in \mathcal{L} \cap [0,1]$ such that $m(S) = 1$. If S is nowhere dense in $[0,1]$, then $[0,1] \setminus \overline{S}$ is nonempty and open so it contains an interval; hence it must have positive Lebesgue measure!

2.4 Sets of Lebesgue Measure Zero

One of the main assets of using a measure theoretic approach to dynamics is that you can ignore "bad" or abnormal behavior on a set of measure zero. These sets are usually much larger than countable sets, as the Cantor examples show, and it is worth understanding zero measure sets for Lebesgue measure before moving on. First, we introduce some terminology.

Definition 2.4.1. Any set $E \in \mathcal{L}$ such that $m(E) = 0$ is called a *null set* (or a set of measure zero). If a statement about points in I or $\mathbb{R}$ is true except for points $x \in E$, where E is a null set, then we say it is true *m-almost everywhere*, abbreviated *m-a.e.*. If there is no ambiguity about the measure under consideration, we say a property holds almost everywhere or for almost every x.

An important feature about Lebesgue measure is that it is complete. This means that any subset of a null set is also measurable. It seems like an obvious property, but we cannot assume it holds in all settings; indeed, not all measures are complete. However, we have that the following proposition holds.

Proposition 2.4.2. *If $E \in \mathcal{L}$ is a null set and $F \subseteq E$ is any subset, then $F \in \mathcal{L}$. That is, Lebesgue measure is complete on $\mathbb{R}$.*

Proof. Assume that $E \in \mathcal{L}$ is a null set, and we may as well assume that $E \subseteq I_n = [n, n+1), n \in \mathbb{Z}$ (if not, we break E into the disjoint union of sets with this property and work on each piece separately). For a set $F \subseteq E$, we have $m^*(F) \leq m^*(E) = 0$; also, $I_n \setminus E \subseteq I_n \setminus F$, so

$$m_*(E) = 0 = 1 - m^*(I_n \setminus E) \geq 1 - m^*(I_n \setminus F) = m_*(F) = 0.$$

Therefore, $m_*(F) = m^*(F)$, which is the definition of Lebesgue measurability. □

Chapter 3

Beginning Symbolic and Topological Dynamics

We focus on the *asymptotic behaviors* exhibited by a dynamical system. More precisely, given $E \subset \mathbb{R}$ and $f : E \to E$ a continuous map, the set E and the map f form a *dynamical system*, that is, a set of possible states (E) and a rule (f) that determines a present state in terms of the past states. Most often E is a compact metric space. Fix $x \in E$ and consider the sequence of forward iterates:

$$x,\ f(x),\ f^2(x) := f(f(x)),\ f^3(x) := f(f(f(x))),\ \dots\ . \tag{3.1}$$

What kind of asymptotic behaviors can such sequences exhibit? We concentrate on the topological, symbolic, and combinatoric behaviors exhibited by such sequences. Our task is to investigate the types of behaviors that are possible, conditions to guarantee existence (or nonexistence) of certain behaviors, frequency of occurrence (both from the topological and measure theoretical viewpoint) of behaviors, and measures of the complexity of a system.

In this chapter we present "beginning pieces" from symbolic and topological dynamics. Additional symbolic and combinatoric tools appear in Chapter 6. We suggest the reader work through the first four sections of this chapter, as they form a foundation for the entire text. The material in Section 3.5 is used in Section 6.2 and Chapters 10 and 13. Parts of Section 3.6 are used in Sections 9.2 and 10.2 and Chapter 11. For further study in symbolic and topological dynamics, see [5, 6, 18, 74, 93, 96, 108, 115, 156, 129].

3.1 Periodic Behavior

We call the sequence in (3.1) the *forward trajectory* or the *forward orbit* of x under the map f. The simplest behavior exhibited by a forward trajectory is periodic, that is, $f^n(x) = x$ for some n, and thus (3.1) becomes

$$x, f(x), f^2(x), \dots, f^n(x) = x, f(x), f^2(x), \dots, f^n(x) = x, \dots ;$$

hence the next definition.

Definition 3.1.1. Let (E, ρ) be a compact metric space and $f : E \to E$ be continuous. We say $x \in E$ is a *periodic point of period n* (or simply *n-periodic*) provided $f^n(x) = x$ and $f^j(x) \neq x$ for $1 \leq j \leq n-1$. If x is n-periodic, we call $\{x, f(x), \dots, f^{n-1}(x)\}$ a *periodic cycle* (or simply an *n-cycle*). If x is 1-periodic, we call x a *fixed point.*

Example 3.1.2. *Let $g_a(x) = ax(1-x)$ for $a \in \mathbb{R}$, referred to as the* quadratic *or* logistic family. *For*

$$a \approx 3.8318740552833155684,$$

the map g_a has a period 3 orbit. Indeed, $x_0 = \frac{1}{2}$ is periodic of period 3; set $x_1 = g_a(x_0)$ and $x_2 = g_a(x_1)$. The graph of the first iterate (g_a) and third iterate(g_a^3) of g_a are shown in Figure 3.1. The period 3 orbit (x_0, x_1, x_2) and the nonzero fixed point (z) are shown. Notice that each x_i is fixed by g_a^3. Thus, the forward trajectory of $x_0 = \frac{1}{2}$ is precisely:

$$x_0, x_1, x_2, x_0, x_1, x_2, x_0, x_1, x_2, \dots .$$

The family of maps given in the next definition, namely, the *family of symmetric tent maps*, appears frequently in the text. Although the family is very simple to describe (piecewise linear), the family is rich in dynamical behaviors. The reader, particularly those new to dynamics, may find it useful to maintain an ongoing investigation of this family with the help of the Java Applets or with other computer analysis. Chapter 10 investigates this family in detail.

Definition 3.1.3. The *symmetric family of tent maps*, $\{T_a\}_{a \in [0,2]}$, is given by $T_a(x) = ax$ for $x \in [0, \frac{1}{2}]$ and $T_a(x) = a(1-x)$ for $x \in [\frac{1}{2}, 1]$. See Figure 3.2.

Exercise 3.1.4. *Find $a \in [0,2]$ such that $\frac{1}{2}$ is periodic of period 3 under T_a. Compare to Example 3.1.2.* **HINT:** *We want:*

$$\frac{1}{2} \overset{T_a}{\mapsto} \frac{a}{2} \overset{T_a}{\mapsto} a(1-\frac{a}{2}) \overset{T_a}{\mapsto} a^2(1-\frac{a}{2}) = \frac{1}{2}.$$

Solve $a^2(1-\frac{a}{2}) = \frac{1}{2}$ for a.

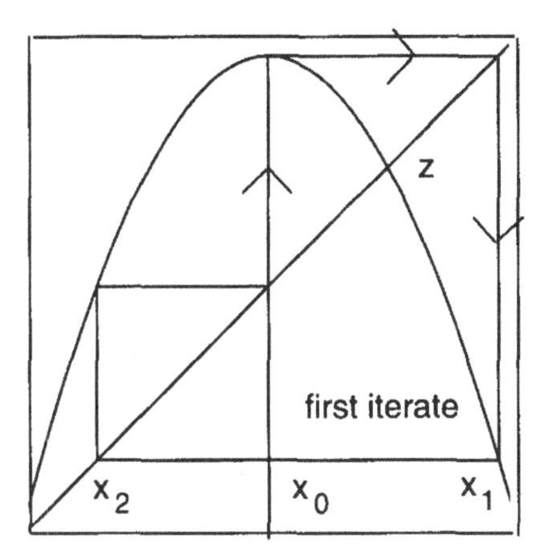

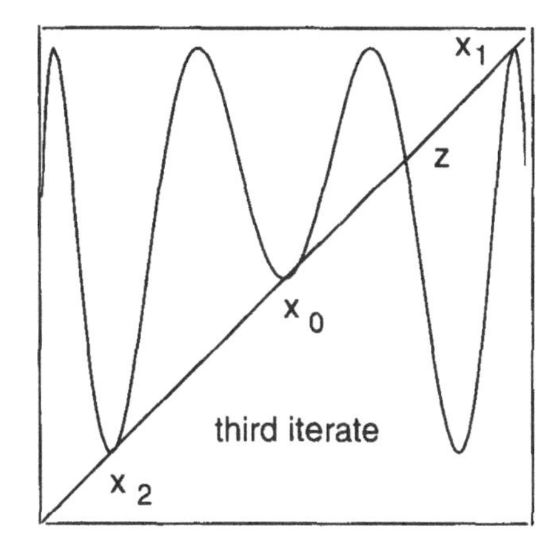

Figure 3.1: Iterates of $g_a(x) = ax(1-x)$ for $a \approx 3.8318740552833155684$

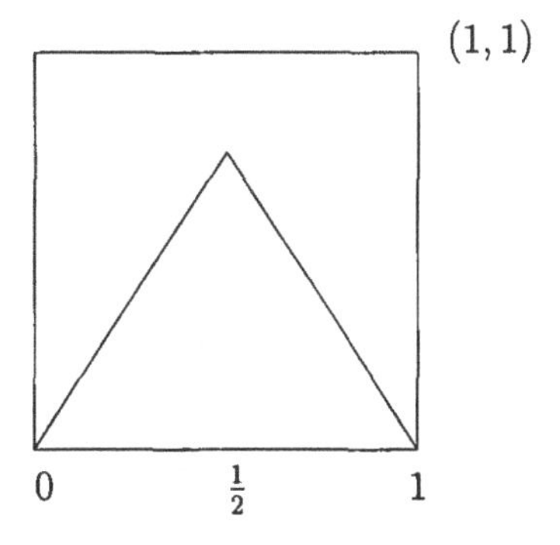

Figure 3.2: Graph of T_a for $a = 1.5$

We say x is *preperiodic* under a map f provided x is not periodic and $f^n(x) = z$ for some periodic point z. The point $x = \frac{1}{2}$ is preperiodic for the symmetric tent map T_2.

Much attention has been paid to periodic orbits because they generally can be "seen" by computers and they provide information as to the complexity of the system. For example, they can be used to estimate the *topological entropy* (see Chapter 9) of a system, the entropy being a measure of complexity.

If the system being studied has periodic orbits, one asks what periods are possible, what permutations are possible, and which periods or permutations force others (in the sense that if a map has one, then it must have the other)? Figure 3.3 provides two period 5 orbits, each with a different permutation. The period 5 orbit for each map in Figure 3.3 is denoted by: x_0, x_1, x_2, x_3, x_4 with $x_i \mapsto x_{i+1}$ for $0 \le i \le 4$. For each orbit, relabel the points in a linearly increasing manner to obtain the associated permutation (the permutation follows the orbit, i.e, $x_2 \mapsto x_3$ gives $1 \mapsto 3$ in the first permutation and $1 \mapsto 2$ in the second permutation). For a study of forcing see [5]; we do not study forcing.

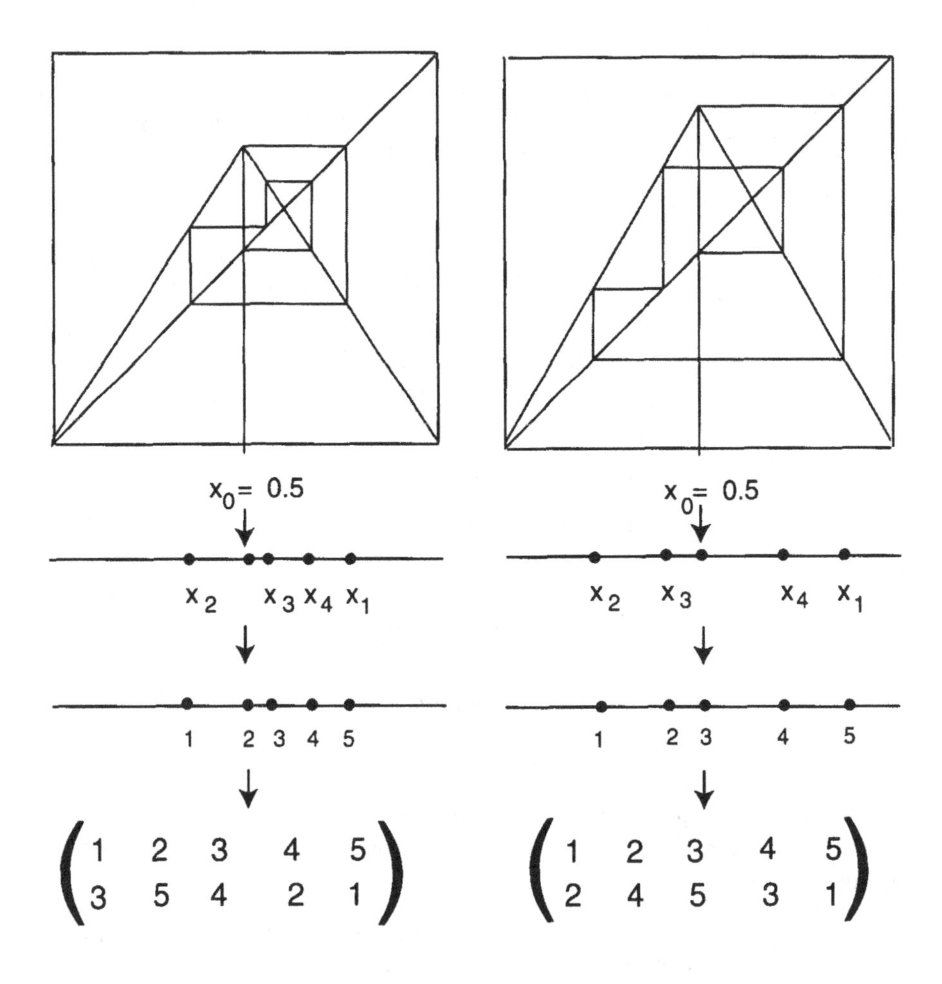

Figure 3.3: Two period 5 orbits and associated permutations

The following theorem of Sharkovsky [155] tells precisely which periods imply which other periods for continuous maps of $\mathbb{R}$. See [20, 64] for a proof of Theorem 3.1.5.

Theorem 3.1.5. *Let $f : \mathbb{R} \to \mathbb{R}$ be continuous. Order the natural numbers* *as follows:*

$$\begin{gathered}3 \succ 5 \succ 7 \succ 9 \cdots \succ 2 \cdot 3 \succ 2 \cdot 5 \succ 2 \cdot 7 \succ 2 \cdot 9 \cdots \succ \\ 2^2 \cdot 3 \succ 2^2 \cdot 5 \succ 2^2 \cdot 7 \succ 2^2 \cdot 9 \cdots \succ \cdots \\ 2^3 \cdot 3 \succ 2^3 \cdot 5 \succ 2^3 \cdot 7 \succ 2^3 \cdot 9 \cdots \cdots \succ 2^3 \succ 2^2 \succ 2 \succ 1.\end{gathered}$$

If f has a periodic orbit of period k, and if $k \succ l$ in the above ordering, then f also has a periodic orbit of period l.

Moreover, given $m \in \mathbb{N}$, there exists a continuous $g : \mathbb{R} \to \mathbb{R}$ such that g has a periodic orbit of period m and no periodic orbit of period n for any $n \succ m$.

Examples for the "moreover" statement from Theorem 3.1.5 can be found in the family $g_a(x) = ax(1-x)$ with $a \in [2,4]$. Next, we look at the basic notions of *attracting* and *repelling* periodic orbits.

Definition 3.1.6. Let $f : E \to E$ be continuous with $E \subset \mathbb{R}^n$, and let p be periodic of period m. We say p is *attracting* provided there is an open set $U \ni p$ such that, for $x \in U \setminus \{p\}$, we have $\lim_{n\to\infty} f^{mn}(x) = p$. We say p is *repelling* provided there is an open set $U \ni p$ such that, for $x \in U \setminus \{p\}$, there is $n_x \in \mathbb{N}$ such that $f^{n_x m}(x) \notin U$.

Exercise 3.1.7. *Let $f : [0,1] \to [0,1]$ be differentiable and suppose $p \in [0,1]$ is periodic of period m. Assume that the derivative of f is continuous. Prove each of the following.*

1. *If $|(f^m)'(p)| < 1$, then p is attracting.*
2. *If $|(f^m)'(p)| > 1$, then p is repelling.*

What about $|(f^m)'(p)| = 1$? **HINT:** *For item 1 choose α such that $|(f^m)'(p)| < \alpha < 1$. Then, for x sufficiently close to p, we have that $|f^m(x) - (f^m(p) = p)| < \alpha|x-p|$. By recursion, we obtain $|f^{jm}(x) - p| < \alpha^j|x-p|$ for $j \geq 1$. Conclude that p is indeed attracting.*

Exercise 3.1.8. *Let T_a be a symmetric tent map with slope $a > 1$. Show no periodic point is attracting.*

Exercise 3.1.9. *Suppose $g_a(x) = ax(1-x)$ is such that $c = 1/2$ is n-periodic for some n. Prove that c is an attracting periodic point.*

The next exercise is an observation used in the proof of Lemma 3.2.12 and is of interest in its own right.

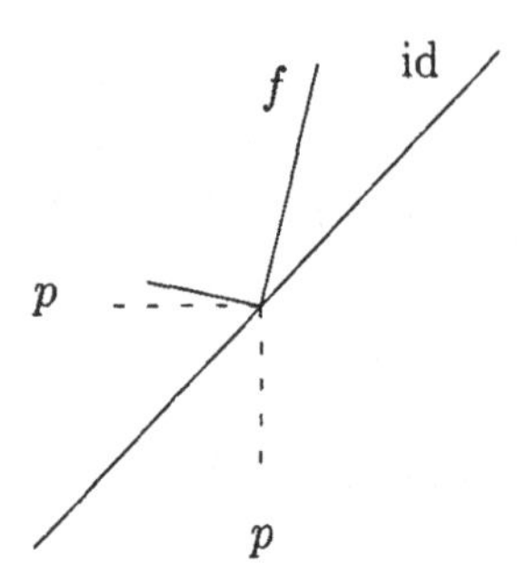

Figure 3.4: Example of repelling fixed point

Exercise 3.1.10. *Let $f : I \to I$ be a continuous map of a closed interval and p a repelling fixed point for f. Prove or disprove that there exists an open interval $W \ni p$ such that $y \in W$ implies $|f(y) - p| > |y - p|$.* **HINT:** *There may not exist such an open interval; see Figure 3.4.*

Remark 3.1.11. Definitions used for the terms *attracting* and *repelling* vary, although all are similar. The definition we use for repelling does not guarantee the existence of an open interval W as described in Exercise 3.1.10. As you read other works, take care as to precisely what definitions are being used.

We close this section with an exercise in which the existence of a period 2 orbit is obtained by analyzing the graph of the second iterate of the map.

Exercise 3.1.12. $\diamondsuit$ *Suppose that f is a continuous map of $\mathbb{R}$, that there is a value c such that $f(f(c)) < c < f(c)$, and that all three points are contained in an interval that maps into itself. Show that f has a period 2 orbit.* **HINT:** *Let $[a, b]$ be the invariant interval. There exists $p \in [a, c)$ such that $f^2(p) = p$. If $f(p) \neq p$, we are done. Otherwise, $f(p) = p$ and there exists $x \in (p, c)$ such that $f(x) = c$. Without loss of generality assume (p, c) contains no other fixed points of f. Consider $f^2|(x, c)$. On the interval (x, c), the graph of f^2 crosses the identity, and hence we obtain a fixed point of f^2 in (x, c). This fixed point is a periodic point of period 2 since (p, c) contains no fixed points of f.*

3.2 Nonwandering and ω-Limit Sets

If x is a periodic (or preperiodic) point for a map f, then the asymptotic behavior of (3.1) is precisely the periodic orbit. More generally, to investigate the asymptotic behavior of (3.1) it is natural to consider the accumulation points of (3.1). We refer to this collection of accumulation points as the ω-limit set for x under f.

Definition 3.2.1. Let (E, ρ) be a compact metric space and $f : E \to E$ be continuous. For $x \in E$ we define the *ω-limit set of x under the map f* as:

$$\omega(x, f) = \omega(x) = \{y \in E \mid \text{ there exists } n_1 < n_2 < \cdots \text{ with } f^{n_i}(x) \to y\}.$$

Intuitively, $y \in \omega(x, f)$ means that points from the forward orbit of x get arbitrarily close to y. We call x *recurrent* provided $x \in \omega(x, f)$, as in this case, points from the orbit of x come back (or recur) close to x. Thus, if x is periodic, then x is recurrent since the forward orbit of x comes back exactly to x infinitely often. Yet it can happen that $x \in \omega(x, f)$ without x being periodic; see Example 3.2.11. Hence, recurrence is a generalization of periodicity, in the sense that the forward orbit of x need not return exactly to x (as in the periodic case) but does return arbitrarily close to x. Recurrence is discussed in Section 3.5. Often $\omega(x, f)$ is a Cantor set; several such examples occur throughout the text.

Exercise 3.2.2. *Let (E, ρ) be a compact metric space, $f : E \to E$ be continuous, and x be n-periodic. Prove $\omega(x) = \{x, f(x), \dots, f^{n-1}(x)\}$. If x is pre-periodic, is $x \in \omega(x, f)$?*

Exercise 3.2.3. *Let $f : [0, 1] \to [0, 1]$ be continuous. Show f has a fixed point and therefore there exists an x with $\omega(x) = \{x\}$.* **HINT:** *If $f(0) = 0$ or $f(1) = 1$, we are done, as either 0 or 1 is a fixed point. Otherwise, $f(0) > 0$ and $f(1) < 1$. As f is continuous, the graph of f must cross the diagonal. This crossing produces a fixed point.*

Definition 3.2.4. Let $f : E \to E$ be a continuous map of a compact metric space. We call $F \subset E$ *invariant* provided $f(F) \subset F$. If $f(F) = F$, we call F *strongly invariant.*

Exercise 3.2.5. *Let (E, ρ) be a compact metric space, $f : E \to E$ continuous, and $x \in E$. Show:*

1. *$\omega(x)$ is not empty.*

2. *$\omega(x)$ is closed and therefore compact (recall Exercise 1.1.28).*

3. *$f(\omega(x)) = \omega(x)$, that is, $\omega(x)$ is strongly invariant.*

Exercise 3.2.6. *Let $f : [0, 1] \to [0, 1]$ be a homeomorphism (not necessarily onto). Prove:*

1. *If f is* orientation preserving, *that is, $x < y \Rightarrow f(x) < f(y)$, then $\omega(z)$ is a fixed point for all $z \in [0,1]$ (not necessarily the same fixed point).*

2. *If f is* orientation reversing, *that is, $x < y \Rightarrow f(x) > f(y)$, then $f^2 = f \circ f$ is orientation preserving and hence (by the above item) $\omega(z, f)$ is either a fixed point or a 2-cycle for all $z \in [0,1]$ (not necessarily the same cycles).*

HINT: *For1 use Exercise 1.1.16 and Exercise 1.1.20. Item 2 is immediate from 1 applied to the homeomorphism f^2.*

A first question would be whether a given $\omega(x, f)$ is finite. Our next discussion shows that, in the setting where $f : E \to E$ is continuous and E is a compact metric space, if $\omega(x, f)$ is finite, then it must consist of one periodic cycle. Hence, there is a sense in which the more interesting ω-limit sets are infinite ones.

Lemma 3.2.7. *Let (E, ρ) be a compact metric space and $f : E \to E$ continuous. Fix $x \in E$ and let F be a nonempty, proper, closed subset of $\omega(x)$. Then $F \cap \overline{f(\omega(x) \setminus F)} \neq \emptyset$.*

Proof. Suppose to the contrary that $F \cap \overline{f(\omega(x) \setminus F)} = \emptyset$. Since F and $\overline{f(\omega(x) \setminus F)}$ are then disjoint closed sets and f is uniformly continuous, we may choose open sets G_1 and G_2 such that (see Figure 3.5)

- $\omega(x) \setminus F \subset G_1$,
- $F \subset G_2$, and
- $\overline{G_2} \cap f(\overline{G_1}) = \emptyset$.

We may then choose $n_1 < n_2 < n_3 < \cdots$ such that

$$f^{n_k}(x) \in G_1 \quad \text{and} \quad f^{n_k+1}(x) \in G_2$$

for all k; we may do this since: $\omega(x) \cap G_1 \neq \emptyset$, $\omega(x) \cap G_2 \neq \emptyset$, and $\omega(x) \subset G_1 \cup G_2$. Let y be a accumulation point of $\{f^{n_k}(x)\}$. Then $y \in \overline{G_1}$ and $f(y) \in \overline{G_2}$, contradicting that $\overline{G_2} \cap f(\overline{G_1}) = \emptyset$. □

Exercise 3.2.8. *Prove that one can make the choice of $n_1 < n_2 < n_3 < \cdots$ in the proof of Lemma 3.2.7.*

Proposition 3.2.9. *Let (E, ρ) be a compact metric space and $f : E \to E$ continuous. Let $x \in E$ and suppose that $\omega(x)$ is a finite set. Then, $\omega(x)$ is a periodic cycle.*

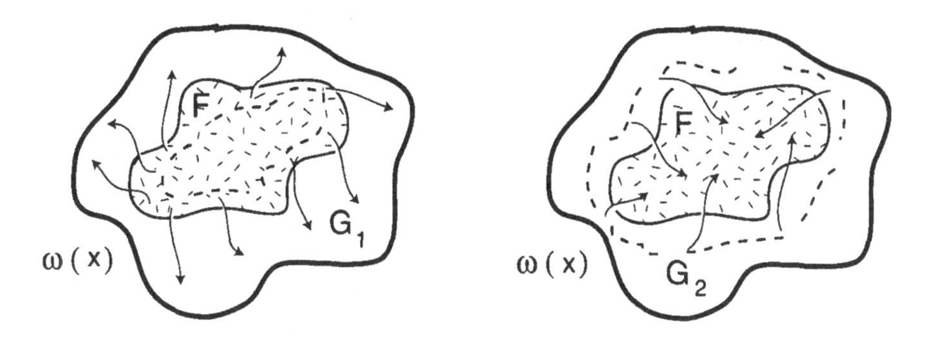

Figure 3.5: Open sets G_1 and G_2

Exercise 3.2.10. *Prove Proposition 3.2.9.* **HINT:** *Use Lemma 3.2.7.*

In contrast to an ω-limit set being a finite set, the next example provides an ω-limit set that is a nondegenerate closed interval. Chapter 5 provides an example where $\omega(x)$ is a Cantor set; several such examples occur throughout the text.

Example 3.2.11. *Consider the symmetric tent map T_2. We show there exists $z \in [0,1]$ with $\omega(z, T_2) = [0,1]$. Indeed, the set $\{z \in [0,1] \mid \omega(z, T_2) = [0,1]\}$ is uncountable.*

Informally code points in $[0, \frac{1}{2})$ with 0, points in $(\frac{1}{2}, 1]$ with 1, and code $\frac{1}{2}$ with $*$. Think of this coding as an address. *Hence, the address of* 0.87 *is* 1, *and the address of* 0.36 *is* 0 *while the address of* 0.5 *is* $*$. *To each $x \in [0,1]$ we associate an element of $\{0, 1, *\}^{\mathbb{N}}$ by listing in a sequence the addresses of the forward orbit of x. This sequence is called the* itinerary *of x and is denoted $I(x, T_2)$. For example, the itinerary of $c = 0.5$ is $I(c, T_2) = \langle *, 1, 0, 0, 0, 0, \ldots \rangle$.*

The map T_2 has two maximal (in length) subintervals of monotonicity, namely, $[0, \frac{1}{2}]$ and $[\frac{1}{2}, 1]$. The second iterate, T_2^2, has four such intervals, namely, $[0, \frac{1}{4}]$, $[\frac{1}{4}, \frac{1}{2}]$, $[\frac{1}{2}, \frac{3}{4}]$, and $[\frac{3}{4}, 1]$. More generally, for each $n \in \mathbb{N}$, the map T_2^n has precisely 2^n maximal subintervals of monotonicity. See Figure 3.6. If x and y belong to the interior *of one such maximal subinterval for T_2^n, then the itinerary of x and y agree in the first n positions (and these n entries are either 0 or 1). Moreover, if u and v belong to the interior of different subintervals of monotonicity, then the itineraries of u and v differ somewhere within the first n terms.*

Note that there are precisely 2^n words of length n with an alphabet of $\{0, 1\}$. Hence, there is a bijection between the maximal subintervals of monotonicity of T_2^n and words of length n on the alphabet $\{0, 1\}$. The bijection is *obtained by identifying a subinterval of monotonicity J of T_2^n with the first n terms of the itinerary of any $x \in J^\circ$.*

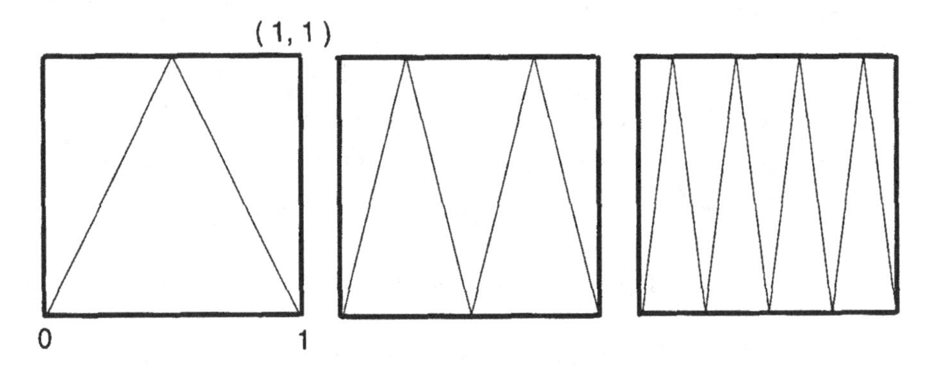

Figure 3.6: Graphs of T_2, T_2^2, and T_2^3

Observe that an $*$ appears in the itinerary of $z \in (0,1)$ if and only if, for some $n \in \mathbb{N}$, z is the endpoint of a subinterval of monotonicity for T_2^n. Hence, a $*$ appears in the itinerary of at most countably many $z \in (0,1)$.

For each $\gamma = \gamma_1, \gamma_2, \gamma_3 \dots \in \{0,1\}^{\mathbb{N}}$ and $n \in \mathbb{N}$ there is precisely one subinterval of monotonicity of T_2^n, denoted $J_{\gamma,n}$, coded by $\gamma_1, \gamma_2, \dots, \gamma_n$, that is, the itinerary of each $z \in J_{\gamma,n}^\circ$ begins with $\gamma_1, \gamma_2, \dots, \gamma_n$. Moreover, $J_{\gamma,n} \supset J_{\gamma,n+1}$ for each n and $\bigcap_n J_{\gamma,n}$ is a single point, say z_γ. Indeed, $I(z_\gamma, T_2)$ is precisely γ.

By concatenating all possible words of length n over the alphabet $\{0,1\}$, for all n, into one infinite sequence, say $\tilde{\gamma} \in \{0,1\}^{\mathbb{N}}$, and choosing $z \in [0,1]$ with $I(z,T_2) = \tilde{\gamma}$, we obtain $\omega(z,T_2) = [0,1]$. Note that each such infinite concatenation (and there are uncountable many) produces a distinct $\tilde{\gamma}$ and z. See Exercise 10.1.8 for a more detailed presentation of this example.

It follows from Exercise 3.2.5 that $(\omega(x,f), f)$ forms a dynamical system. In Chapter 13 we are interested in determining when $f|\omega(x,f)$ is a homeomorphism. Lemma 3.2.12 tells us that this cannot happen for interval maps if $\omega(x,f)$ properly contains a repelling periodic point satisfying condition 1 of Lemma 3.2.12. Recall from Exercise 3.1.10 that condition 1 is not automatic for a repelling periodic point.

Lemma 3.2.12 ([34, Lemma 1]). *Let $f : I \to I$ be a continuous map of a closed interval in $\mathbb{R}$, and let p be a repelling m-periodic point of f. Suppose*

1. *there exists an open interval $W \ni p$ such that $y \in W$ implies $|f^m(y) - p| > |y - p|$,*

2. *there exists $x \in I$ such that the orbit of p is properly contained in $\omega(x, f)$.*

Then $f|\omega(x, f)$ is not one-to-one.

Proof. We break the proof into two parts. We first do the case when $m = 1$, that is, p is a fixed point. Fix $\epsilon > 0$ such that $|y - p| < \epsilon$ implies $|f(y) - p| > |y - p|$. Set $\Delta = (p - \epsilon, p + \epsilon)$. Choose $\epsilon_i > 0$ such that $\epsilon > \epsilon_1 > \epsilon_2 > \cdots \downarrow 0$. As p is properly contained in $\omega(x, f)$, we may choose $n_1 \in \mathbb{N}$ such that $f^{n_1}(x) \in \Delta_1 = (p - \epsilon_1, p + \epsilon_1)$ and $f^{n_1 - 1}(x) \notin \Delta_1$. Set $z_1 = f^{n_1 - 1}(x)$. Choose $n_2 > n_1$ such that $f^{n_2}(x) \in \Delta_2 = (p - \epsilon_2, p + \epsilon_2)$ and $z_2 = f^{n_2 - 1}(x) \notin \Delta_2$. Then $z_2 \notin \Delta$. For otherwise

$$|(f(z_2) = f^{n_2}(x)) - p| > |z_2 - p|; \tag{3.2}$$

however, $f^{n_2}(x) \in \Delta_2 = (p - \epsilon_2, p + \epsilon_2)$ and $z_2 \notin \Delta_2$ contradict (3.2). See Figure 3.7. Continuing, we obtain a sequence $\{z_n\}$ and a point z such that (passing to subsequence if needed) $\lim_{n\to\infty} z_n = z \neq p$ and $\lim_{n\to\infty} f(z_n) = p$. Thus, $f(z) = p$ with $z \neq p$ and $z, p \in \omega(x, f)$. Hence, $f|\omega(x, f)$ is not one-to-one.

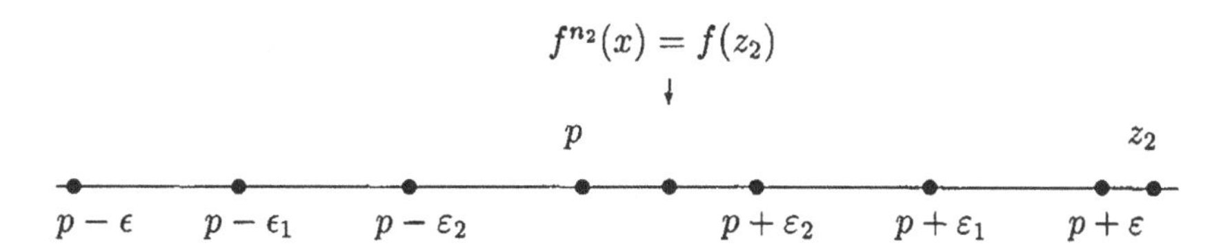

Figure 3.7: Placement of z_2

Next assume $m > 1$ and set $g = f^m$. Then p is a repelling fixed point for g. Let $\{n_k\}$ be a strictly increasing sequence of positive integers such that $\lim_{k\to\infty} f^{n_k}(x) = p$. Express each n_k as $n_k = t_k m + r_k$, where $0 \leq r_k \leq m-1$; passing to subsequence if needed, assume the sequence $\{t_k\}$ is increasing and $r_k = r$ for all k. Then, $f^{n_k}(x) = g^{t_k}(f^r(x))$ and hence $\omega(f^r(x), g) \ni p$. If $\omega(f^r(x), g)$ $\{p\}$, then $p \in orb_g(f^r(x)) \subset orb_f(x)$ and thus $\omega(x, f) = orb_f(p)$, a contradiction. Hence, $\omega(f^r(x), g)$ properly contains $\{p\}$ and therefore (by the first part of the proof) $g|\omega(f^r(x), g)$ is not one-to-one. As $\omega(f^r(x), g) \subset \omega(x, f)$, we are done. □

Exercise 3.2.13. ◇ *Does Lemma 3.2.12 hold if condition 1 is dropped?*

Remark 3.2.14. It follows from Exercise 3.1.8 and Lemma 3.2.12 that if T_a is a symmetric tent map with slope $a > 1$, $p \neq c$ a periodic point of T_a, and the orbit of p is properly contained in $\omega(x, T_a)$, then $T_a|\omega(x, T_a)$ is not one-to-one. More generally, Chapter 13 investigates the question of whether $T_a|\omega(c, f)$ is one-to-one when $\omega(c, T_a)$ is a Cantor set.

If an ω-limit set is infinite, one might ask whether the ω-limit set can be broken into a finite disjoint union of closed invariant sets; if yes, we would investigate the subsystems given by these invariant sets. Exercise 3.2.15 says this is not possible. What about an infinite disjoint union of closed invariant sets?

Exercise 3.2.15. *Let (E, ρ) be a compact metric space, $f : E \to E$ continuous, and $x \in E$. Prove that $\omega(x)$ cannot be written as a finite disjoint union of closed invariant sets.* **HINT:** *Use Lemma 3.2.7.*

For a compact metric space E and continuous map $f : E \to E$, the next proposition tells us that, if there is some $x \in E$ with $\overline{\{f^j(x)\}_{j\geq 0}} = E$ (i.e., the orbit of x is dense in E), then for any $z \in E$ either $\omega(z, f) = E$ or $\omega(z, f)$ has an empty interior (i.e., is nowhere dense). Notice in Proposition 3.2.16 that $\overline{\{f^n(x)\}_{n\geq 1}}$ is used, rather than $\overline{\{f^n(x)\}_{n\geq 0}}$.

Proposition 3.2.16. *Let E be a compact metric space and $f : E \to E$ be continuous. Then there exists a proper subset A of E such that A is closed, $f(A) \subset A$, and $A^\circ \neq \emptyset$ if and only if $\overline{\{f^n(x)\}_{n\geq 1}} \neq E$ for all $x \in E$.*

Proof. Let A be a proper subset of E such that A is closed, $f(A) \subset A$, and $A^\circ \neq \emptyset$. Suppose to the contrary that $\{f^n(x)\}_{n\geq 1}$ is dense in E for some $x \in E$. First notice that $x \notin A$, for otherwise $A = E$. As $A^\circ \neq \emptyset$ and $\overline{\{f^n(x)\}_{n\geq 1}} = E$, there exists $m = \min\{n \mid f^n(x) \in A\}$. Then, since $f(A) \subset A$, $f^j(x) \in A$ for all $j \geq m$. However, then $E \setminus A = \{x, f(x), \ldots, f^{m-1}(x)\}$ and $\{x\}$ is an open set (as $E \setminus A$ is a finite open set). Lastly, $\{x\}$ open and $\{f^n(x)\}_{n\geq 1}$ dense imply that $f^k(x) = x \notin A$ for some k, contradicting $f^j(x) \in A$ for all $j \geq m$.

Suppose there does not exist a proper subset A of E with $f(A) \subset A$, A closed, and $A^\circ \neq \emptyset$. We show there exists $x \in E$ with $\{f^n(x)\}_{n\geq 1}$ dense in E. Let $\mathcal{U} = \{U_1, U_2, \ldots\}$ be a collection of open sets as in Exercise 1.1.46. Then for each $j \geq 1$ the set $\bigcup_{n\geq 1} f^{-n}(U_j)$ is dense in E since $E \setminus \bigcup_{n\geq 1} f^{-n}(U_j)$ is either empty or invariant and closed (and hence has empty interior by assumption). Set $\mathcal{D} = \bigcap_{j\geq 1} \bigcup_{n\geq 1} f^{-n}(U_j)$. It follows from the Baire Category Theorem (Theorem 1.1.54) that $D \neq \emptyset$. However, if $x \in D$, then $\{f^n(x)\}_{n\geq 1}$ is dense in E. □

A proof of the next theorem is beyond the text; however, note that it provides a characterization of what kinds of sets can be ω-limit sets for continuous self-maps of an interval. Compare to Proposition 3.2.16.

Theorem 3.2.17. *[29] Let I be a nondegenerate closed interval in $\mathbb{R}$ and let C denote the class of continuous self-maps of I. A necessary and sufficient condition that a set $M \subset I$ be an ω-limit set for some $f \in C$ is that M be either a nonempty nowhere dense set or that M be a finite union of closed intervals.*

Thus, for f a continuous self-map of an interval and $M = \omega(x, f)$ for some x with M not a finite union of closed intervals, we have that M is nowhere dense. As M is a subset of $\mathbb{R}$ and M is nowhere dense, M cannot contain an open interval and hence must be totally disconnected. Thus, M is compact (as it is closed) and is totally disconnected, and therefore two of the three properties needed for a Cantor set are present. If M is finite, we know it consists of precisely a periodic cycle. In case M is not finite, it can happen that M is countable (see Exercise 10.2.3) or that M is uncountable (see Examples 10.2.2 and 10.2.4). In the uncountable case, is M a Cantor set? The answer is not necessarily (again, see Example 10.2.4). If x has the property of *strong recurrence*, which will be discussed in the next section, then indeed M is a Cantor set.

We can say a bit more for piecewise strictly monotone maps with a finite number of intervals of monotonicity, such as a symmetric tent map. For such maps, we have a uniform bound on the cardinality of any preimage set $f^{-1}(x)$. The next two exercises address what can happen in this setting when an ω-limit set is uncountable and totally disconnected.

Exercise 3.2.18. *Let (E, ρ) be a compact metric space and $f : E \to E$ continuous. Suppose there exists $L \in \mathbb{N}$ such that for all $x \in E$, $f^{-1}(x)$ has no more than L elements. If $\omega(x)$ is uncountable and totally disconnected, prove $\omega(x)$ is a Cantor set or $\omega(x)$ can be expressed uniquely as a Cantor set K and a countable set $\omega(x) \setminus K$ such that:*

- *$f(K) = K$, and*
- *$\omega(x) \setminus K$ is not closed.*

HINT: *Recall Proposition 1.1.52 and Exercises 1.1.53 and 3.2.15.*

Exercise 3.2.19. $\diamondsuit$ *Construct an example where (E, ρ), f, x, and $\omega(x)$ are as in Exercise 3.2.18 and $\omega(x) \setminus K$ is nonempty. (See Example 10.2.4.)*

Continuing our investigation of the asymptotic behavior of (3.1), one can ask what happens to an open neighborhood of x under iteration of the map f? Let $U \ni x$ be open. Does U revisit itself with iteration, that is, does there exist $j \in \mathbb{N}$ such that $f^j(U) \cap U \neq \emptyset$? Notice that if indeed an open neighborhood of x revisits itself with iteration, it need not be the case that x is periodic (or recurrent); can you provide an example?

Definition 3.2.20. Let (E, ρ) be a compact metric space and $f : E \to E$ continuous. We call $x \in E$ *nonwandering* provided that, for every open set U containing x, there is an integer $n > 0$ such that $f^n(U) \cap U \neq \emptyset$. Set $\Omega(f) = \{x \in E \mid x \text{ is nonwandering}\}$. We call $\Omega(f)$ the *nonwandering* set for f.

For a large class of maps, the asymptotic behavior of (3.1) for $x \notin \Omega(f)$ is limited and understood; see [162, Section 2]. Hence, one investigates the set $\Omega(f)$ and the dynamical system restricted to $\Omega(f)$.

Proposition 3.2.21. *Let (E, ρ) be a compact metric space and $f : E \to E$ continuous. Then $\Omega(f) \subset \cap_{i=0}^{\infty} f^i(E)$.*

Proof. Let $x \in E \setminus \cap_{i=0}^{\infty} f^i(E)$ and m be minimal such that $x \notin f^m(E)$. As E is compact and f continuous, we have that $f^m(E)$ is compact (recall Exercise 1.1.27). Hence, the distance between x and $f^m(E)$ is positive (recall Exercise 1.1.30), and thus we may choose disjoint open sets U and V with $x \in U$ and $f^m(E) \subset V$. Set $W = U \cap (\cap_{i=1}^{m} f^{-i}(V))$. Then W is open, $W \ni x$, and $f^i(W) \subset V$ for all $i \geq 1$. Thus $x \notin \Omega(f)$. □

Exercise 3.2.22. *Let (E, ρ) be a compact metric space and $f : E \to E$ continuous. Prove the following.*

1. *If $x \in \omega(x, f)$, then x is nonwandering.*

2. *If x is nonwandering, it need not be that $x \in \omega(x, f)$.* **HINT:** *See Example 3.2.24 below and consider $x = \frac{3}{4}$. Also consider $x = 1$ and the map T_2.*

Exercise 3.2.23. *Let (E, ρ) be a compact metric space and $f : E \to E$ continuous. Prove:*

1. *$\Omega(f)$ is closed.*

2. *$\cup_{x \in E} \omega(x) \subset \Omega(f)$.*

3. *Periodic points are contained in $\Omega(f)$.*

4. *$f(\Omega(f)) \subset \Omega(f)$.*

Can you give an example where one has strict set containment in (2) and (4)?

Example 3.2.24. *Let g be as in Figure 3.8. Then, $\frac{3}{4} \notin g(\Omega(g))$, $\frac{3}{4} \notin \cup_{x \in [0,1]} \omega(x)$, and $\frac{3}{4} \in \Omega(g)$.*

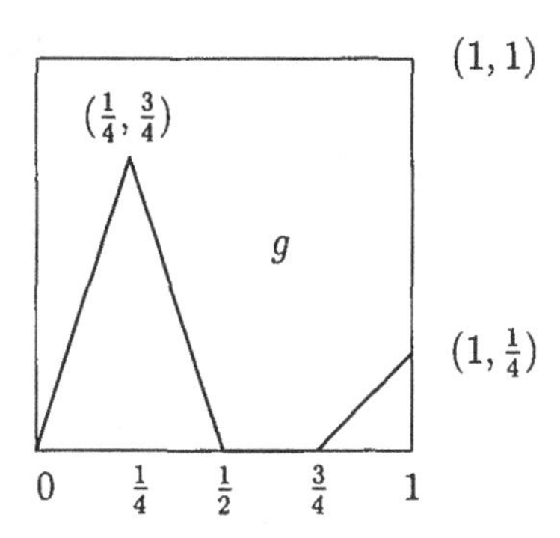

Figure 3.8: Graph of g

In Section 10.7 we observe that if $\sqrt{2} < a^m < 2$ for some $m \in \{1, 2, 2^2, 2^3, 2^4, \dots\}$, then the nonwandering set of T_a (symmetric tent map) consists of m disjoint closed intervals and a finite number of periodic points. For $a \in [\sqrt{2}, 2]$, the nonwandering set of T_a is precisely $[T_a^2(\frac{1}{2}), T_a(\frac{1}{2})]$. For further discussion see [162, page 78].

Lastly, we raise the question of the *depth* of an ω-limit set.

Definition 3.2.25. Let $E \subset \mathbb{R}$ and $e \in E$. We call e an *isolated point of* E provided there is an open set $U \ni e$ such that $U \cap E = \{e\}$.

Definition 3.2.26. Let $E \subset \mathbb{R}$ and let $I(E)$ denote the isolated points of E. Set $d(E) = E \setminus I(E)$. Define the *depth of* E, denoted $D(E)$, to be $\min\{n \mid d^n(E) = \emptyset\}$. If no such n exists, we set $D(E) = \infty$.

Finite subsets of $\mathbb{R}$ have depth 1. Cantor sets and nondegenerate intervals have depth ∞. Sets such as $\{0, 1, 1/2, 1/3, 1/4, \dots\}$ have depth 2. Thus, ω-limit sets can have depth 1, 2, and ∞. What other possibilities are there?

Question 3.2.27. What depths are possible for $\omega(x, f)$ where $E \subset \mathbb{R}$ and $f : E \to E$ is continuous?

3.3 Topological Conjugacy

Often in mathematics one uses tools (e.g., change of variables or change of coordinates) to place a given problem into a setting more easily understood. One such tool in dynamics is *topological conjugacy*. We will see (Theorem 3.4.27) that to study topological dynamical behaviors for certain low-dimensional systems, it suffices (via a topological conjugacy) to study the family of symmetric tent maps. As these maps are piecewise linear maps, they can be more easily understood. Recall that the symmetric family of tent maps, $\{T_a\}_{a\in[0,2]}$, is given by $T_a(x) = ax$ for $x \in [0, \frac{1}{2}]$ and $T_a(x) = a(1-x)$ for $x \in [\frac{1}{2}, 1]$. Chapter 10 investigates this family in more detail.

Definition 3.3.1. Let $f : A \to A$ and $g : B \to B$ be given. We say f and g are *topologically conjugate* provided there is an onto homeomorphism $h : A \to B$ such that $h \circ f = g \circ h$. (Remember that an onto homeomorphism is a bijection that is continuous in both directions.) We refer to h as the *conjugacy* between f and g. See Figure 3.9.

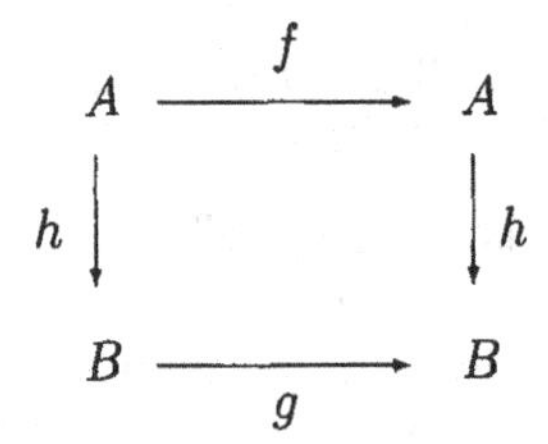

Figure 3.9: Topological conjugacy

If $f : A \to A$ and $g : B \to B$ are topologically conjugate, many dynamical behaviors of the system (f, A) are preserved by the conjugacy, that is, the system (g, B) also has the behaviors; several such examples are provided in the next section. Hence there is a sense in which topologically conjugate systems are dynamically the same and thus, depending on what dynamical behavior you are investigating, it might be useful to use one or the other representation of the system. However, *not* all dynamical behaviors are preserved via a conjugacy. Typically, "topological behaviors," such as periodicity and recurrence, are preserved by conjugacies, whereas "metric or analytic behaviors" are not necessarily preserved. For example, if the map f is differentiable, properties such as C^r *smooth* (the map f is C^r smooth provided f is r times differentiable and each of the r derivatives is continuous), *Collet-Eckmann* (Definition 10.5.2), or *negative Schwarzian derivative* (Definition 10.5.1) are not necessarily preserved by a conjugacy; see Section 10.5 for a brief discussion on Collet-Eckmann and negative Schwarzian and see [115] for more detail. Naturally, one investigates conditions to put on a conjugacy in order to preserve a desired dynamical behavior. However, we are getting ahead of our discussion; see Chapters 9 and 10 for further uses of conjugacies.

Remark 3.3.2. If in Definition 3.3.1 the map h satisfies $h \circ f = g \circ h$, is continuous, and onto, then we call (B, g) a *factor* of (A, f). Note it is no longer required that h be one-to-one. In this setting we say f and g are *semiconjugate* and refer to h as the *semiconjugacy*.

Exercise 3.3.3. *Suppose that f is topologically conjugate to g via a homeomorphism h. Prove that for each $n > 0$ the map f^n is topologically conjugate to g^n via h.*

Exercise 3.3.4. *Let f and g be topologically conjugate with conjugacy h, and suppose that $f(p) = p$. Show that $g \circ h(p) = h(p)$, that is, $h(p)$ is a fixed point for the map g. Thus fixed points are preserved by a conjugacy. Generalize to k-periodic points.*

Exercise 3.3.5. *Let $f : E \to E$ and $g : H \to H$ be topologically conjugate with conjugacy h, and suppose $\overline{\{f^i(x)\}_{i\geq 0}} = E$. Prove $\overline{\{g^i(h(x))\}_{i\geq 0}} = H$.*

Exercise 3.3.6. *Let $f : E \to E$ and $g : H \to H$ be topologically conjugate with conjugacy h, and suppose x is recurrent, that is, $x \in \omega(x, f)$ (Section 3.2). Prove that $h(x)$ is recurrent under g, that is, prove that $h(x) \in \omega(h(x), g)$.*

Remark 3.3.7. For each $b \in (1, \sqrt{2})$ there exists a $n_b \in \{2, 4, 8, 16, \dots\}$, an interval $J_b \ni c$, and a unique $a \in [\sqrt{2}, 2]$ such that $T_b^{n_b}|J_b$ is affinely topologically conjugate (i.e., the conjugacy h can be expressed as $h(x) = \alpha x + \beta$ for some $\alpha, \beta \in \mathbb{R}$) to T_a. See Section 10.7 and Example 10.7.2 for further discussion.

3.4 Transitive Behavior

In this section we investigate when a dynamical system has a dense orbit; such systems are called *topologically transitive*. For interval maps, there is a handy property called *locally eventually onto* that implies topological transitivity. On the other hand, there are obstructions to the existence of dense orbits. One such obstruction is the occurrence of attracting periodic orbits. We identify two more obstructions: the occurrences of wandering intervals (Definition 3.4.7) and of restrictive intervals, that is, renormalization (Definition 3.4.24).

We see that a symmetric tent map with slope $> \sqrt{2}$ is transitive on its core (see Remark 3.4.17) and that a unimodal map (Definition 3.4.11) without the three obstructions mentioned above is topologically conjugate to a symmetric tent map with slope in $(\sqrt{2}, 2]$ (Theorem 3.4.27). This fact is an aid in our study of ω-limit sets. A detailed study of the dynamics of the symmetric tent family occurs in Chapter 10.

Definition 3.4.1. Let E be a compact metric space and $f : E \to E$ be continuous. We say f is *topologically transitive* provided there exists $x \in E$ with $\{f^n(x) \mid n \geq 0\}$ dense in E, that is, $\overline{\{f^n(x)\}_{n\geq 0}} = E$.

Proposition 3.4.2. *Let E be a compact metric space, $f : E \to E$ be continuous, and $f(E) = E$. The following are equivalent.*

1. *The map f is topologically transitive.*
2. *If $A \subset E$ is closed with $f(A) \subset A$, then either $A = E$ or A is nowhere dense.*
3. *If U is open with $f^{-1}(U) \subset U$, then $U = \emptyset$ or U is dense in E.*
4. *If U, V are nonempty open sets, then there exists $n \geq 1$ such that $f^{-n}(U) \cap V \neq \emptyset$.*
5. *The set $\{x \in E \mid \overline{\{f^n(x)\}_{n\geq 0}} = E\}$ is dense in E.*

Proof. Assume (1) holds; we show (2). Let $y \in E$ be such that $\{f^n(y)\}_{n\geq 0}$ is dense in E, and let $A \subset E$ be closed and such that $f(A) \subset A$. If A has an empty interior, that is, if A is nowhere dense, we are done. Thus, suppose A has a nonempty interior, and thus let U be open with $U \subset A$. Then $f^p(y) \in U$ for some p and hence

$$\{y, f(y), \dots, f^{p-1}(y)\} \cup A = E. \tag{3.3}$$

Applying f to (3.3) we have $\{f(y), \dots, f^{p-1}(y)\} \cup A = E$, and hence, by applying f a finite number of times to (3.3), we have that $A = E$. Hence (2) holds.

Assume (2) holds; we show (3). Let U be a nonempty open set with $f^{-1}(U) \subset U$. Set $A = E \setminus U$; then A is closed and $f(A) \subset A$. Since $U \neq \emptyset$, we have $A \neq E$ and hence, by (2), A is nowhere dense. It then follows that U is dense in E.

Assume (3) holds; we show (4). Let U, V be nonempty open sets. Then $\cup_{n=1}^{\infty} f^{-n}(U)$ is open and $f^{-1}(\cup_{n=1}^{\infty} f^{-n}(U)) \subset \cup_{n=1}^{\infty} f^{-n}(U)$, and therefore $\cup_{n=1}^{\infty} f^{-n}(U)$ is dense by (3). Hence $f^{-n}(U) \cap V \neq \emptyset$ for some n.

Assume (4) holds; we show (5). Let $\mathcal{U} = \{U_1, U_2, U_3, \dots\}$ be a countable collection of open sets with the property that, given $x \in E$ and $W \ni x$ open in E, there exists $U \in \mathcal{U}$ with $x \in U \subset W$; recall Exercises 1.1.46 and 1.1.47. From (4) we have that each $\cup_{m=0}^{\infty} f^{-m}(U_n)$ is dense in E. We have (5) by noting that

$$\{x \mid \overline{\{f^n(x)\}_{n\geq 0}} = E\} = \cap_{n=1}^{\infty} \cup_{m=0}^{\infty} f^{-m}(U_n) \tag{3.4}$$

and that the right-hand side of (3.4) is dense by the Baire Category Theorem.

That (5) implies (1) is clear. □

It follows from Propositions 3.4.2 and 3.2.16 that if $f : E \to E$ is transitive (with f continuous and E compact), then for any $z \in E$ we have that $\omega(z, f) = E$ or $\omega(z, f)$ is nowhere dense. Shortly we will apply this fact to symmetric tent maps.

Exercise 3.4.3. *Let $E, \tilde{E}$ be a compact metric spaces; $f : E \to E$ and $g : \tilde{E} \to \tilde{E}$ be continuous maps; $f(E) = E$; and $g(\tilde{E}) = \tilde{E}$. Suppose f is topologically transitive and that f and g are topologically conjugate. Prove g is topologically transitive.* **HINT:** *Recall Exercises 3.3.3 and 3.3.5.*

Exercise 3.4.4. *Let $f : I \to I$ be continuous with $I \subset \mathbb{R}$ a closed interval. Assume f is topologically transitive. Let U, V be nonempty open sets. Prove there exists $k > 0$ such that $f^k(U) \cap V \neq \emptyset$.*

Exercise 3.4.5. ♣ *[19] Let $I \subset \mathbb{R}$ be a nondegenerate closed interval and $f : I \to I$ continuous with $f(I) = I$. Prove f is transitive if and only if, for every closed interval $J \subset I$, we have that $\cup_{n \geq 0} f^n(J)$ is dense in I.*

Definition 3.4.6. Let $f : I \to I$ be a continuous map of a closed interval. We say f is *strongly transitive* provided that, for every nondegenerate interval $J \subset I$, there exists M such that $\cup_{k=0}^{M} f^k(J) = I$ [135].

Next we discuss wandering sets; again, the nonexistence of such sets is required for Theorem 3.4.27.

Definition 3.4.7. Let $f : E \to E$ be a continuous map of a compact metric space. We say a nondegenerate open set $U \subset E$ is *wandering* provided

- $f^n(U) \cap U = \emptyset$ for all $n > 0$, and
- U does not tend towards a periodic orbit, that is, $\cup_{x \in U} \omega(x, f)$ is not a single periodic orbit.

In the event that $E \subset \mathbb{R}$ and U is an open interval, we refer to U as a *wandering interval.* If an open set is not wandering, it is called *nonwandering.*

Exercise 3.4.8. *Let $f : E \to E$ be a continuous map with $E \subset \mathbb{R}$ compact. Suppose $x \in \Omega(f)$ (recall Definition 3.2.20) and $U \ni x$ is open. Show U is nonwandering.*

Exercise 3.4.9. *Let $f : E \to E$ be a continuous map of a compact metric space. Suppose $U \subset E$ is a wandering set. Prove f is not transitive.*

Remark 3.4.10. There are conditions to put on f to guarantee there are no wandering intervals. For example, let $f : [0,1] \to [0,1]$ be a C^2 map that is nonflat (see Definition 6.2.4) at each critical point; then there are no wandering intervals. See [115] for further discussion on the existence of wandering intervals. The often-studied family of maps $\{g_a(x) = ax(1-x)\}_{a\in[0,4]}$ is such that each system $([c_2, c_1], g_a)$ has no wandering intervals; here, the critical point $c = 1/2$ is nonflat.

We have used symmetric tent maps for examples; these maps are examples of *unimodal maps*, and unimodal maps play a significant role in the text.

Definition 3.4.11. A continuous map $f : [0,1] \to [0,1]$ is called *unimodal* if there exists a unique *turning* or *critical point*, c, such that $f|_{[0,c)}$ is increasing, $f|_{(c,1]}$ is decreasing, and $f(0) = f(1) = 0$. For each $x \neq c$, let $\hat{x} \neq x$ be the unique point such that $f(x) = f(\hat{x})$. Set $c_i = f^i(c)$ for $i \geq 0$. We call f *strictly unimodal* provided $f|_{(c,1]}$ is strictly decreasing and $f|_{[0,c)}$ is strictly increasing.

For a unimodal map $f : I \to I$ with $c_2 < c < c_1$ and $c_2 \leq c_3$, we call the interval $[c_2, c_1]$ the *core* of the system (I, f). Notice that the core maps into itself. As we are interested in the asymptotic dynamics of the system, we generally restrict our attention to f on the core. Often we refer to $f|[c_2, c_1]$ as unimodal, even though the endpoints are not tied down at zero as stated in Definition 3.4.11. The dynamics of the system (I, f) are trivial if it is not the case that $c_2 < c < c_1$ and $c_2 \leq c_3$.

Definition 3.4.12. We say a unimodal map f is *locally eventually onto, leo,* provided that for every $\epsilon > 0$ there exists $M \in \mathbb{N}$ such that, if U is an interval with $|U| > \epsilon$ and if $n \geq M$, then $f^n(U) = [c_2, c_1]$.

Exercise 3.4.13. $\diamondsuit$ *Let $f : I \to I$ be unimodal. Prove f is leo if and only if for each open interval U there exists a positive integer m such that $[c_2, c_1] \subset f^m(U)$.* **HINT:** *Prove the "if" direction by contradiction. Suppose f is not leo. Then there exists $\epsilon > 0$; open intervals $U_1, U_2, U_3, \ldots$; and positive integers $n_1 < n_2 < n_3 < \ldots$ such that for all i we have*

- $|U_i| \geq \epsilon$
- $f^{n_i}(U_i) \supseteq [c_2, c_1]$
- $f^j(U_i) \not\supseteq [c_2, c_1]$ *for $j < n_i$.*

Next show that if $\{W_1, W_2, W_3, \dots\} \subset \{U_1, U_2, U_3, \dots\}$*, then* $\bigcap_i W_i$ *does not contain a nondegenerate open interval. Lastly, take a finite partition of* I *by intervals of length less than* $\frac{\epsilon}{4}$*, say* $P = \{P_1, P_2, \dots, P_m\}$*. Then there exists* j *such that* P_j *contains the left endpoint of* U_i *for infinitely many* i*, and thus either* P_j *or* P_{j+1} *sits in infinitely many* U_i*, a contradiction.*

Exercise 3.4.14. *Let* $f : I \to I$ *be unimodal with core* I*. Suppose* f *is leo. Prove* f *is topologically transitive.* **HINT:** *Use leo to prove that for each open interval* U *we have* $\overline{\{f^{-n}(U)\}_{n \geq 0}} = I$ *and hence condition (3) of Proposition 3.4.2 holds.*

Exercise 3.4.15. *Let* f *be unimodal and leo. Prove that indeed* f *is strongly transitive. Is unimodal needed? Can you give an example of a transitive map that is not strongly transitive?*

Exercise 3.4.16. ♣ *Prove that* T_a *is leo (on the core) for* $a \in (\sqrt{2}, 2]$*. What about* $a \in [1, \sqrt{2}]$ *(see Proposition 3.4.26)?* **HINT:** *See either [32, Lemma 2] or [162, Proposition 2.5.5].*

Remark 3.4.17. Let T_a be a symmetric tent map with $a \in (\sqrt{2}, 2]$ and set $E = [c_2, c_1]$. Since T_a is leo, we have that it is transitive. Hence, by Proposition 3.2.16, if B is a proper subset of $[c_2, c_1]$ that is closed and invariant (i.e., $T_a(B) \subset B$), then $B^\circ = \emptyset$. Thus all proper, closed, invariant subsets of $[c_2, c_1]$ are nowhere dense. Therefore, either $\omega(x, T_a) = [c_2, c_1]$ or $\omega(x, T_a)$ is nowhere dense.

Exercise 3.4.18. *Let* T_a *be some symmetric tent map with* $a \in (\sqrt{2}, 2]$*. Show that the system* $([c_2, c_1], T_a)$ *has no wandering intervals.* **HINT:** *Use the fact that* T_a *is leo for* $a \in [\sqrt{2}, 2]$*.*

Exercise 3.4.19. *Let* $f : I \to I$ *and* $g : J \to J$ *be unimodal maps with cores* I *and* J*, respectively. . Assume* f *is leo. Suppose* f *and* g *are topologically conjugate. Prove* g *is leo.* **HINT:** *Use Exercises 3.3.3 and 3.4.13.*

Exercise 3.4.20. ♣ *Let* f *be strictly unimodal. Prove* f *is leo if and only if* f^2 *has a dense orbit.* **HINT:** *See, for example, [31, Lemma 1.2].*

We briefly discuss renormalization; see [115] for more detail.

Definition 3.4.21. Let $f : [0,1] \to [0,1]$ be onto and continuous. An interval $J \subset [0,1]$ is called *restrictive*, or *periodic*, if $J \neq [0,1]$ and there exists $n \in \mathbb{N}$ with minimal such n greater than 1, such that $f^n(J) \subset J$.

Exercise 3.4.22. *Show that a unimodal map is not leo if it has a restrictive interval.*

Exercise 3.4.23. *Show that a unimodal map with an attracting periodic point has a restrictive interval.*

There exist unimodal maps with restrictive intervals and no attracting periodic points. One such example is the 2^∞ map investigated in Chapter 5. Of course, as we will see below, other such examples exist within the symmetric tent family.

Definition 3.4.24. We say a unimodal map f (with turning point c) is *renormalizable* provided there exists a restrictive interval $J \ni c$ and $n > 2$ such that $f^n(J) \subset J$ and $f^n|J$ is again a unimodal map. (We relax the definition of unimodal to allow for a map to be decreasing to the left of the turning point and increasing to the right.)

In Definition 3.4.21, the interval J is called a *restrictive interval* because the orbit of J traps orbits. Indeed, J is periodic (i.e., $f^n(J) \subset J$), and hence once the orbit of an x enters J, it remains in $J \cup f(J) \cdots \cup f^{n-1}(J)$. Thus, f will not have a point x with a dense orbit, with the notable exception of the map $f = T_{\sqrt{2}}$. This map is renormalizable of period 2, but the restrictive interval $J = [\sqrt{2}-1, 2-\sqrt{2}]$ and its image $T_{\sqrt{2}}(J) = [2-\sqrt{2}, \frac{1}{2}\sqrt{2}]$ are adjacent, and together they form the core of $T_{\sqrt{2}}$.

Figure 3.10 is an example of renormalization. Here $a = 1+\sqrt{5} \approx 3.2360679$ and $c = 1/2$ is periodic of period 2 under g_a. The map g_a is renormalizable with restrictive interval J and $n = 2$. In fact, the maps $g_a^2|J$ and $g_a^2|E$ are each topologically conjugate to g_2. Note that $c = 1/2$ is periodic of period 1 under g_2 and this is also the case for $g_a^2|E$ and $g_a^2|E$.

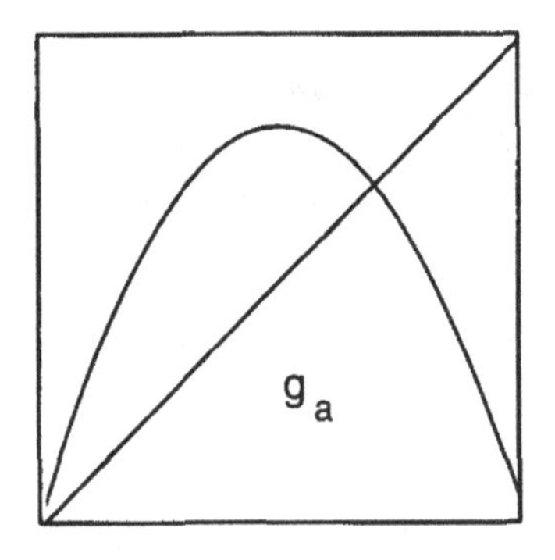

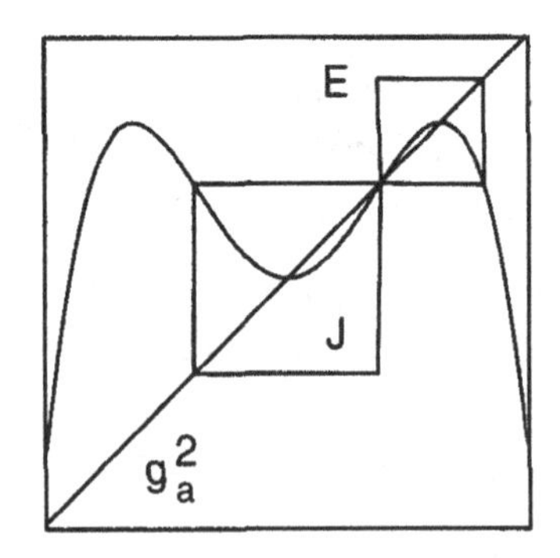

Figure 3.10: Once renormalizable map g_a

Remark 3.4.25. Let f be a renormalizable unimodal map with $n > 2$ and restrictive interval J (as in Definition 3.4.24); say $f^n|J$ is again unimodal.

Define $g := f^n|J$. Then g is again a unimodal map. If g is renormalizable, we call f *twice renormalizable.* Similarly we speak of *three times renormalizable* $\cdots$ *infinitely renormalizable* [115].

We close this section with a brief discussion of renormalizability in the symmetric tent family and the quadratic family. The next proposition precisely classifies the renormalization behavior for the symmetric tent family. Item (2) can be proved by induction. We suggest the reader work through Section 10.7 before attempting a proof. Also see [56, 162].

Proposition 3.4.26. *[56] Three observations.*

1. *Each symmetric tent map T_a with $a \in (1, \sqrt{2}]$ is renormalizable.*
2. *If $\sqrt{2} < a^m \leq 2$ for some $m \in \{2, 2^2, 2^3, \dots\}$, then the symmetric tent map T_a is m times renormalizable. No symmetric tent map is infinitely renormalizable.*
3. *A symmetric tent map T_a with $a \in (\sqrt{2}, 2]$ is not renormalizable.*

Many examples of infinitely renormalizable maps are found in the family $g_a(x) = ax(1 - x)$, $a \in [0, 4]$. One such example is investigated in Chapter 5. For this example, the ω-limit set for the turning point is a Cantor set. Indeed, if f is unimodal, infinitely renormalizable, and sufficiently smooth, then $\omega(c, f)$ is a Cantor set. There are no infinitely renormalizable maps within the symmetric tent family. However, we are again getting beyond the scope of the text; see [115].

The combinatorics associated with renormalization in the unimodal setting is completely understood and is discussed in Section 10.7. For interval maps with higher modality (i.e., two or more turning points), the combinatorics is not completely understood; see [33].

The reader should not get distracted or slowed by the concept of renormalization. Our purpose in introducing renormalization is to state Theorem 3.4.27. Often, we use symmetric tent maps with slopes $> \sqrt{2}$ (i.e., not renormalizable). Due to Theorem 3.4.27, we see that indeed results for such symmetric tent maps follow over for a broader class of unimodal maps (namely, those satisfying the hypotheses of Theorem 3.4.27). In addition, one can pull a result for a nonrenormalizable symmetric tent map over to the renormalizable case via Remark 3.3.7. Theorem 3.4.27 is a corollary of a much more general result given in Theorem 9.5.1.

Theorem 3.4.27. *[122, 115] Let f be a unimodal map that is not renormalizable, has no attracting periodic points, and has no wandering intervals. Then f is topologically conjugate to a symmetric tent map with slope in $(\sqrt{2}, 2]$.*

Consider the one-parameter family of unimodal maps given by $g_a(x) = ax(1-x)$, where $a \in [0,4]$. We saw in Remark 3.4.10 that for each a, the system $([c_2, c_1], g_a)$ has no wandering intervals. In fact, the set of parameters a for which

1. g_a is not renormalizable,

2. there are no wandering intervals, and

3. there are no attracting periodic points

is of positive Lebesgue measure. However, this set of parameters is not dense in any nondegenerate closed subinterval of $[0,4]$. Proofs of these facts are beyond the scope of the text. See [115] for details.

Let $\mathcal{T} \subset [0,4]$ be those parameters a such that g_a is topologically conjugate to some symmetric tent map. Theorem 3.4.27 and the above paragraph give that $\mathcal{T}$ has positive Lebesgue measure. Examples of parameters in $\mathcal{T}$ include all a's for which g_a is leo (on its core), as in this case g_a is topologically conjugate to a symmetric tent map.

One might then ask, given a symmetric tent map T_b, when is there a quadratic map g_a topologically conjugate to it? In fact, if T_b is not renormalizable and $c = 1/2$ is not periodic for T_b, then there is exactly one quadratic map g_a topologically conjugate to T_b. These results are discussed further in Chapter 10. In the periodic or renormalizable case we have factors but not conjugacies.

3.5 Recurrence

In this section we introduce recurrence, uniform recurrence, and persistent recurrence.

Definition 3.5.1. Let $f : E \to E$ be a continuous map of a compact metric space. Let $x \in E$. We say x is *recurrent* provided that, for every open set U containing x, there is some $m \in \mathbb{N}$ with $f^m(x) \in U$.

Exercise 3.5.2. *Let $f : E \to E$ be a continuous map of a compact metric space. Prove $x \in \omega(x, f)$ if and only if x is recurrent.*

Remark 3.5.3. Let $f : E \to E$ be a continuous map of a compact metric space and $x \in E$. Suppose that $\omega(x, f)$ is not finite and that x is recurrent. Then $\omega(x, f)$ is uncountable (recall Remark 1.1.44). If in addition $\omega(x, f)$ is totally disconnected, then (recall Proposition 1.1.52) $\omega(x, f)$ is a Cantor set or is a Cantor set modulo a countable set.

Exercise 3.5.4. *Let f and g be topologically conjugate and suppose that x is recurrent under f. Show that $h(x)$ is recurrent under g.*

Definition 3.5.5. Let $f : E \to E$ be a continuous map of a compact metric space. Let $x \in E$. We say x is *uniformly recurrent* provided that, for every open set U containing x, there exists $M \in \mathbb{N}$ such that $f^j(x) \in U$, $j \geq 0$, implies $f^{j+k}(x) \in U$ for some $0 < k \leq M$.

Exercise 3.5.6. *Let $f : I \to I$ be unimodal with turning point c. Assume c is recurrent and not uniformly recurrent. Prove there are arbitrarily small open intervals about c that allow arbitrarily large return times.* **HINT:** *Let $W \ni c$ be an open interval that allows arbitrarily large return times, that is, for every $M \in \mathbb{N}$ there exists $j \geq 0$ such that $f^j(c) \in W$ and $f^{j+k}(c) \notin W$ for all $0 < k \leq M$. Let V be an open interval with $c \in V \subset W$. Prove V also allows arbitrarily large return times. Suppose not, that is, suppose $\tilde{M} \in \mathbb{N}$ is such that, if $f^j(c) \in V$, then $f^{j+k}(c) \in V$ for some $0 < k \leq \tilde{M}$. As c is recurrent, there is a first time the forward orbit of c returns to V, that is, we have $\tilde{m} = \min\{j > 0 \mid f^j(c) \in V\}$. Set $M = \max\{\tilde{m}, \tilde{M}\}$. Again, as c is recurent, we have $n_1 < n_2 < n_3 < \ldots$ such that $f^j(c) \in V$ for $j > 0$ if and only if $j \in \{n_i \mid i \geq 1\}$. Then $|n_{j+1} - n_j| \leq M$ for all j. Conclude that $f^j(c) \in W$ implies $f^{j+k}(c) \in W$ for some $k \leq M$, a contradiction.*

Definition 3.5.7. Let $f : E \to E$ be a continuous map of a compact metric space. We say $F \subset E$ is *minimal* provided $F \neq \emptyset$, F is closed, $f(F) \subset F$ (i.e., F is *invariant*), and no proper subset of F has these three properties.

Lemma 3.5.8. *Let $f : E \to E$ be a continuous map of a compact metric space. Then a nonempty subset F of E is minimal if and only if $\omega(x) = F$ for all $x \in F$.*

Proof. If F is minimal, then $\omega(x) = F$ for all $x \in F$ since any $\omega(x)$ is nonempty, closed, and invariant (recall exercise 3.2.5).

Assume $\omega(x) = F$ for all $x \in F$. Suppose that $\emptyset \neq M \subset F$ is closed and invariant. Let $y \in M$. Then $F = \omega(y) \subset M$ implies that $F = M$. □

It follows from Lemma 3.5.8 and Exercise 3.2.5(3), that a minimal set is strongly invariant.

Exercise 3.5.9. ♣ *[18, page 92] Let $f : I \to I$ be a continuous map of a closed interval from $\mathbb{R}$. Let $F \subset I$ be infinite and minimal. Show F is a Cantor set.*

Exercise 3.5.10. ◊ *[15, 74] Let $f : E \to E$ be a continuous map of a compact metric space and $x \in E$. Suppose that $x \in \omega(x)$. Prove $\omega(x, f)$ is minimal if and only if x is uniformly recurrent.*

Recall the definition of a unimodal map and of $\hat{x}$ given in Definition 3.4.11.

Definition 3.5.11. [114] Let $f : I \to I$ be unimodal. We call a point $x \in I$ *nice* provided $f^n(x)$ is not in the open interval $\langle x; \hat{x} \rangle$ for all $n \geq 0$.

Lemma 3.5.12. *Let $f : I \to I$ be unimodal with turning point c. Suppose c is recurrent but not uniformly recurrent. Then there exists an open interval $U \ni c$ such that U allows arbitrarily large return times and such that the boundary of U consists of nice points.*

Proof. Since c is not uniformly recurrent, we may choose an open interval $W \ni c$ such that W allows arbitrarily large return times, say $W = (a, b)$. Thus, for every $M \in \mathbb{N}$ there exists $j \geq 0$ such that $f^j(c) \in W$ and $f^{j+k}(c) \notin W$ for all $0 < k \leq M$. Without loss of generality, we may assume that $f(a) = f(b)$, that is, we may assume that $W = (a, \hat{a})$ (recall Exercise 3.5.6).

Suppose for every $w \in [a, c]$ there exists $m_w > 0$ such that $f^{m_w}(w) \in W$. Then, by continuity of f, for each $w \in [a, c]$, let $U_w \ni w$ be such that $f^{m_w}(U_w) \subset W$. As $[a, c]$ is compact, there exists $w_1, \dots, w_k$ such that $\bigcup_{i=1}^{i=k} U_{w_i} \supseteq [a, c]$. Hence, $[a, c]$ does not allow arbitrarily large return times. The argument is similar on the other side of the turning point.

Thus, without loss of generality we may choose $w \in W$ with $c < w < b$ and such that $f^j(w) \notin W$ for all $j > 0$. Set $U = (\hat{w}, w)$. □

Definition 3.5.13. [42, 111] Let $f : I \to I$ be unimodal with turning point c. For $x \in I$, let $H_n(x) = (a, b) \ni x$ be the maximal interval containing x on which f^n is monotone. Set

$$r_n(x) = \min\{|f^n(x) - f^n(a)|,\ |f^n(x) - f^n(b)|\}.$$

We say the turning point c is *persistently recurrent* provided c is recurrent and $\lim_{n\to\infty} r_n(f(c)) = 0$. See Figure 3.11.

The notion of persistently recurrent first appears in [24] and then later in [111]. For further discussion on persistent recurrence, see [42].

Exercise 3.5.14. ♣ *[42, Section 3] Let $f : I \to I$ be unimodal with turning point c. Prove that c is persistently recurrent if and only if*

- *c is recurrent,*
- *it is not the case that there is an open set $U \ni c$, $n_i \to \infty$, and $V_i \ni c_1$ such that f^{n_i} maps V_i monotonically onto U for all i.*

Lemma 3.5.15 ([39, Lemma 3.4]). *Let $f : I \to I$ be unimodal with turning point c. If c is persistently recurrent, then c is uniformly recurrent.*

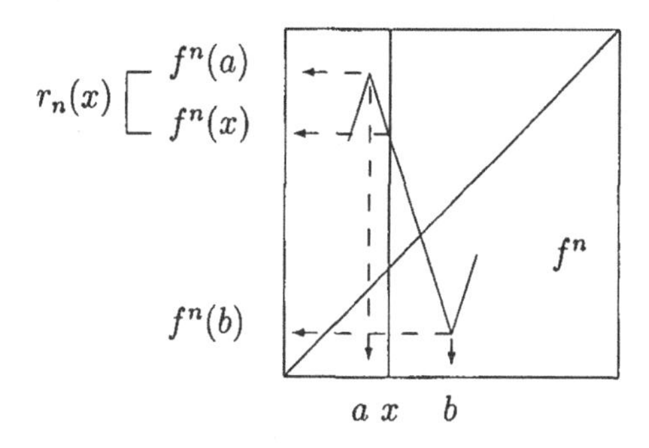

Figure 3.11: Construction of $r_n(x)$

Proof. Assume that c is persistently recurrent but not uniformly recurrent. Let U be an open interval such that $U \ni c$, U allows arbitrarily large return times and $f^n(\partial U) \notin U$ for all $n \geq 1$. For ease of notation, set $c_n = f^n(c)$ for $n \geq 1$.

For each $c_n \in U$, let $b_n = \min\{j \geq 1 \mid f^{n+j}(c) \in U\}$. We may choose for each $c_n \in U$ a smallest open interval $W_n \ni c_n$ such that $f^{b_n}(\partial W_n) \subset \partial U$ and $f^{b_n}(W_n) \subset U$ (use continuity of f, $c_{n+b_n} \in U$ and $f^i(\partial U) \notin U$ for all $i \geq 0$).

Claim 1: For each $c_n \in U$ with $c \notin W_n$, we have that $f^{b_n}|W_n$ is monotone.

Claim 2: For each c_n, $c_m \in U$, if $b_n \neq b_m$, then $W_n \cap W_m = \emptyset$.

From Claim 1 we have that $c_n \in U$ and $c \notin W_n$ imply $f^{b_n}(W_n) = U$. For such n, let $V_n \ni c_1$ be such that $f^{n-1}(V_n) = W_n$ (recall that $f^{n-1}(c_1) = c_n \in W_n$).

Claim 3: Fix n such that $c_n \in U$. Let $n' = \min\{j \leq n \mid c_j \in W_n\}$. If $n' < n$, then

- $b_{n'} = b_n$.
- $W_{n'} = W_n$.

Since U allows arbitrarily large return times, there exist infinitely many distinct sets W_n. By Claim 3 we may assume we have labeled these infinitely many distinct W_n's such that n is the smallest positive integer with $c_n \in W_n$. Hence, there are infinitely many distinct sets V_n each of which is mapped monotonically onto U by f^{n+b_n-1}. Lastly, as there exist arbitrarily large return times, we may assume that $\lim_{n\to\infty} n + b_n - 1 = \infty$, contradicting c being persistently recurrent. $\square$

Exercise 3.5.16. *Prove Claims 1–3 from the proof of Lemma 3.5.15.*

Exercise 3.5.17. ◇ *Provide an example of symmetric tent map where the turning point c is uniformly recurrent but not persistently recurrent.*

3.6 Shift Spaces

A finite set $\mathcal{A} = \{0, 1, \ldots, n-1\}$ is a metric space, where the distance between distinct points is taken to be 1 and the distance from a point to itself is 0. Let ρ denote this metric. Notice that each point is both open and closed and that $(\mathcal{A}, \rho)$ is a compact metric space. Set

$$X_n = \{\langle x_1, x_2, x_3, \ldots, \rangle \mid x_i \in \mathcal{A} \text{ for all } i \in \mathbb{N}\}$$

and define a metric on X_n by

$$d(x, y) = \sum_{i \geq 1} \frac{\rho(x_i, y_i)}{2^i}.$$

A few observations. First, note that $X_n = \mathcal{A}^{\mathbb{N}}$. Next, suppose $\{x^k\}_{k\geq 1}$ is a sequence in (X_n, d) converging to $y = \langle y_1, y_2, \ldots \rangle$. Say each x^k is of the form: $x^k = \langle x_1^k, x_2^k, x_3^k, \ldots \rangle$. Then, (due to the metric ρ) for each $j \in \mathbb{N}$, the sequence $\{x_j^k\}_{k\geq 1}$ is eventually constant at entry y_j (note that $y \in X_n$). Lastly, observe that (X_n, d) is a compact metric space; see Exercise 3.6.1.

Exercise 3.6.1. *Prove that (X_n, d) is a compact metric space.* **HINT:** *It is easy to check that d is indeed a metric. To check compactness, let $\{x^k\}_{k\geq 1}$ be a sequence from X_n. Thus each x^k is of the form: $x^k = \langle x_1^k, x_2^k, x_3^k, \ldots \rangle$. Then for each $j \in \mathbb{N}$, $\{x_j^k\}_{k\geq 1}$ is a sequence from the compact metric space $(\mathcal{A}, \rho)$. Choose $n_1^1 < n_2^1 < n_3^1 < n_4^1 < \ldots$ such that $\{x_1^{n_j^1}\}_{j\geq 1}$ converges, say to $y_1 \in \mathcal{A}$. Next, let $n_1^2 < n_2^2 < n_3^2 < n_4^2 < \ldots$ be a subsequence of $\{n_j^1\}_{j\geq 1}$, with $n_1^1 < n_1^2$, such that $\{x_2^{n_j^2}\}_{j\geq 1}$ converges, say to $y_2 \in \mathcal{A}$. Continue this process to obtain a sequence $\{n_1^j\}_{j\geq 1}$ and $y = \langle y_i \rangle \in \mathcal{A}^{\mathbb{N}}$ such that $\{x^{n_1^j}\}_{j\geq 1}$ converges to $y \in \mathcal{A}^{\mathbb{N}}$.*

Exercise 3.6.2. *Prove each of the following.*

1. *(X_n, d) is perfect.*

2. *(X_n, d) is totally disconnected.* **HINT:** *Suppose H is a nonempty connected subset of X_n. We want to show that H is a single point. For each $j \in \mathbb{N}$, let $\pi_j : X_n \to \mathcal{A}$ be the projection map given by $\pi_j(\langle x_1, x_2, \ldots \rangle) = x_j$. Then each π_j is continuous and hence $\pi_j(H)$ is a connected subset of $\mathcal{A}$ (recall Exercise 1.1.24). However, the only connected subsets of $\mathcal{A}$ are single points. Hence, for each j, $\pi_j(H)$ is a single point and therefore H is a single point.*

We now have that (X_n, d) is compact metric, perfect, and totally disconnected and hence is a Cantor set (Definition 1.1.39). Recall that every Cantor set is homeomorphic to every other Cantor set (Exercise 1.1.40), and hence (X_n, d) is homeomorphic to (X_m, d) for all $n, m \in \mathbb{N}$.

Let $\sigma : X_n \to X_n$ be the shift map, that is,

$$\sigma(\langle x_1, x_2, x_3, \dots \rangle) = \langle x_2, x_3, \dots \rangle.$$

The map σ is an onto homeomorphism. We will see in Chapter 9 that (X_n, σ) is not topologically conjugate to (X_m, σ) for $n \neq m$ as their topological entropies differ (Exercise 9.2.2 and Theorem 9.1.9). We refer to (X_n, d) as the *full n-shift* with *shift map* σ. We call $S \subset X_n$ *shift-invariant* provided $\sigma(S) \subset S$. Of course, X_n is shift-invariant. A set $S \subset X_n$ is called a *shift space* provided S is compact and shift-invariant. Recall (Exercise 1.1.28) that a subset of a compact metric space is compact if and only if it is closed. Hence, a shift space is a closed and shift-invariant subset of X_n (for some n). As noted above, X_n (with the metric d) is a Cantor set. It is not the case that every shift space is a Cantor set; take, for example, $S = \{\langle 0, 0, 0 ... \rangle\}$. However, every shift space is compact and totally disconnected, and hence one need check whether the space is perfect to determine if it is indeed Cantor.

Next, consider the finite set $\mathcal{A} = \{0, 1, 2, \dots, n-1\}$ as an alphabet and for $j \in \mathbb{N}$ set $\mathcal{A}^j = \{\langle x_1, x_2, \dots, x_j \rangle \mid x_i \in \mathcal{A} \text{ for all } i\}$. An element from $\cup_{j \geq 1} \mathcal{A}^j$ is referred to as a *word* over the alphabet $\mathcal{A}$. We say $x = \langle x_1, x_2, \dots \rangle \in X_n$ contains a word $w = w_1, w_2, \dots, w_m$ provided there exists i with

$$x_{i+1}, x_{i+2}, \dots, x_{i+m} = w_1, w_2, \dots, w_m.$$

Let $\mathcal{F}$ be a collection (not necessarily finite and possibly empty) of words over the alphabet $\mathcal{A}$. Define $X_{\mathcal{F}}$ to be the subset of elements from X_n that do *not* contain any words from $\mathcal{F}$. We view the words from $\mathcal{F}$ as forbidden. The next theorem gives a characterization of shift spaces as precisely the subsets of X_n (for some n) of the form $X_{\mathcal{F}}$.

Theorem 3.6.3 ([108, Theorem 6.1.21]). *A subset S of X_n is a shift space if and only if there is a collection of words $\mathcal{F}$ such that $S = X_{\mathcal{F}}$.*

Proof. For any choice of $\mathcal{F}$, the set $X_{\mathcal{F}}$ is closed and shift-invariant and hence is a shift space.

Assume S is a shift space. We need to find a collection of words $\mathcal{F}$ such that $S = X_{\mathcal{F}}$. If $S = X_n$, then, letting $\mathcal{F}$ be the empty collection, we have $S = X_{\mathcal{F}}$. Thus, assume $S \neq X_n$. We have that $X_n \setminus S$ is open. Hence, for each $y \in X_n \setminus S$ there is $k(y) \in \mathbb{N}$ such that, if $z = \langle z_1, z_2, \dots \rangle \in X_n$ and $z_1, z_2, \dots, z_{k(y)} = y_1, y_2, \dots, y_{k(y)}$, then $Z \in X_n \setminus S$. Set $\mathcal{F} = \{y_1, y_2, \dots, y_{k(y)} \mid y \in X_n \setminus S\}$. Then $S = X_{\mathcal{F}}$. □

Exercise 3.6.4. *1. Prove that, for any choice of $\mathcal{F}$, the set $X_{\mathcal{F}}$ is closed and shift-invariant and hence is a shift space.*

2. Suppose $S \subset X_n$ is a shift space and that $S \neq X_n$. Let $\mathcal{F}$ be as in the proof of Theorem 3.6.3. Prove that indeed we have $S = X_{\mathcal{F}}$.

Example 3.6.5. *Consider $X_2 = \{0,1\}^{\mathbb{N}}$. Let S be the collection of all sequences in X_2 with no two 1's next to each other. Thus, $S = X_{\mathcal{F}}$, where $\mathcal{F} = \{11\}$. This shift space is called the* golden mean shift *as its topological entropy (see Exercise 9.2.8) is precisely* $\log(\frac{1+\sqrt{5}}{2})$. *This shift space is also called the* Fibonacci shift.

Exercise 3.6.6. *Let $\mathcal{F}_1$ and $\mathcal{F}_2$ be collections of words over $\mathcal{A}$. Prove that $X_{\mathcal{F}_1} \cap X_{\mathcal{F}_2} = X_{\mathcal{F}_2 \cup \mathcal{F}_2}$; hence the intersection of two shift spaces over the same alphabet is again a shift space. What about unions?*

Exercise 3.6.7. *Let $S \subset X_n$ be a shift space and $M \in \mathbb{N}$. Prove there is a collection of words $\mathcal{F}$ all of which have length at least M, with $X_{\mathcal{F}} = S$.*

In the event that the collection of forbidden words, $\mathcal{F}$, is a finite collection, we refer to $X_{\mathcal{F}}$ as a *shift of finite type*, denoted SFT. There is a natural way to represent a shift of finite type via a *transition matrix.*

Let B be a square $\{0,1\}$ matrix (i.e., entries from $\{0,1\}$) with its rows and columns indexed by $\{0,1,\dots,n-1\}$; thus B is a $n \times n$ matrix. The entry of B in row i and column j is denoted by $B_{i,j}$. Set

$$X_B = \{x = \langle x_1, x_2, \dots \rangle \in X_n \mid B_{x_i, x_{i+1}} = 1 \text{ for all } i \in \mathbb{N}\}.$$

Exercise 3.6.8. *Prove that X_B is an SFT. We call X_B the* shift of finite type defined by the matrix *B and refer to the matrix B as the* transition matrix.

Associated to the SFT X_B is a *transition graph* illustrating the allowed transitions from one symbol (i.e., a letter from the finite alphabet $\{0,1,\dots,n-1\}$) to another. Thus X_B has associated to it a transition matrix B and a transition graph G_B. Figure 3.12 gives three examples (for matrices A, B, and C). Note that B^l counts the numbers of paths in G_B of length l, that is, between vertices i and j there are exactly $B^l_{i,j}$ paths in G_B of length l.

We have just described how to form a shift of finite type from a square $\{0,1\}$ matrix. On the other hand, given a shift of finite type $X_{\mathcal{F}}$, there is a natural way to define a square $\{0,1\}$ matrix B such that $X_{\mathcal{F}} = X_B$. We leave this to the next exercise.

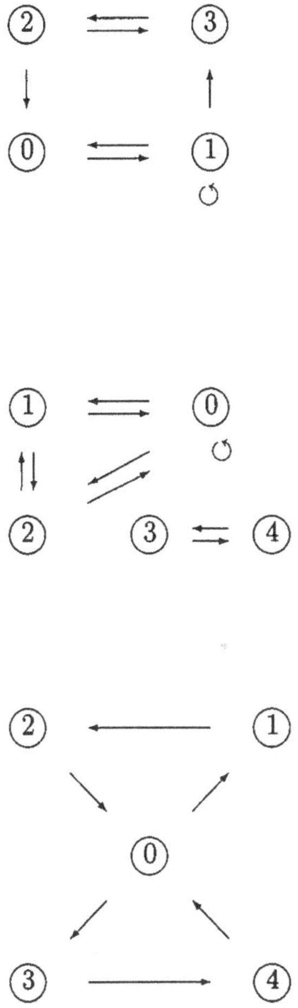

$$A = \begin{pmatrix} 0 & 1 & 0 & 0 \\ 1 & 1 & 0 & 1 \\ 1 & 0 & 0 & 1 \\ 0 & 0 & 1 & 0 \end{pmatrix}$$

$$B = \begin{pmatrix} 1 & 1 & 1 & 0 & 0 \\ 1 & 0 & 1 & 0 & 0 \\ 1 & 1 & 0 & 0 & 0 \\ 0 & 0 & 0 & 0 & 1 \\ 0 & 0 & 0 & 1 & 0 \end{pmatrix}$$

$$C = \begin{pmatrix} 0 & 1 & 0 & 1 & 0 \\ 0 & 0 & 1 & 0 & 0 \\ 1 & 0 & 0 & 0 & 0 \\ 0 & 0 & 0 & 0 & 1 \\ 1 & 0 & 0 & 0 & 0 \end{pmatrix}$$

Figure 3.12: Some transition graphs with transition matrices

Exercise 3.6.9. *Let $X_{\mathcal{F}}$ be a shift of finite type. Define a square $\{0,1\}$ matrix B such that $X_{\mathcal{F}} = X_B$. We refer to the matrix B as the transition matrix for the shift of finite type.* **HINT:** *Let $\mathcal{A}$ be the alphabet. First, one may assme that words in $\mathcal{F}$ have the same length, say the length is M [108, Proposition 2.1.7]. Let G be a graph with vertices labeled by words from $\mathcal{A}^M \setminus \mathcal{F}$. The forbidden words of $\mathcal{F}$ dictate the allowed edges in the graph G. The matrix B is precisely the transition matrix for the graph G. An example is provided in Figure 3.13 for $\mathcal{F} = \{11\}$. The path v_3, v_1, v_2, v_3, v_1 in the graph of Figure 3.13 corresponds to an element from $X_{\mathcal{F}}$ beginning with* 01001.

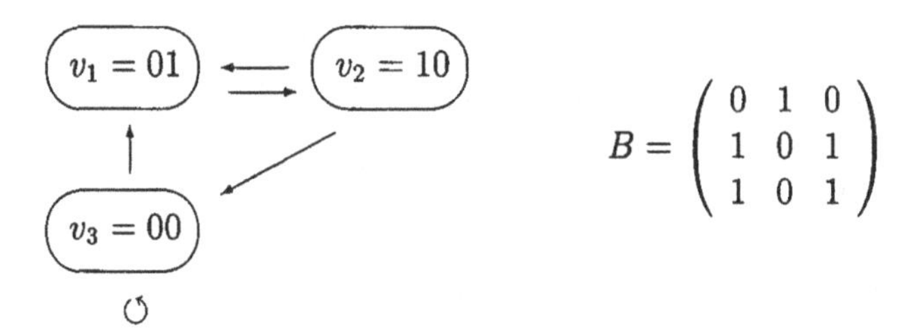

Figure 3.13: Transition matrix B for $X_{\mathcal{F}=\{11\}}$

Definition 3.6.10. We call a $\{0, 1\}$ matrix A, with rows and columns indexed by $\{0, 1, \ldots, n-1\}$, *irreducible* provided that, for each $0 \leq i, j \leq n-1$, there exists $l \geq 1$ such that $A^l_{i,j} > 0$. More strongly, the matrix is called *primitive* if there exists $l_0 \geq 1$ such that $A^l_{i,j} > 0$ for all $1 \leq i, j \leq n$ and $l \geq l_0$.

Similarly, we have irreducibel/primitive transition graphs; that is, if matrix $A = A_G$ belongs to the transition graph $G = G_A$ and if A is irreducible/primitive, then we call the graph G irreducible/primitive.

In the event that a square $\{0, 1\}$ matrix A is irreducible (primitive), we call the associated shift of finite type X_A irreducible (primitive).

If B is irreducible, then for every pair of vertices i and j in $G = G_B$ there is a path in G from i to j; if B is primitive, there are paths between any two vertices of G, of any length $l \geq l_0$. In Figure 3.12, A is primitive, B is not irreducible, and C is irreducible but not primitive. We say C has "period" 3, because every loop in the graph has a multiple of 3 as its length.

Exercise 3.6.11. *If the transition graph $G = G_A$ is disconnected, show that A_G is not irreducible.*

Exercise 3.6.12. *Let $\pi : \{0, 1, \ldots, n-1\} \to \{0, 1, \ldots, n-1\}$ be a permutation. You can associate a graph G and transition matrix A_G to π. When is A_G irreducible? When is it primitive?*

Exercise 3.6.13. *Suppose G is an irreducible transition graph. Suppose also that some vertex i has two loops of lengths n and m. If the greatest common divisor of n and m is 1, show that G is primitive.*

Let $S \subset X_n$ be a shift space and $x \in S$. We call x *periodic* provided $\sigma^j(x) = x$ for some j; the minimal $j > 0$ with $\sigma^j(x) = x$ is referred to as the *period*. A point $x \in S$ is called *transitive* provided $\overline{\{\sigma^j(x)\}}_{j \geq 0} = S$, that is, provided the forward orbit of x under the shift map is dense in S.

Exercise 3.6.14. *Let A be an irreducible $\{0,1\}$ matrix, with rows and columns indexed by $\{0,1,\ldots,n-1\}$. Prove each of the following.*

1. *X_A is either a finite set (consisting of a periodic orbit) or is a Cantor set.*
2. *The periodic points are dense in X_A.*
3. *The transitive points are dense in X_A and are a G_δ set (i.e., countable intersection of open sets).*

Chapter 4

Beginning Measurable Dynamics

We frequently consider the long-term behavior of a dynamical system on a large set of orbits without knowing the behavior of *every* orbit. In particular, Poincaré proved a type of recurrence for all orbits *except for those lying in a set of measure 0* for dynamical systems. This result is presented below and contrasts with the recurrence definitions and results of Section 3.5. This is referred to as the qualitative theory of dynamics since frequently, even though one can predict what the orbits will do on a set of full measure, it is not known precisely what will happen at even a single point! We introduce the ideas behind measurable dynamics and ergodic transformations in this chapter but refer to texts on ergodic theory such as [80, 137, 168].

The reader should be familiar with the material in Chapter 2 before beginning this chapter. The notion of a *measurable isomorphism*, introduced in Section 4.2, is used in Chapter 12; otherwise, this chapter is not used elsewhere in the text.

4.1 Preliminaries

We restrict our setting to the examples of interest in this book. As before, $f : I \to I$ denotes a map of a compact interval, that is, $I = [a, b]$. We assume in addition that I is endowed with the Lebesgue measure structure restricted to I, so we denote the measurable sets as $\mathcal{L}_I$ and normalized Lebesgue measure as m_I; that is, if $I = [a, b]$, then $m(I) = b - a$ and $m_I(A) = m(A)/(b - a)$ for all $A \in \mathcal{L}_I$. If no ambiguity arises, we will drop the subscript and write $(I, \mathcal{L}, m)$ when we want to emphasize the Lebesgue measurable structure of I. We also have the usual metric $\rho(x, y) = |x - y|$ on

I; the connection between the metric structure and the Lebesgue measure structure is that, for any points $x \leq y \in I$, $\rho(x,y) = m([x,y])$.

4.2 Measurable Maps on I

While many of the transformations $f : I \to I$ under consideration are continuous, we only require a weaker condition in keeping with the philosophy that we do not insist that the behavior of f be nice at every point.

Definition 4.2.1. The map $f : (I, \mathcal{L}_I, m_I) \to (I, \mathcal{L}_I, m_I)$ is *a (Borel) measurable transformation of I* if, for every open set C, $f^{-1}C = \{x \in I : f(x) \in C\} \in \mathcal{L}$. A function $g : (I, \mathcal{L}_I, m_I) \to (\mathbb{R}, m)$ is *a (Borel) measurable function on I* if, for every open set C, $g^{-1}C \in \mathcal{L}_I$.

Definition 4.2.2. Let $\mathcal{B}$ denote the smallest collection of sets in $\mathbb{R}$ that includes all open sets and is closed under complementation and countable unions. That is, if U is open, then $U \in \mathcal{B}$; if $A \in \mathcal{B}$, then $\mathbb{R} \setminus A \in \mathcal{B}$; and if $A_1, \ldots, A_n, \ldots \in \mathcal{B}$, then $\cup_{n=1}^{\infty} A_j \in \mathcal{B}$. The sets in $\mathcal{B}$ are called the *Borel sets* of $\mathbb{R}$. In an analogous fashion we define the Borel sets of any interval I and denote them $\mathcal{B}$ as well.

Exercise 4.2.3.

1. *Show that every continuous function on I is Borel measurable.*

2. *Prove that f is Borel measurable if and only if, for every $B \in \mathcal{B}$, $f^{-1}B \in \mathcal{L}$.*

3. *Prove that we have the following relations between measurable sets on $\mathbb{R}$: $\mathcal{B} \subseteq \mathcal{L}$, that is, all Borel sets are Lebesgue measurable (this is easy); however, $\mathcal{L} \not\subseteq \mathcal{B}$, that is, there exist sets in $\mathcal{L}$ not in $\mathcal{B}$.* **HINT:** *An outline of this is as follows: Consider first the* Cantor function f, *a continuous function that maps the Middle Third Cantor set K of measure 0 onto $[0,1]$. Now define the function $g(x) = f(x) + x$ and show it is a homeomorphism from $[0,1]$ onto $[0,2]$ and that $m(g(K)) = 1$. Use the nonmeasurable sets S and S_r constructed in Proposition 2.2.1 to show that if $B \subset [0,1]$ is any Lebesgue set of positive measure, then $B = \cup_{r \in \mathbb{Q}}(B \cap S_r)$. Hence B contains a Lebesgue nonmeasurable set. Find a Lebesgue nonmeasurable subset $A \subset g(K)$; then $B = g^{-1}A$ is Lebesgue but not Borel measurable.*

Since we are only interested in what happens on sets of positive measure, even the notion of invertibility is defined up to sets of measure 0. However, we need to dispense with some annoying formalities regarding measurable sets. We are always interested in Lebesgue measure. We have proved that Lebesgue measure is a complete measure; furthermore, it is well known that if we start with the collection of Borel sets on $\mathbb{R}$ and complete it with respect to Lebesgue measure (which means we add to $\mathcal{B}$ all subsets of sets of Lebesgue measure 0 and nothing more), then we get $\mathcal{L}$ back. Due to the asymmetry in the definition of a measurable transformation, henceforth we will use Borel sets as our collection of measurable sets unless the extra measure 0 sets are necessary. Furthermore, no matter what interval we are using, we will denote the collection of Borel sets by $\mathcal{B}$.

Definition 4.2.4. A measurable transformation $f : (I, \mathcal{B}, m) \to (I, \mathcal{B}, m)$ is *invertible* if there exists a measurable set $I' \subseteq I$, $m(I') = m(I)$, and $f(I') = f^{-1}(I') = I'$ such that the restriction of f to I' is one-to-one and onto. We say that f is *n-to-one* if there exists a set I' as above and each point $x \in I'$ has precisely n preimages. The transformation f is *bounded-to-one* if there exists a set I' as above and each point $x \in I'$ has precisely $n(x)$ preimages, and $n(x) \leq M$ for some integer M and every $x \in I'$.

Example 4.2.5. *The map $f(x) = x^2$ on $\mathbb{R}$ is two-to-one with respect to m. We can ignore the fact that the point $x = 0$ only has one preimage, since it is a set of measure* 0.

Recall that the notion of topological conjugacy was defined in Section 3.3 to show when two continuous maps f and g are dynamically the same. There is a measure theoretic analogue that is sometimes useful (see for example Section 12.4). It is important to realize that, even though continuous maps $f : I_1 \to I_1$ and $g : I_2 \to I_2$ are measurable, the notion of measurable isomorphism neither implies nor is implied by topological conjugacy.

Definition 4.2.6. Assume $f : (I_1, \mathcal{B}, m_1) \to (I_1, \mathcal{B}, m_1)$ and $g : (I_2, \mathcal{B}, m_2) \to (I_2, \mathcal{B}, m_2)$ are measurable transformations. We say f and g are *(measure theoretically) isomorphic* if there exists a measurable invertible (in the measurable sense) map $h : (I_1, \mathcal{B}, m_1) \to (I_2, \mathcal{B}, m_2)$ such that $h \circ f = g \circ h$ m-almost everywhere and $m_{I_2}(B) = m_{I_1}(h^{-1}(B))$ for every $B \in \mathcal{B}$.

Exercise 4.2.7. *Find f and g with $h \circ f = g \circ h$ and such that*

1. *h is an isomorphism but not a topological conjugacy.* **HINT:** *Isomorphisms need not be continuous.*

2. *h is a topological conjugacy but not an isomorphism.* **HINT:** *Continuous maps need not preserve measure.*

From here on (in this chapter), we consider only measurable transformations $f : I \to I$ and use the notation $\mathcal{L}$ and $\mathcal{B}$ to denote the Lebesgue and Borel sets, respectively (in either I and $\mathbb{R}$). Many of the measurable transformations we consider are also measure preserving, that is, the inverse image of any set in $\mathcal{B}$ under f has the same measure as the original set. We now give the formal definition.

Definition 4.2.8. Let I denote an interval, bounded or unbounded, and consider a measurable transformation $f : I \to I$. For $A \in \mathcal{B}$ set $f^{-1}A = \{x : f(x) \in A\}$. The transformation $f : I \to I$ is called *measure preserving* provided that, for each $A \in \mathcal{B}$, we have $m(f^{-1}A) = m(A)$.

Using Definition 4.2.8 to determine whether a given tranformation f is measure preserving could be difficult. The next theorem provides an elegant way to determine whether a transformation is measure preserving. The proof is not very difficult and hence is left as an exercise.

Theorem 4.2.9. *Suppose that $f : (I, \mathcal{B}, m_I) \to (I, \mathcal{B}, m_I)$ is a transformation. Consider the collection $\mathcal{I}$ of all subintervals contained in I. If for every $A \in \mathcal{I}$ we have $f^{-1}A \in \mathcal{B}$ and $m(f^{-1}(A)) = m(A)$, then f is a measure preserving transformation.*

Definition 4.2.10. An arbitrary Borel probability measure on $(I, \mathcal{B})$ is a function

$$\nu : \mathcal{B} \to [0, 1]$$

such that the following hold:

1. $\nu(\emptyset) = 0$ and $\nu(I) = 1$.

2. If $\{A_j\}_{j \in N}$ is a disjoint collection of sets from $\mathcal{B}$, then $m(\cup_{j=1}^{\infty} A_j) = \sum_{j=1}^{\infty} m(A_j)$.

As described in Section 2.4, we can always complete ν by adding subsets of measure 0 sets. Hence, we assume throughout that all measures are complete Borel probability measures. We say a measure ν is *absolutely continuous* with respect to m on $\mathcal{L}$ and write $\nu \ll m$ if all sets of measure 0 for m have measure 0 for ν, that is, $m(A) = 0 \Rightarrow \nu(A) = 0$. Similarly, we say that ν and μ are *equivalent measures* and write $\nu \sim m$ if $\nu \ll m$ and $m \ll \nu$.

Exercise 4.2.11. *Prove the following:*

1. *$f(x) = 2x$ on $\mathbb{R}$ is not measure preserving.*

2. *$f(x) = 2x \pmod 1$ on $[0, 1)$ is measure preserving.*

3. $g_b(x) = x + b$ *is measure preserving on* $\mathbb{R}$ *for any* $b \in \mathbb{R}$. **HINT:** *This property was built into the construction of Lebesgue measure.*

4. $f_b(x) = x + b \pmod 1$ *on* $[0, 1)$ *is measure preserving.* **HINT:** *This follows immediately from the previous example. Note, however, that* f_b *preserves a finite measure* m_I *on* $I = [0, 1)$, *while* g_b *preserves the infinite measure* m.

4.3 Poincaré Recurrence

In Section 3.5 we discussed recurrence, topological recurrence, and persistent recurrence. Here we discuss the measure theoretic analog of recurrence that was introduced by Poincaré in 1912. It is a property shared by all finite measure preserving transformations and, as we shall see, by many other measurable transformations as well. We assume that we are in the setting outlined above.

Definition 4.3.1. If f is a measurable transformation on I and A is any measurable subset of I, a point $x \in A$ is called *Poincaré recurrent* with respect to f and A if $f^n x \in A$ for at least one positive integer n. The transformation f is called *Poincaré recurrent* if, for every set $A \in \mathcal{B}$, m-almost every point is recurrent.

Theorem 4.3.2 (Poincaré recurrence theorem). *If* f *is any measure preserving transformation of* $(I, \mathcal{B}, m)$ *(not necessarily invertible) and* $m(I) < \infty$, *and if* $A \in \mathcal{B}$, *then* m*-almost every point of* A *is Poincaré recurrent. That is,* f *is Poincaré recurrent.*

Proof. Let $B \subset A$ denote the set of points that never returns to A and suppose $m(B) > 0$. We note that B is measurable, since

$$B = A \cap f^{-1}(I \setminus A) \cap f^{-2}(I \setminus A) \cap \cdots.$$

If $x \in B$, then B does not intersect $f(x), f^2(x), \ldots, f^j(x)$; this means that B is disjoint from $f^{-n}B$ for all positive n.

Furthermore, the sets $B, f^{-1}B, \ldots, f^{-j}B, \ldots$ are all pairwise disjoint, since

$$f^{-i}B \cap f^{-(i+j)}B = f^{-i}(B \cap f^{-j}B).$$

Since f is measure preserving, and $m(I) = 1$, this is a contradiction. Therefore, no such set B exists. □

Actually, a stronger statement can be made, namely, that almost every point in each A returns to A infinitely many times.

Corollary 4.3.3. *If f is any measure preserving transformation of $(I, \mathcal{B}, m)$ (not necessarily invertible) and $m(I) < \infty$, and if $A \in \mathcal{B}$, then m-almost every point of A returns to A infinitely often. We call this property the infinite Poincaré recurrence.*

Proof. Given a set A of positive measure, we apply Theorem 4.3.2 and let C_1 be the set of points that return to A eventually under some iterate of f. Similarly, for each integer $j \geq 1$, f^j is measure preserving; so, applying Theorem 4.3.2 to f^j, let C_j be the set of points that return to A under some iterate of f^j. Then $m(C_j) = m(A)$, and we consider $C = \cap_{j=1}^{\infty} C_j \subset A$; clearly $m(C) = m(A)$. It is easy to see that any $x \in C$ must return to A infinitely often, otherwise there would be a "last visit" and say $f^n(x) \in A$, but no higher iterate is in A, which is a contradiction. □

Example 4.3.4. *We remark that Poincaré recurrence fails if $m(I) = \infty$. An easy example is obtained by setting $I = \mathbb{R}$ and $f(x) = x + 1$. Then, if $A = (0, 1)$, the sets $f^{-n}A = (-n, -n + 1)$ are all pairwise disjoint.*

However, we can weaken the hypotheses on the recurrence theorem 4.3.2 so that the finite measure need not be preserved. We give some properties similar to recurrence and prove that they are in fact equivalent.

Definition 4.3.5. Assume f is a measurable transformation of $(I, \mathcal{B}, m)$ (not necessarily invertible) and $m(I) < \infty$.

1. f is *nonsingular* if, for every measurable set A, $m(A) = 0 \Leftrightarrow m(f^{-1}A) = 0$.

   Assume now that f is nonsingular.

2. f is *conservative* if, for every A of positive measure,

$$m(A \cap f^{-n}A) > 0 \quad \text{for some} \quad n.$$

   If f is not conservative, then there exists a set W of positive measure such that $\{f^{-k}W\}_{k \in \mathbb{N}}$ are disjoint. W is called a *wandering set* .

3. f is *incompressible* if $f^{-1}B \subset B \Rightarrow m(B \setminus f^{-1}B) = 0$.

The next proposition is a classical result that can be found for example in [103]. The proof of Proposition 4.3.6 is fairly straightforward, and hence we give an outline of it and leave any missing details as an exercise.

Proposition 4.3.6. *Let f be a nonsingular transformation of $(I, \mathcal{B}, m)$ and $m(I) < \infty$. Then the following are equivalent.*

1. *f is conservative.*

2. *f is Poincaré recurrent.*

3. *f is infinitely Poincaré recurrent.*

4. *f is incompressible.*

Proof. $1 \Rightarrow 2$ is obvious.

$2 \Rightarrow 3$ is proved in Corollary 4.3.3.

$3 \Rightarrow 2$ is trivial.

$2 \Rightarrow 4$: If $f^{-1}B \subseteq B$, then

$$\cdots \subseteq f^{-2}B \subseteq f^{-1}B \subseteq B,$$

so we see that

$$\begin{aligned} B^* &= B \setminus f^{-1}B = B \setminus \cup_{k \in N} f^{-k}B \\ &= \{x \in B \mid x \text{ never returns to } B\}. \end{aligned}$$

However, recurrence implies that $m(B^*) = 0$, so 4 is proved.

$4 \Rightarrow 1$: Suppose W is a wandering set. If we define $B = \cup_{k=0}^{\infty} f^{-k}W$, then we have that $B \subseteq f^{-1}B$, and we can write $W = B \setminus f^{-1}B$. By (4), $m(W) = 0$. □

4.4 Ergodicity

Suppose that $f : (I, \mathcal{B}, m) \to (I, \mathcal{B}, m)$ is a nonsingular transformation. All finite measure preserving maps are automatically conservative and nonsingular. Any C^1 map of I can be shown to be nonsingular, but not all smooth maps are conservative. An example of a unimodal map of the interval that is C^∞ but not conservative is given by the 2^∞ map and is studied in Chapter 5. It is known to be ergodic with respect to m.

An ergodic transformation is an indecomposable one from the measure theoretic point of view. Unlike topological notions of irreducibility, the property of f being ergodic depends on the measure. Here is the precise definition.

Definition 4.4.1. If f is a nonsingular transformation and $m(I) = 1$, then f is *ergodic* if the only sets $B \in \mathcal{B}$ with $f^{-1}B = B$ satisfy $m(B) = 0$ or $m(B) = 1$. (If $m(I) = \infty$, the definition is the same, except we say that $m(B) = 0$ or $m(I \setminus B) = 0$).

We remark that changing to an equivalent measure does not change the property of being ergodic. However, if f is ergodic with respect to m and ν is another measure on the same collection of sets that is not absolutely continuous, then nothing about ergodicity with respect to ν is known. Therefore, when the dynamical system f is clearly understood, we sometimes say m is ergodic (instead of f). There are many equivalent characterizations of ergodicity. The property of being ergodic does not depend on invertibility or conservativity, but we state the equivalent properties for a finite measure preserving transformation. The proofs of these equivalences appear in almost every book on ergodic theory and are useful exercises.

Theorem 4.4.2. *If $f : (I, \mathcal{B}, m) \to (I, \mathcal{B}, m)$ is a measure preserving transformation with $m(I) = 1$, then the following are equivalent:*

1. *f is ergodic.*
2. *If $B \in \mathcal{B}$ with $m(B \,\triangle\, f^{-1}B) = 0$, then $m(B) = 0$ or $m(B) = 1$.*
3. *For every measurable set A of positive measure, $m(\bigcup_{n=1}^{\infty} f^{-n}A) = 1$.*
4. *For every $A, B \in \mathcal{B}$, $m(A) > 0$, $m(B) > 0$, there exists $n > 0$ such that $m(f^{-n}A \cap B) > 0$.*
5. *For every measurable set A of positive measure and any positive integer N, $m(\cup_{n=N}^{\infty} f^{-n}A) = 1$.*
6. *For every $A, B \in \mathcal{B}$, $m(A) > 0$, $m(B) > 0$, and any positive integer N, there exists $n > N$ such that $m(f^{-n}A \cap B) > 0$.*
7. *For every $A, B \in \mathcal{B}$,*
$$\frac{1}{n}\sum_{i=0}^{n-1} m(f^{-i}A \cap B) \to m(A)m(B).$$

These are the set theoretic properties that say that, for an ergodic measure preserving transformation, every set of positive measure has a measure theoretically dense orbit. We now continue the equivalent statements, turning to measurable functions.

8. *If φ is a measurable function and $\varphi \circ f(x) = f(x)$ for every $x \in I$, then f is constant m-a.e.*

9. *If φ is a measurable function and $\varphi \circ f(x) = f(x)$ for m-a.e. $x \in I$, then f is constant m-a.e.*

Exercise 4.4.3. *Prove as many implications in the theorem above as possible using the definitions. Check your solutions by looking in an ergodic theory text.*

4.4.1 Integration of Measurable Functions

As mentioned in Chapter 2, one of Lebesgue's primary goals in defining Lebesgue measure was to enlarge the class of functions that can be integrated. We give a brief overview of Lebesgue integration and refer to a measure theory text such as Folland [70] for the complete picture. We first describe how to integrate nonnegative-valued measurable functions. Then we outline the minor adjustments needed to pass to complex-valued measurable functions. We start by defining the basic functions that can be integrated.

Definition 4.4.4. Let A be a measurable subset of I. We define the *indicator function* or *characteristic function*

$$\chi_A(x) = \begin{cases} 1 & \text{if } x \in A, \\ 0 & \text{if } x \notin A. \end{cases}$$

Since A is a measurable set, χ_A is a measurable function.

Indicator functions and the simple functions made from them replace the rectangular regions used to obtain a Riemann integral.

Definition 4.4.5. A function $\varphi : I \to \mathbb{R}$ is a *simple function* if it is of the form: $\varphi(x) = \sum_{i=1}^n a_i \chi_{A_i}(x)$, with $a_i \in \mathbb{R}, A_i \in \mathcal{B}$, and the sets A_i are disjoint. We define the integral of a simple function φ to be

$$\int_I \varphi dm = \sum_{i=1}^n a_i m(A_i).$$

The integral is independent of the choice of A_i's. This is the basic building block for Lebesgue integration. If φ is a measurable function such that $\varphi \geq 0$, then we have the following result.

Theorem 4.4.6. *If $\varphi : I \to [0, \infty]$ is a measurable function, there exists a sequence of simple functions, $\{\varphi_n\} \to \varphi$ pointwise, with $0 \leq \varphi_1 \leq \ldots \leq \varphi_n \leq \ldots \leq \varphi$.*

Proof. For every $n \in \mathbb{N}$ and $0 \leq k \leq 2^{2n} - 1$, we define

$$A_n^k = \varphi^{-1}((k2^{-n}, (k+1)2^{-n}]) \quad \text{and} \quad B_n = \varphi^{-1}((2^n, \infty])$$

and

$$\varphi_n = \sum_{k=0}^{2^{2n}-1} k2^{-n}\chi_{A_n^k} + 2^n\chi_{B_n}.$$

We leave it as an exercise to show that this sequence of φ_n's satisfies the conclusion of the theorem. □

We now extend the definition of the Lebesgue integral to all measurable nonnegative functions using the theorem.

Definition 4.4.7. If $\varphi : I \to [0, \infty]$ is a measurable function, then we define the *Lebesgue integral* of φ to be

$$\int_I \varphi dm = \sup\{\int_I \psi dm \mid 0 \leq \psi \leq \varphi \text{ and } \psi \quad \text{simple}\}.$$

We remark that the integral could be infinite. Extending this definition to a real-valued measurable function φ is straightforward; we define the positive and negative parts of φ by $\varphi^+(x) = \max(\varphi(x), 0)$ and $\varphi^-(x) = \max(-\varphi(x), 0)$. Next we integrate the positive and negative parts separately. Then, assuming at least one of those two integrals is finite, we define

$$\int_I \varphi dm = \int_I \varphi^+ dm - \int_I \varphi^- dm.$$

We say that φ is *integrable* if $\int_I \varphi^+ dm, \int_I \varphi^- dm < \infty$; φ is integrable if and only if $|\varphi|$ is integrable. If $\varphi : I \to \mathbb{C}$ is a measurable function, we say φ is integrable if, when we write $\varphi = \varphi_1 + i\varphi_2$, the real-valued functions φ_1 and φ_2 are integrable. In this case,

$$\int_I \varphi dm = \int_I \varphi_1 dm + i \int_I \varphi_2 dm.$$

For integrable functions g and h, if $g = h$ m-a.e., then

$$\int_I g dm = \int_I h dm.$$

There are three fundamental theorems that allow us to compute Lebesgue integrals. We refer to a measure theory book such as [70] for proofs of these results. Although we do not make use of these three theorems, we include them to perhaps motivate the reader to further investigate measure theory.

Theorem 4.4.8 (Monotone convergence theorem). *If $\{g_n\}$ is a sequence of nonnegative measurable functions on I such that $g_n \leq g_{n+1}$ for all n, and $g(x) = \lim_{n\to\infty} g_n$, then*

$$\int_I g dm = \lim_{n\to\infty} \int_I g_n dm.$$

Lemma 4.4.9 (Fatou's lemma). *If $\{g_n\}$ is a sequence of nonnegative measurable functions on I, then*

$$\int_I (\liminf g_n) dm \leq \liminf \int_I g_n dm.$$

Theorem 4.4.10 (Dominated convergence theorem). *If $\varphi : I \to \mathbb{R}$ is nonnegative and integrable, and if $\{g_n\}$ is a sequence of measurable functions on I with $|g_n| \leq \varphi$ m-a.e. for each n, and if $g(x) = \lim_{n\to\infty} g_n$ a.e., then g is integrable and*

$$\lim_{n\to\infty} \int_I g_n dm = \int_I g dm.$$

We denote by $L^1(I, \mathcal{B}, m)$ or $L^1(I, m)$ the space of all integrable complex-valued functions on I, where two such functions are identified if they are equal a.e.

Exercise 4.4.11.

1. *If g is a bounded real-valued function on $I = [a, b]$ that is Riemann integrable, then g is Lebesgue measurable and integrable and*

$$\int_a^b g(x)dx = \int_I g dm.$$

2. *Consider the function $\chi_{\mathbb{Q}} : [0, 1] \to \mathbb{R}$, the indicator function of the rational numbers on $[0, 1]$. Show that $\chi_{\mathbb{Q}}$ is not Riemann integrable, but it is Lebesgue integrable and*

$$\int_I \chi_{\mathbb{Q}} dm = 0.$$

4.4.2 Averaging Measurable Functions Along Orbits

The Birkhoff Ergodic Theorem, proven in 1931, is one of the earliest results in ergodic theory. In order to understand this theorem, it is helpful to have some understanding of integrating measurable functions as outlined in the previous section. However, it is acceptable to think about the usual Riemann integral with dm denoting the usual dx to get a feel for this theorem.

Recall that a measurable function φ is integrable if $\int_I |\varphi| dm < \infty$, and we write $\varphi \in L^1(I, m)$. We say that m is a σ-finite measure on I if I can be written as the countable union of sets of finite measure.

Theorem 4.4.12 (Birkhoff Ergodic Theorem). *Suppose that $f : (I, \mathcal{B}, m) \to (I, \mathcal{B}, m)$ is a measure preserving transformation and m is σ-finite. If $\varphi \in L^1(I, m)$, then*

$$\frac{1}{n}\sum_{i=0}^{n-1} \varphi(f^i x) \longrightarrow \varphi^* \in L^1(I, m) \quad m\text{-a.e.}$$

Furthermore, $\varphi^* \circ f = \varphi^*$ a.e., and if $m(I) < \infty$, then $\int \varphi^* dm = \int \varphi dm$.

In particular, if f is a finite measure preserving transformation, then the following corollary holds.

Corollary 4.4.13. *If f is ergodic and finite measure preserving, then φ^* is constant m-a.e., so*

$$\frac{1}{n}\sum_{i=0}^{n-1} \varphi(f^i x) \longrightarrow \int \varphi dm \quad m\text{-a.e.}$$

There are many proofs of the Birkhoff Ergodic Theorem in the literature. For example, Keane [93], Katok and Hasselblatt [91], Walters [168], Garsia [73], Petersen [137], and Krengel [103] offer a sampling of the different styles of proofs.

Suppose that m is an invariant and ergodic measure for a transformation f. The usefulness of the Birkhoff Ergodic Theorem is that, even though we cannot describe orbits of f precisely, we can at least say what most orbits "do" on the average.

For example, if A is any measurable subset of I, then we apply the Birkhoff Ergodic Theorem to the indicator function χ_A to see that m-a.e. point of A spends $\int_I \chi_A(x) dm = m(A)$ of its time in A.

We now give some concrete examples, based on Exercise 4.2.11.

Example 4.4.14. *Let $f(x) = 2x \pmod 1$ on $[0, 1)$. Then Lebesgue measure m is invariant and ergodic. (Despite Theorem 4.4.2, in general it is hard to establish the ergodicity of a transformation, and the ergodicity of m*

for this example is beyond the scope of this book.) If $A = [0, \frac{1}{2})$, then we get for m-a.e. x:

$$\begin{aligned} \lim_n \tfrac{1}{n}\#\{0 \leq i < n; f^i(x) \in A\} &= \lim_n \tfrac{1}{n}\sum_{i=0}^{n-1} \chi_A(x) \\ &= \int \chi_A dm = \tfrac{1}{2}. \end{aligned} \tag{4.1}$$

So almost all points spend half of their time in $[0, \frac{1}{2})$, which agrees with our intuition.

Exercise 4.4.15. *Find some points in $[0, 1)$ for which (4.1) does not hold.*

Exercise 4.4.16. *Extend Example 4.4.14 to arbitrary dyadic intervals $A = A_{n,a} = [\frac{a}{2^n}, \frac{a+1}{2^n})$. Show that m-a.e. point x visits each $A_{a,n}$ with frequency 2^{-n}.*

Points with this property are called normal numbers, *because in their dyadic expansion $x = \sum_{i=1}^{\infty} x_i 2^{-i}$, with $x_i \in \{0, 1\}$, each word $y_1 \dots y_N \in \{0, 1\}^N$ appears in $x_1 x_2 x_3 \dots$ with frequency 2^{-N}.*

Example 4.4.17. *Let $f(x) = x + \alpha \pmod 1$ on $[0, 1)$ and some irrational α. Again Lebesgue measure m is invariant and ergodic. In Exercise 8.1.3 we outline a topological proof that each x has a dense orbit, but the Birkhoff Ergodic Theorem shows that at least m-a.e. (i.e., a set of full measure) x has a dense orbit. Indeed, if $A \subset [0, 1)$ is any interval (other than a point), then m-a.e. x visits A with frequency $m(A) > 0$. Hence, m-a.e. point visits every interval (infinitely often). (See the next section also.)*

In fact, we can replace m-a.e. x in the last statement in Example 4.4.17 with *every* x. The reason is that f has only one invariant probability measure, m. In this case f is called *uniquely ergodic.* For a proof that f is uniquely ergodic and why this implies that every point x visits A with frequency $m(A)$, we refer the reader to Walters [168]. Since in this example the argument is not very difficult, it is outlined in Exercise 4.4.19.

Exercise 4.4.18. *Why do the arguments in Example 4.4.17 break down if $\alpha \in \mathbb{Q}$?*

Exercise 4.4.19. *Show that, for the map $f(x) = x + \alpha$ on the circle $\mathbb{S}^1$, every point visits intervals A with frequency $m(A)$.* **HINT:** *The statement is true for m-a.e. x, and in particular for a dense set X of points. If $y \notin X$, then there are points in X arbitrarily close to y. Now use the fact that f preserves distances.*

4.4.3 A Connection to Topological Dynamics

For an arbitrary metric space (X, d), we define the σ-algebra of Borel sets to be the smallest collection of sets in X containing all the open sets, and closed under complementation and countable unions. As before, we denote the Borel σ-algebra of sets by $\mathcal{B}$. The next result is one example of many that combine the topological and the measure theoretic dynamics of a transformation.

Theorem 4.4.20. *[168] Let (X, d) be a compact metric space, let $\mathcal{B}$ be the Borel sets of X, and assume that m is a probability measure on $(X, \mathcal{B})$ with the property that $m(U) > 0$ for every nonempty open set U. Assume that f is a continuous transformation on X that preserves m and is ergodic. Then m-almost every point $x \in X$ has a dense orbit in X.*

Proof. Let $\{U_n\}_{n \in N}$ be a countable collection of open sets as in Exercise 1.1.46. Then a point x has a dense orbit in X if and only if $x \in \bigcap_{n=1}^{\infty} \bigcup_{k=0}^{\infty} f^{-k}U_n$. For each $n \geq 1$ set $B_n^* = \bigcup_{k=0}^{\infty} f^{-k}U_n$; by hypothesis, $m(B_n^*) > 0$ for each n. Since $f^{-1}(B_n^*) \subseteq B_n^*$ and f is both measure preserving and ergodic, we have that $m(B_n^*) = 1$. This proves the result. □

Chapter 5

A First Example: The 2^∞ Map

In this chapter, we look more closely at the logistic family, that is, the family $\{g_a(x) = ax(1-x)\}_{a\in[0,4]}$, and tell a story showing how to construct an interesting dynamical system within this family. Indeed, we obtain a dynamically constructed Cantor set such that the action of the map restricted to this Cantor set is the *dyadic adding machine*. More precisely, we investigate the map within this family that has a periodic point of period 2^n for each $n \geq 0$ but no other periodic points (there is only one such map within this family). We refer to this map as the 2^∞ map; it is also referred to as a "Coullet Tresser" or the "Feigenbaum" map. This and related maps have received much attention due to the many interesting properties exhibited; see [12, 58, 68, 69, 115] and [4, Chapter 12]. Here, we focus on the fact that this map restricted to the ω-limit set of the turning point is topologically conjugate to the dyadic adding machine (Theorem 5.4.5).

The 2^∞ map appears as an example in Sections 6.1, 10.6, and 10.7 and in Chapter 13. We use the notion of recurrence from Section 3.5 in this chapter.

5.1 Logistic Family

We begin with some coding.

Definition 5.1.1. We recursively define words $H_n \in \{0,1\}^{2^n-1}$, for $n \geq 1$. Set $H_1 = 1$. For $n \geq 2$ define

$$H_n = \begin{cases} H_{n-1}\ 0\ H_{n-1} & \text{if the number of 1's in } H_{n-1} \text{ is odd} \\ H_{n-1}\ 1\ H_{n-1} & \text{if the number of 1's in } H_{n-1} \text{ is even.} \end{cases}$$

Set $H_\infty = \lim_{n\to\infty} H_n$; see Example 5.1.2 below. Thus $H_\infty \in \{0,1\}^{\mathbb{N}}$.

Example 5.1.2. *Examples of H_n's:*

$$\begin{aligned}
H_1 &= 1\\
H_2 &= 1\,0\,1\\
H_3 &= 1\,0\,1\,1\,1\,0\,1\\
H_4 &= 1\,0\,1\,1\,1\,0\,1\,0\,1\,0\,1\,1\,1\,0\,1\\
H_5 &= 1\,0\,1\,1\,1\,0\,1\,0\,1\,0\,1\,1\,1\,0\,1\,1\,1\,0\,1\,1\,1\,0\,1\,0\,1\,0\,1\,1\,1\,0\,1.
\end{aligned}$$

Notice that H_{n+1} extends H_n.

Definition 5.1.3. Let $g_a(x) = 4ax(1-x)$ be given. For each $x \in [0,1]$ we define the *itinerary of x under the map g_a*, denoted

$$I(x, g_a) = \langle I_0(x), I_1(x), I_2(x), \dots \rangle,$$

as follows:

$$I_j(x) = \begin{cases} 0 & \text{if } g_a^j(x) < c \\ * & \text{if } g_a^j(x) = c \\ 1 & \text{if } g_a^j(x) > c. \end{cases}$$

We make the convention that if, for some j, we have $I_j(x) = *$, then we stop the sequence, that is, the itinerary is a finite string. Hence, if $g_a^n(x) \neq c$ for all n, then $I(x, g_a) \in \{0,1\}^{\mathbb{N}}$.

Exercise 5.1.4. *1. Show there exists a unique $a_1 \in [0,4]$ such that*

$$I(\frac{a_1}{4}, g_{a_1}) = H_1 \ *.$$

2. Show there exists a unique $a_2 \in [0,4]$ such that

$$I(\frac{a_2}{4}, g_{a_2}) = H_2 \ *.$$

3. Show that $0 < a_1 < a_2 < 4$.

Theorem 5.1.5. *For each $n \geq 1$ there exists a unique $a_n \in [0,4]$ such that*

$$I(\frac{a_n}{4}, g_{a_n}) = H_n \ *.$$

Thus, for each $n \geq 1$ the point $x = \frac{1}{2}$ is periodic of period 2^n for the map g_{a_n}. Moreover, $0 < a_1 < a_2 < a_3 < \cdots < 4$.

Definition 5.1.6. For each $n \in \mathbb{N}$, let a_n be as in Theorem 5.1.5. Set $a_* = \lim_{n\to\infty} a_n$. We know a_* exists since the sequence $\{a_n\}$ is increasing and is bounded (recall Exercise 1.1.16). Note that $a_* \in [0,4]$. We call g_{a_*} the 2^∞ map. A numerical estimate for a_* is 3.569945668. For ease of notation, throughout the rest of this chapter we set $\mathbf{g} = g_{a_*}$, that is, $\mathbf{g}$ denotes the 2^∞ map.

Theorem 5.1.7. *The 2^∞ map has periodic points of periods 2^n for all n and no other periodic points.*

Proofs of Theorems 5.1.5 and 5.1.7 are left to the reader.

Exercise 5.1.8. ◇ *Prove that* $\mathbf{g}$ *has periodic points of period 2^n for each* $n \geq 0$.

Exercise 5.1.9. ◇ *Show that* $c = \frac{1}{2}$ *is recurrent for* $\mathbf{g}$. *Is c uniformly recurrent?*

Exercise 5.1.10. ◇ *Prove that* $\mathbf{g}$ *is infinitely renormalizable.*

5.2 A Bit of Combinatorics

Recall that, for $a \in [0,4]$ and $x \in [0,1]$, the itinerary of x under the map g_a, denoted $I(x, g_a)$, is either a finite string of 0's and 1's followed by an $*$ or an infinite string of 0's and 1's. We need to put an ordering on this collection of itineraries. The ordering we use is a slight variation of the usual lexicographical ordering and is called the *parity-lexicographical ordering* (plo for short). The plo works as follows: Let $v \neq w$ be itineraries and find the first position where v, w differ; compare in that position using the usual ordering $0 < * < 1$ if the number of 1's preceding this position is even and use the ordering $0 > * > 1$ otherwise.

Example 5.2.1.
1. *Let* $v = 0\ 0\ 1\ 0\ 1\ 0\ *$ *and* $w = 0\ 0\ 0\ 1\ 1\ *$. *Then* $v \succeq w$.
2. *Let* $v = 1\ 0\ 0\ 1\ 1\ 1\ *$ *and* $w = 1\ \overline{0}$. *Then* $w \succeq v$.

Exercise 5.2.2. *Fix* $a \in [0,4]$ *and* $x, y \in [0,1]$ *with* $x < y$. *Then* $I(y, g_a) \succeq I(x, g_a)$. *(See Exercise 10.1.1.)*

5.3 Construction of the Cantor Set $\omega(c, \mathbf{g})$

In this section we see that $\omega(c, \mathbf{g})$ is a Cantor set. Let $c_0 = \frac{1}{2}$ and for $n \geq 1$ set

$$c_n = \mathbf{g}^n(\tfrac{1}{2}).$$

Recall that $I(c_1, \mathbf{g})$ begins as:

$$1\,0\,1\,1\,1\,0\,1\,0\,1\,0\,1\,1\,1\,0\,1\,1\,1\,0\,1\,1\,1\,0\,1\,0\,1\,0\,1\,1\,1\,0\,1\ \ldots. \tag{5.1}$$

From (5.1) we obtain:

$$
\begin{aligned}
I(c_1, \mathbf{g}) &= 1\,0\,1\,1\,1\,0\,1\,0\,1\,0\,1\,1\,1\,0\,1\,1\,1\,0\,1\,1\,1\,0\,1\,0\,1\,0\,1\,1\,1\,0\,1 \ldots \\
I(c_2, \mathbf{g}) &= 0\,1\,1\,1\,0\,1\,0\,1\,0\,1\,1\,1\,0\,1\,1\,1\,0\,1\,1\,1\,0\,1\,0\,1\,0\,1\,1\,1\,0\,1 \ldots \\
I(c_3, \mathbf{g}) &= 1\,1\,1\,0\,1\,0\,1\,0\,1\,1\,1\,0\,1\,1\,1\,0\,1\,1\,1\,0\,1\,0\,1\,0\,1\,1\,1\,0\,1 \ldots \\
I(c_4, \mathbf{g}) &= 1\,1\,0\,1\,0\,1\,0\,1\,1\,1\,0\,1\,1\,1\,0\,1\,1\,1\,0\,1\,0\,1\,0\,1\,1\,1\,0\,1 \ldots \\
I(c_5, \mathbf{g}) &= 1\,0\,1\,0\,1\,0\,1\,1\,1\,0\,1\,1\,1\,0\,1\,1\,1\,0\,1\,0\,1\,0\,1\,1\,1\,0\,1 \ldots \\
I(c_6, \mathbf{g}) &= 0\,1\,0\,1\,0\,1\,1\,1\,0\,1\,1\,1\,0\,1\,1\,1\,0\,1\,0\,1\,0\,1\,1\,1\,0\,1 \ldots \\
I(c_7, \mathbf{g}) &= 1\,0\,1\,0\,1\,1\,1\,0\,1\,1\,1\,0\,1\,1\,1\,0\,1\,0\,1\,0\,1\,1\,1\,0\,1 \ldots \\
I(c_8, \mathbf{g}) &= 0\,1\,0\,1\,1\,1\,0\,1\,1\,1\,0\,1\,1\,1\,0\,1\,0\,1\,0\,1\,1\,1\,0\,1 \ldots \\
\ldots &= \ldots .
\end{aligned}
$$

From the above and Exercise 5.2.2 we have the following:

$$c_2 < c_6 < c_8 < c_0 < c_4 < c_3 < c_7 < c_5 < c_1.$$

Definition 5.3.1. Define

- $I_n = \langle c_{2^n}; c_{2^{n+1}} \rangle$ for $n \geq 1$,
- $F_n = \cup_{j=0}^{2^n-1} \mathbf{g}^j(I_n)$ for $n \geq 1$,
- $\Omega_\infty = \bigcap_{n \geq 1} F_n$.

See Figure 5.1; the figure is not drawn to scale in the sense that $|c_j - c_i|$ is not to scale.

Exercise 5.3.2. *Prove each of the following.*

1. *Let $x \in \Omega_\infty$. Then there exists a unique code $\langle k_1(x), k_2(x), k_3(x), \ldots \rangle$ such that*
$$0 \leq k_n(x) < 2^n \quad \textit{and} \quad x \in \mathbf{g}^{k_n(x)}(I_n),$$
for $n \geq 1$.
2. *$\mathbf{g}^j(I_{n+1}) \subset \mathbf{g}^k(I_n)$ iff $j = k \mod 2^n$.*

Theorem 5.3.3. *The set Ω_∞ is a Cantor set. Moreover, $\omega(\frac{1}{2}, \mathbf{g}) = \Omega_\infty$.*

Exercise 5.3.4. $\diamondsuit$ *Prove Theorem 5.3.3.*

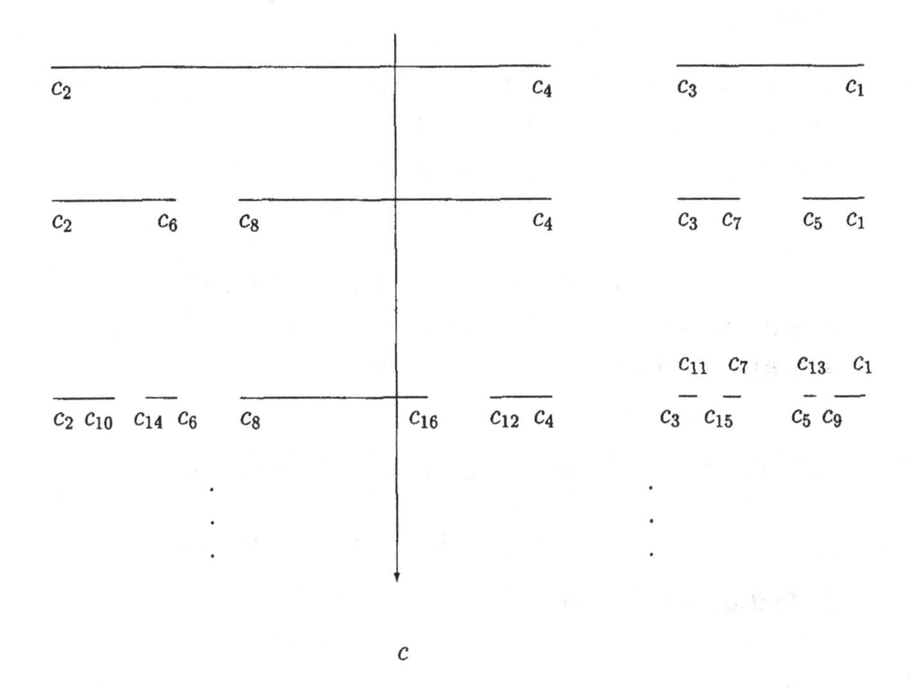

Figure 5.1: Construction of $\omega(c, \mathbf{g})$

5.4 Cantor Set and Adding Machines

In the previous section we saw that Ω_∞ is a Cantor set. We now study the dynamical system:

$$\mathbf{g} : \Omega_\infty \to \Omega_\infty.$$

What are the dynamics of this system? We see in Theorem 5.4.5 that the system is an adding machine.

Example 5.4.1. *Set*

$$\Sigma = \{\langle m_1, m_2, m_3, \dots, \rangle \mid 0 \le m_i < 2^i \ \ \textit{and} \ \ m_{i+1} = m_i \mod 2^i \ \textit{for all} \ \ i\}.$$

Let $\tau : \Sigma \to \Sigma$ be defined by:

$$\tau(\langle m_1, m_2, \dots, m_i, \dots \rangle) =$$

$$\langle (m_1 + 1) \mod 2,\ (m_2 + 1) \mod 2^2,\ \dots, (m_i + 1) \mod 2^i, \dots \rangle.$$

The space Σ along with the self-map τ give a dynamical system, denoted (Σ, τ); $(\Sigma\tau)$ is an example of an adding machine.

See Chapter 13 for references and material on other adding machines.

Example 5.4.2. *Let $\eta : \{0,1\}^{\mathbb{N}} \to \{0,1\}^{\mathbb{N}}$ be addition by 1 (with carry). Examples are*

$$\begin{aligned} \eta(\overline{1\,1\,0}) &= 0\,0\,1\,\overline{1\,1\,0} \\ \eta(\overline{1}) &= \overline{0}. \end{aligned}$$

Again, the dynamical system $(\{0,1\}^{\mathbb{N}}, \eta)$ is an example of an adding machine. This adding machine is called the dyadic adding machine.

We need to define two maps. First, define $\varphi : \Omega_\infty \to \Sigma$ by

$$\varphi(x) = \langle k_1(x), k_2(x), k_3(x), \ldots \rangle.$$

Next, define $\Psi : \Sigma \to \{0,1\}^{\mathbb{N}}$ by

$$\Psi(\langle m_1, m_2, m_3, \ldots \rangle) = \langle \alpha_1, \alpha_2, \alpha_3, \ldots \rangle,$$

where $m_0 = 0$ and

$$\alpha_{i+1} = \begin{cases} 1 & \text{if } m_{i+1} = m_i + 2^i \\ 0 & \text{if } m_{i+1} = m_i. \end{cases}$$

Example 5.4.3. *Suppose that $z \in \Omega_\infty$ is such that*

$$\langle k_1(z) = 1, k_2(z) = 3, k_3(z) = 7, k_4(z) = 7, k_5(z) = 23, k_6(z) = 55, \ldots \rangle.$$

Then,

$$\begin{aligned} \langle k_1(\mathbf{g}(z)) &= 0, k_2(\mathbf{g}(z)) = 0, k_3(\mathbf{g}(z)) = 0, k_4(\mathbf{g}(z)) = 8, \\ k_5(\mathbf{g}(z)) &= 24, k_6(\mathbf{g}(z)) = 56, \ldots \rangle, \end{aligned}$$

and

$$\begin{aligned} \varphi(z) &= \langle 1,\ 3,\ 7,\ 7,\ 23,\ 55, \ldots \rangle \\ \varphi(\mathbf{g}(z)) &= \langle 0,\ 0,\ 0,\ 8,\ 24,\ 56, \ldots \rangle \\ \tau(\varphi(z)) &= \langle 0,\ 0,\ 0,\ 8,\ 24,\ 56, \ldots \rangle \\ \Psi(\varphi(z)) &= \langle 1,\ 1,\ 1,\ 0,\ 1,\ 1,\ \ldots \rangle \\ \eta(\Psi(\varphi(z))) &= \langle 0,\ 0,\ 0,\ 1,\ 1,\ 1,\ \ldots \rangle \\ \Psi(\tau(\varphi(z))) &= \langle 0,\ 0,\ 0,\ 1,\ 1,\ 1,\ \ldots \rangle \\ \Psi(\varphi(\mathbf{g}(z))) &= \langle 0,\ 0,\ 0,\ 1,\ 1,\ 1,\ \ldots \rangle. \end{aligned}$$

Thus,

$$\begin{aligned}\eta(\Psi(\varphi(z))) &= \Psi(\varphi(\mathbf{g}(z)))\\ \tau(\varphi(z)) &= \varphi(\mathbf{g}(z)).\end{aligned}$$

Hence, we have Figure 5.2 and more generally Theorem 5.4.5. Theorem 5.4.5 tells us that studying the system $(\Omega_\infty, \mathbf{g})$ is "equivalent" to studying either of the adding machines (Σ, τ) or $(\{0,1\}^{\mathbb{N}}, \eta)$.

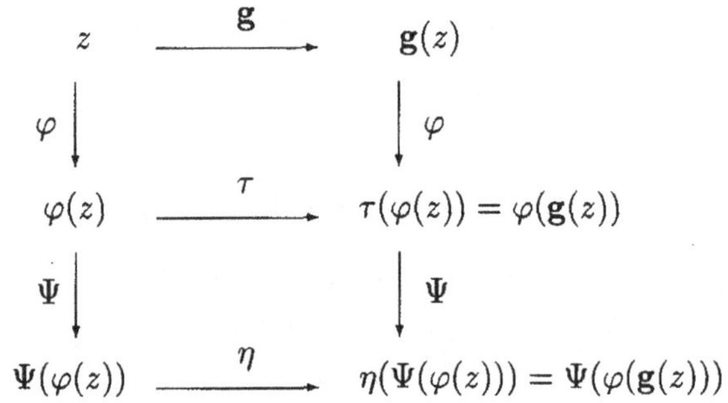

$$\begin{array}{ccc} z & \xrightarrow{\mathbf{g}} & \mathbf{g}(z) \\ \varphi \downarrow & & \downarrow \varphi \\ \varphi(z) & \xrightarrow{\tau} & \tau(\varphi(z)) = \varphi(\mathbf{g}(z)) \\ \Psi \downarrow & & \downarrow \Psi \\ \Psi(\varphi(z)) & \xrightarrow{\eta} & \eta(\Psi(\varphi(z))) = \Psi(\varphi(\mathbf{g}(z))) \end{array}$$

Figure 5.2: Commuting diagram

Exercise 5.4.4. *Show that φ and Ψ are one-to-one maps. Are they onto?*

Theorem 5.4.5. *The following diagram commutes.*

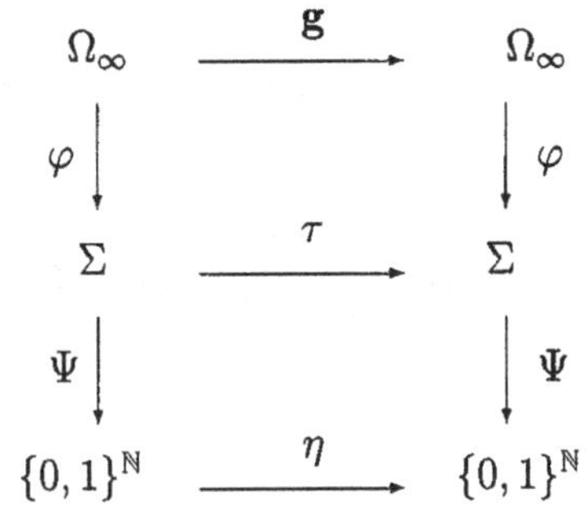

$$\begin{array}{ccc} \Omega_\infty & \xrightarrow{\mathbf{g}} & \Omega_\infty \\ \varphi \downarrow & & \downarrow \varphi \\ \Sigma & \xrightarrow{\tau} & \Sigma \\ \Psi \downarrow & & \downarrow \Psi \\ \{0,1\}^{\mathbb{N}} & \xrightarrow{\eta} & \{0,1\}^{\mathbb{N}} \end{array}$$

Exercise 5.4.6. *Suppose that $x \in [0,1] \setminus \Omega_\infty$ and is not periodic under $\mathbf{g}$. Prove or disprove $\omega(x, \mathbf{g}) = \Omega_\infty$.*

Exercise 5.4.7. *Prove that $\omega(c, \mathbf{g})$ is a minimal set. Moreover, prove that $g|\omega(c, \mathbf{g})$ is a homeomorphism.* **HINT:** *To establish that $\mathbf{g}|\omega(c, \mathbf{g})$ is a homeomorphism, one need only check that $\mathbf{g}$ is one-to-one on $\omega(c, \mathbf{g})$. To do this, use the fact the $(\mathbf{g}, \omega(c, \mathbf{g}))$ is topologically conjugate to $(\{0,1\}^{\mathbb{N}}, \eta)$.*

5.5 A Toeplitz Sequence

The sequence given in formula (5.1) is an example of a *Toeplitz* sequence. An element $x = \langle x_1, x_2, \dots \rangle \in \{0,1\}^{\mathbb{N}}$ is called a *Toeplitz* sequence provided that, for every $i \in \mathbb{N}$, there exists a period $p_i > 1$ such that $x_i = x_{i+np_i}$ for all $n \geq 1$, that is, $\mathbb{N}$ can be decomposed into arithmetic progressions such that x_i is constant on each arithmetic progression. See [108] for a further discussion of Toeplitz sequences and shifts.

Exercise 5.5.1. *Construct $s = \langle s_1, s_2, \dots \rangle \in \{0,1\}^{\mathbb{N}}$ as follows:*

- *Put a 1 at every odd entry, leaving the even entries blank.*
- *Put a 0 at entries $2, 6, 10, 18, \dots$, leaving entries $4, 8, 12, 16, \dots$ blank.*
- *Put a 1 at entries $4, 12, 20, 28, \dots$, leaving entries $8, 16, 24, 32, \dots$ blank.*
- *Put a 0 at entries $8, 24, 40, 56, \dots$, leaving entries $16, 32, 48, 64, \dots$ blank.*

Continue this process until the sequence s is determined. Show that s coincides with the sequence in formula (5.1). This sequence is referred to as the Feigenbaum sequence. By construction, s is a Toeplitz sequence.

Exercise 5.5.2. *Let $s \in \{0,1\}^{\mathbb{N}}$ be a Toeplitz sequence. Prove that s is uniformly recurrent.* **HINT:** *Let $s_{m+1}, \dots s_n$ be a word from s. Set N equal to the least common multiple of $\{p_{m+1}, \cdots, p_n\}$. Prove that the block $s_{m+1}, \cdots, s_n$ reoccurs within N positions.*

Chapter 6

Kneading Maps

In this chapter we present two combinatoric tools, Hofbauer towers and kneading maps, developed by Hofbauer and Keller [87]. These tools allow combinatoric characterizations (Section 6.2) for certain dynamical behaviors of unimodal maps that will prove useful in the remaining chapters. We next investigate *shadowing* for symmetric tent maps and identify, using these tools, a combinatoric characterization for shadowing in this family of maps. Lastly, we use these tools to construct examples of unimodal maps where $\omega(c, f) = [c_2, c_1]$ or $\omega(c, f)$ is a Cantor set.

The reader should be familiar with the material from Sections 3.1 through 3.5 before working in this chapter. Section 6.1 is needed for Section 9.3 and Chapter 11. Both Sections 6.1 and 6.2 are required for Chapters 10 and 13. Sections 6.3 and 6.4 are not used elsewhere in the text.

6.1 Hofbauer Towers and Kneading Maps

Recall that a continuous map $f : [0, 1] \to [0, 1]$ is called *unimodal* if there exists a unique *turning* or *critical point*, c, such that $f|_{[0,c)}$ is increasing, $f|_{(c,1]}$ is decreasing, and $f(0) = f(1) = 0$. As before, $c_i = f^i(c)$ for $i \geq 0$.

We assume $c_2 < c < c_1$ and $c_2 \leq c_3$; otherwise, the asymptotic dynamics are uninteresting. Note that the interval $[c_2, c_1]$ is invariant, that is, f maps $[c_2, c_1]$ onto itself. Hence, to study the asymptotic dynamics of the system, it suffices to restrict our attention to $[c_2, c_1]$. We call $[c_2, c_1]$ the *core* of the map f. Throughout this section we assume that the forward orbit of c is infinite and that c is not attracted to a periodic orbit (meaning that there does not exist an m-periodic point x such that $\lim_{k\to\infty} f^{km}(c) = x$).

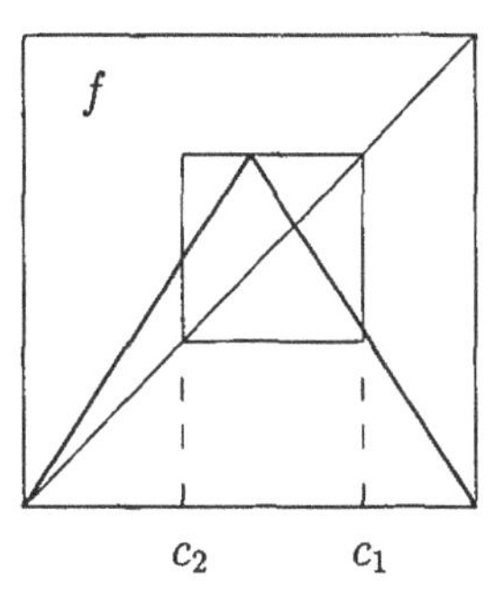

Figure 6.1: Core of f

Let f^n be some iterate of f and let J be any maximal subinterval on which $f^n|_J$ is monotone. Then $f^n : J \to [0,1]$ is called a *branch* of f^n. A branch $f^n : J \to [0,1]$ is called a *central branch* if $c \in \partial J$. Hence there are always two central branches, the images of which are the same when f is symmetric about c. An iterate n is called a *cutting time* if the image of the central branch of f^n contains c. The cutting times are denoted by $S_0, S_1, S_2, \dots (S_0 = 1$ and $S_1 = 2)$. If $f^{S_k} : J \to [0,1]$ is the left central branch of f^{S_k}, then there is a unique point $z_k \in J$ such that

$$f^{S_k}(z_k) = c. \tag{6.1}$$

By construction, z_k has the property that $\cup_{0<j\le S_k} f^{-j}(c) \cap (z_k, c) = \emptyset$ and is therefore called a *closest precritical point.* The point $\hat{z}_k$, defined analogously for the right central branch of f^{S_k}, is also a closest precritical point.

It can be proven (see Exercise 6.1.6) that the difference between two consecutive cutting times is again a cutting time. Hence we can write

$$S_k - S_{k-1} = S_{Q(k)}, \tag{6.2}$$

where $Q : \mathbb{N} \to \mathbb{N} \cup \{0\}$ is an integer function, called the *kneading map.* The kneading map (or cutting times) determines the combinatorics of f completely. It follows from (6.2) that

$$Q(k) < k. \tag{6.3}$$

Given a unimodal map f, the associated Hofbauer tower [87] is the disjoint union of intervals $\{D_n\}_{n\ge1}$, where $D_1 = [0, c_1]$ and, for $n \ge 2$,

$$D_{n+1} = \begin{cases} f(D_n) & \text{if} \quad c \notin D_n \\ [c_{n+1}, c_1] & \text{if} \quad c \in D_n. \end{cases} \tag{6.4}$$

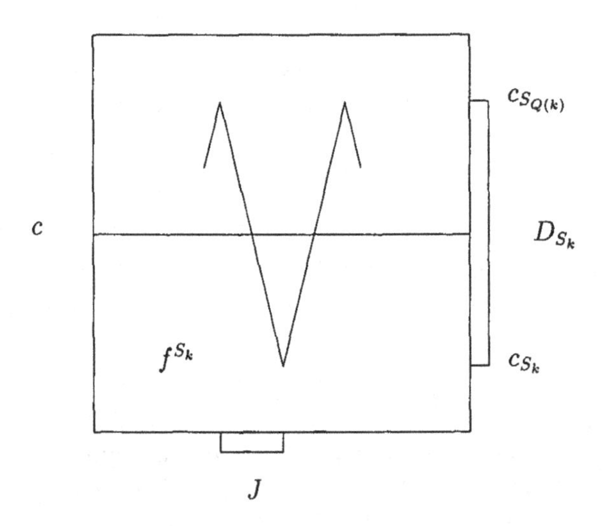

Figure 6.2: Central branches

See Figure 6.3 for the first 13 levels in the Hofbauer tower for the Fibonacci combinatorics. Most often we draw the tower starting with level D_2.

Notice that the image of either central branch $f^n : J \to [0,1]$ is such that $f^n(J) = D_n$. In fact,

$$D_{S_k} = \langle c_{S_k}; c_{S_{Q(k)}} \rangle.$$

Exercise 6.1.1. *Let f be unimodal and fix $j \in \mathbb{N}$. Let J be the maximal closed interval with c_1 as a right-hand endpoint such that f^j is monotone on J (note that J may not be a branch of f but is contained in a branch). Prove that $f^j(J) = D_{j+1}$.* **HINT:** *Induct on j.*

Exercise 6.1.2. *Let f be unimodal with cutting times given by the Fibonacci sequence, that is,*

$$S_0 = 1,\ S_1 = 2,\ S_2 = 3,\ S_3 = 5,\ S_4 = 8,\ S_5 = 13, \dots .$$

We say f has the Fibonacci combinatorics. *Show that the itinerary of c_1 begins as*

$$1\,0\,0\,1\,1\,1\,0\,1\,1\,0\,0\,1\,0.$$

What is $Q(k)$? Figure 6.3 consists of the first few levels of the Hofbauer tower for a unimodal map with the Fibonacci combinatorics. A numerical estimate for the parameter value for the symmetric tent map T_a with the Fibonacci combinatorics is $a = 1.7292119317\dots$.

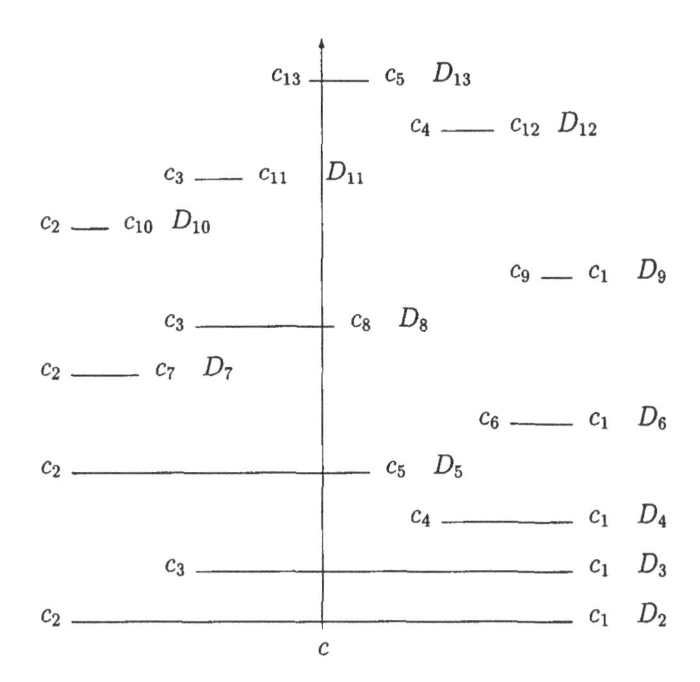

Figure 6.3: Hofbauer tower for Fibonacci combinatorics

Definition 6.1.3. Let f be unimodal with turning point c and $c_n = f^n(c)$. We call $I(c_1, f)$ the *kneading sequence* of the map f. (The itinerary $I(c_1, f)$ is as in Definition 5.1.3, with f replacing g_a and c_1 replacing x; see also Section 10.1.)

Let f be unimodal with the kneading sequence $e_1, e_2, \ldots$. For $j \geq 1$ set

$$B_j = e_{S_{j-1}+1}, \ldots, e_{S_j}.$$

It follows from the Hofbauer tower that

$$B_j = e_1, \ldots, e_{S_{Q(j)}-1}, \overline{e}_{S_{Q(j)}},$$

where $\overline{1} = 0$ and $\overline{0} = 1$.

Exercise 6.1.4. ♣ *[84] Let f be unimodal and $I(c_1, f) = e_1, e_2, \ldots$. Prove that, for each k, the string $e_1, \ldots, e_{S_k}$ contains an odd number of 1's.*

Exercise 6.1.5. *Draw the first eight levels in the Hofbauer tower for the 2^∞ map (recall Definition 5.1.6). Prove that $Q(k) = k - 1$. Can you see the Cantor set, Ω_∞, in the tower (recall Definition 5.3.1)?*

Recall that $\langle a; b\rangle$ denotes $[a, b]$ when $a \leq b$ and $[b, a]$ when $b \leq a$.

Exercise 6.1.6. *Prove each of the following.*

1. *For each* $k \in \mathbb{N}$ *we have:* $D_{S_k+1} = \langle c_1; c_{S_k+1}\rangle$.
2. *For* $S_k < n \leq S_{k+1}$ *we have that* $D_n = \langle c_n; c_{n-S_k}\rangle$ *and* $D_n \subset D_{n-S_k}$ *with the common endpoint* c_{n-S_k}.
3. *If* $n = S_{k+1}$, *then* $c \in D_n = \langle c_{S_{k+1}}; c_{S_{k+1}-S_k}\rangle \subset D_{S_{k+1}-S_k}$.

Notice that it follows from item (3) of Exercise 6.1.6 that $S_{k+1} - S_k$ is indeed a cutting time.

Exercise 6.1.7. *Let* f *be unimodal and* $Q(k)$ *its associated kneading map. Prove* $c_{S_{k-1}} \in (z_{Q(k)-1}, z_{Q(k)}) \cup (\hat{z}_{Q(k)}, \hat{z}_{Q(k)-1})$.

Exercise 6.1.8. *Fix* $a > \sqrt{2}$. *Let* $Q(k)$ *be the kneading map for* T_a. *Prove that* $\lim_{k\to\infty} Q(k) = \infty$ *if and only if* $\lim_{n\to\infty} |D_n| = 0$. **HINT:** *Note that* $\lim_{k\to\infty} Q(k) = \infty$ *implies that* $\lim_{k\to\infty} |c_{S_k} - c| = \lim_{k\to\infty} |c_{S_{Q(k)}} - c| = 0$.

Exercise 6.1.9. ◊ *Let* $a > \sqrt{2}$ *be such that* $Q_a(k) = \max\{0, k-2\}$ *is the kneading map for* T_a, *that is, we have the Fibonacci combinatorics. Prove that*

$$\omega(c, T_a) = \cap_{k\geq 0} \cup_{j=S_k}^{S_{k+1}-1} D_j. \tag{6.5}$$

Exercise 6.1.10. *Can you find other kneading maps* $Q(k)$, *with* $\lim_{k\to\infty} Q(k) = \infty$, *such that formula (6.5) holds?*

Exercise 6.1.11. ♣ *[40, page 1346] Let* f *be a renormalizable unimodal map with* $J \ni c$, $f^n(J) \subset J$, *and* $f^n|J$ *unimodal. Assume* J *is chosen maximal and* n *chosen minimal. Prove that* $n = S_k$ *for some* k *and* $Q(m) \geq k$ *for all* $m > k$.

As we have seen, given a unimodal map f, there is an associated kneading map $Q : \mathbb{N} \to \mathbb{N} \cup \{0\}$. One then asks, given an arbitrary map $Q : \mathbb{N} \to \mathbb{N} \cup \{0\}$, does there exist a unimodal map f with kneading map precisely Q? Definition 6.1.12 provides a characterization of those maps $Q : \mathbb{N} \to \mathbb{N} \cup \{0\}$ for which the answer is yes; see Exercise 6.1.13 and Remark 6.1.14. Admissibility at the level of kneading sequences is discussed in Section 10.1; more precisely, see Definition 10.1.11, Exercise 10.1.12, and Remark 10.1.13.

Definition 6.1.12. Let $Q : \mathbb{N} \to \mathbb{N} \cup \{0\}$ and set $Q(0) = 0$. We say Q is *admissible* provided

$$\{Q(k+j)\}_{j\geq 1} \succeq_L \{Q(Q^2(k)+j)\}_{j\geq 1} \quad \text{and} \quad Q(k) < k$$

for all $k > 0$, where $\succeq_L$ is the lexicographical order.

Exercise 6.1.13. ◇ *Let f be unimodal and $Q(k)$ its associated kneading map. Prove that Q is admissible.*

HINT for Exercise 6.1.13: It follows from Exercise 6.1.6(2) that $c_{S_k} \in \langle c; c_{Q^2(k)} \rangle$. See Figure 6.4. Thus, $c_{S_{k+j}} \in \langle c; c_{S_{Q^2(k)+j}} \rangle$ for $j = 0$. If $c_{S_{k+j}} \in \langle c; c_{S_{Q^2(k)+j}} \rangle$ for some j, then (use Exercise 6.1.7) $Q(k+j+1) \geq Q(Q^2(k)+j+1)$, that is,

$$c_{S_{k+j}} \in \langle c; c_{S_{Q^2(k)+j}} \rangle \quad \Rightarrow \quad Q(k+j+1) \geq Q(Q^2(k)+j+1)$$

for $j \geq 0$.

Next, $c_{S_{k+j}} \in \langle c; c_{S_{Q^2(k)+j}} \rangle$ and $Q(k+j+1) = Q(Q^2(k)+j+1)$ implies that $c_{S_{k+j+1}} \in \langle c; c_{S_{Q^2(k)+j+1}} \rangle$. This completes the hint.

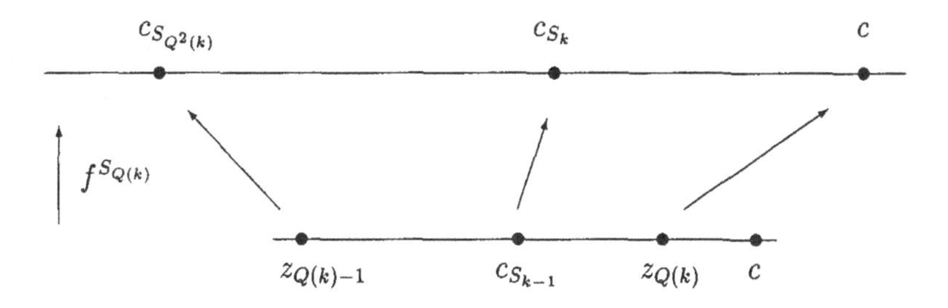

Figure 6.4: Geometry of admissibility condition

Remark 6.1.14. Indeed it is the case that a map $Q : \mathbb{N} \to \mathbb{N} \cup \{0\}$ is admissible if and only if there exists a unimodal map f with kneading map Q. Exercise 6.1.13 provides a proof in one direction; the other direction is beyond the scope of the text (see [87]).

We close this section by defining an *action* on the Hofbauer tower, and then we use this action to construct a graph with vertices precisely the levels in the Hofbauer tower. These ideas are used in Section 9.3.

Definition 6.1.15. Let $f : [0,1] \to [0,1]$ be unimodal and recall the Hofbauer tower from equation (6.4). We define an *action*, denoted $(\hat{I}, \hat{f})$, on the tower as follows: For $x \in D_n$, set

$$\hat{f}(x) = f(x) \in \begin{cases} D_{n+1} & \text{if } c \notin [c_n, x], \\ D_{S_{Q(k)}+1} & \text{if } n = S_k \text{ and } c \in [c_n, x]. \end{cases}$$

Here $\hat{I}$ denotes the tower and $\hat{f}$ the action or map on the tower.

We can now construct a graph, G_f, with vertices the levels D_i, $i \geq 2$, of the tower by placing an arrow $D_i \to D_j$ if and only if $\hat{f}(D_i) \cap D_j \neq \emptyset$.

Exercise 6.1.16. *Show that $D_i \to D_{i+1}$, for all i, $D_{S_k} \to D_{S_{Q(k)}+1}$ for all cutting times S_k, and that there are no other arrows in G_f, that is, the action on the Hofbauer tower describes the graph G_f.*

6.2 First Uses of Kneading Maps

In this section we work with unimodal maps that are not *renormalizable*, have no *periodic attractors*, and have no *wandering intervals*. As such, the maps are topologically conjugate to a symmetric tent map (see Theorems 3.4.27 and 9.5.1).

Exercise 6.2.1. *Let f be a unimodal map that is not renormalizable, has no attracting periodic points, and has no wandering intervals. Prove that* $\lim_{k\to\infty} z_k = c$. *(Recall Exercises 3.4.19 and 3.4.16 and Theorem 3.4.27.)*

Exercise 6.2.2. *Let f be a unimodal map that is not renormalizable, has no attracting periodic points, and has no wandering intervals. Let Q be the kneading map of f and suppose that* $\lim_{k\to\infty} Q(k) = \infty$. *Prove that*

- *the turning point c is persistently recurrent (recall Definition 3.5.13),*
- *$\omega(c, f)$ is minimal (recall definition 3.5.7),*
- *$\omega(c, f)$ is a Cantor set.*

Indeed, the result holds for f with no attracting periodic orbits or wandering intervals. **HINT:** *Show c is persistently recurrent via Exercises 6.2.1, 6.1.7, and 3.5.14. That $\omega(c, f)$ is minimal follows from persistent recurrence, Lemma 3.5.15, and Exercise 3.5.10. To show $\omega(c, f)$ is a Cantor set, recall Exercise 3.5.9 and Proposition 3.2.9.*

Definition 6.2.3. Let f be unimodal with turning point c. We call f *longbranched* provided there exists $\delta > 0$ such that $|D_n| > \delta$ for all $n \geq 1$. (Remember the D_n's are the intervals from the Hofbauer tower.)

The proof of Proposition 6.2.5 is beyond the scope of the text; however, the proposition is easily understood and very useful. First we need a technical condition: *nonflatness*. This condition is easily satisfied for the logistic family $g_a(x) = 4ax(1 - x)$.

Definition 6.2.4. Let f be unimodal with turning point c. We say f is *nonflat* at c provided some derivative of c does not vanish.

Proposition 6.2.5. *[115] Let f be unimodal with a nonflat turning point, no periodic attractors, and no wandering intervals. Then, for every $\epsilon > 0$, there exists $\delta > 0$ such that $|f^n(U)| > \delta$ whenever $|U| > \epsilon$ and $f^n|U$ is monotone.*

Proposition 6.2.6. *[39] Let f be unimodal with a nonflat turning point c and kneading map Q. Assume there are no wandering intervals or periodic attractors. Then, f is longbranched iff $\{Q(k)\}_{k\geq 1}$ is bounded.*

Proof. Assume f is longbranched. Claim 1 and Exercise 6.1.7 give that Q is bounded.

Claim 1: There exists $\epsilon > 0$ such that $|c_{S_k} - c| > \epsilon$ for all k.

Proof: Suppose such an ϵ does not exist. Let δ be the constant from longbranchness. Since f is continuous on a compact space, f is uniformly continuous. Thus, choose $\gamma > 0$ such that $|x-y| < \gamma$ implies $|f(x)-f(y)| < \frac{\delta}{2}$. Choose k such that $|c_{S_k} - c| < \gamma$. Then, $|f(c) - f(c_{S_k})| = |D_{S_k+1}| < \frac{\delta}{2}$; contradicting the choice of δ. This completes the proof of Claim 1.

Assume that Q is bounded. From Exercise 6.1.7 we may choose $\epsilon > 0$ such that $|c - c_{S_k}| > \epsilon$ for all k. Thus, $|\langle c_{S_k}; c_{S_{Q(k)}}\rangle| = |D_{S_k}| > 2\epsilon$ for all k. From Proposition 6.2.5 we may choose $\gamma > 0$ such that $|D_j| > \gamma$ for all $j \notin \{S_0, S_1, S_2, S_3, \dots\}$. Set $\delta = \min\{\gamma, \epsilon\}$. Then f is longbranched with constant δ. □

Let f be unimodal with turning point c. If c is not recurrent or if c is periodic, then f is longbranched. For, suppose c is periodic of period m; then, for any n and any $[x, y]$ a subinterval of monotonicity of f^n, we have $f^n(x), f^n(y) \in \{c, c_1, \dots, c_{m-1}\}$. Thus f is longbranched. If c is not recurrent, then its forward orbit is bounded away from c (again, the image of any endpoint of a subinterval of monotonicity is in the forward orbit of c), and hence f is longbranched. Can f be longbranched and c be recurrent? Exercise 6.2.7 says yes.

Exercise 6.2.7. ◇ *Construct a unimodal map f with kneading map Q such that Q is bounded but the turning point c is recurrent and not periodic.*

Proposition 6.2.8. *[39] Let f be a unimodal map that is not renormalizable, has no attracting periodic points, and has no wandering intervals; let Q be the associated kneading map. If Q is bounded, then $\omega(c, f)$ is nowhere dense (nwd).*

Proof. From Theorem 3.2.17 we know that $\omega(c, f)$ is either nwd or is a finite union of closed intervals. From Theorem 3.4.27 and Exercises 3.4.19 and

3.4.16, we have that f is leo. Thus either $\omega(c, f)$ is nwd or $\omega(c, f) = [c_2, c_1]$. Hence, to prove $\omega(c, f)$ is nwd, it suffices to exhibit an open interval $V \subset [c_2, c_1]$ such that $V \cap \{c_i\}_{i\geq 0} = \emptyset$.

Assume that $Q(k) \leq B$ for all k. If $S_{k-1} < n < S_k$ for some n, k, then $n - S_{k-1} < S_k - S_{k-1} = S_{Q(k)} \leq S_B$. Hence for $S_{k-1} < n < S_k$ we have:

$$D_n = \langle c_n; c_{n-S_{k-1}} \rangle \quad \text{with} \quad n - S_{k-1} \leq S_B.$$

Choose $r > B + 1$ such that $c_1, c_2, c_3, \dots, c_{S_B} \notin (z_{r-1}, \hat{z}_{r-1})$. Let $V \subset (z_{r-1}, z_r)$ be such that $f^{S_{r-1}}(V) = [z_{r-1}, c]$ or $[c, \hat{z}_{r-1}]$; see Figure 6.5.

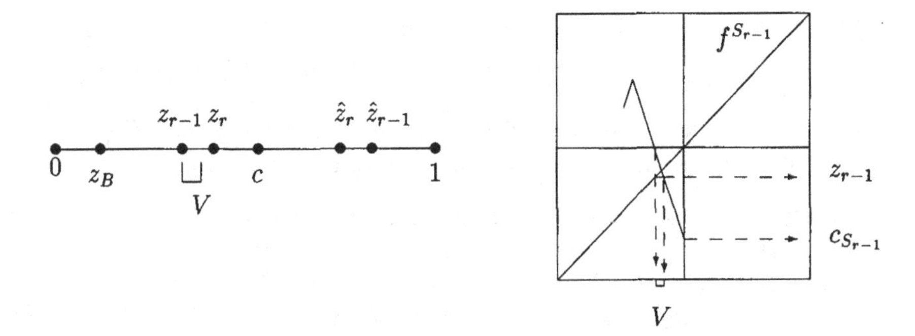

Figure 6.5: Choice of B and construction of V

We now show that the forward orbit of c is disjoint from V. First, for each $k \geq 0$ we have $c_{S_k} \in (z_{Q(k+1)-1}, z_{Q(k+1)}) \cup (\hat{z}_{Q(k+1)}, \hat{z}_{Q(k+1)-1})$ and $Q(k+1) \leq B$. Hence, for each $k \geq 0$, we have $c_{S_k} \notin V$.

Next suppose that $c_n \in V$ for some $S_{k-1} < n < S_k$. Then $D_n = \langle c_n; c_{n-S_{k-1}} \rangle$ with $n - S_{k-1} \leq S_B$, and therefore $c_{n-S_{k-1}} \notin (z_{r-1}, \hat{z}_{r-1})$ (recall our choice of r: $r > B + 1$ such that $c_1, c_2, \dots, c_{S_B} \notin (z_{r-1}, \hat{z}_{r-1})$). Hence, as n is *not* a cutting time, it must be that $D_n = [c_{n-S_{k-1}}, c_n]$ with $c_{n-S_{k-1}} < z_{r-1}$. Thus D_n is such that $c_{n-S_{k-1}} < c_n$, $c_n \in V = (z_{r-1}, z_r)$ and $c_{n-S_{k-1}} < z_{r-1}$. Thus, $z_{r-1} \in D_n$, and therefore $n + S_{r-1}$ is a cutting time, say S_m, with $c_{S_m} = c_{n+S_{r-1}}$. Our choice of V gives that $c_{S_m} \in \langle z_{r-1}; c \rangle$. Thus, $Q(m+1) \geq r > B$, a contradiction. □

Definition 6.2.9. Let $f : I \to I$ be unimodal. We call $J \subset I$ a *homterval* provided $f^n|J$ is monotone for all n.

Exercise 6.2.10. *Let $f : I \to I$ be unimodal with turning point c. Assume there are no homtervals. Prove that $\lim_{k\to\infty} z_k = c$. (Recall the definition of z_k from formula (6.1) and see Exercise 6.2.1.)*

Exercise 6.2.11. *Let f be unimodal and leo. Prove f has no homtervals in the core.*

Proposition 6.2.12. *[39] Let f be a unimodal map with no attracting periodic points or homtervals; let Q be the associated kneading map. If*

$$\liminf_{k \geq 0} Q(k) \geq 2,$$

then $\omega(c, f) \neq [c_2, c_1]$.

Proof. Suppose $\liminf_{k \geq 0} Q(k) \geq 2$ and let $k_0 \in \mathbb{N}$ be such that $Q(k) \geq 2$ for all $k \geq k_0$. Let p be the nonzero fixed point of f; we may assume that p is repelling. Let $U \ni p$ be an open interval such that $c_i \notin U$ for $0 \leq i \leq S_{k_0}$. Set $m = \min\{j \mid f^j(U) \ni c\}$; since there are no homtervals, positive integers with $f^j(U) \ni c$ exist. Let $c_{-m} = f^{-m}(c) \cap U$. Then c_{-m} is a closest preimage of c to p; that is, if $c_{-l} \in \langle c_{-m}; p \rangle$, then $l > m$. Since p is a repelling fixed point, $c_{-m-1}, c_{-m-2}, c_{-m-3}, \ldots$ can be chosen to be the closest preimages of c to p; see Figure 6.6.

Without loss of generality, assume $p < c_{-m}$. Hence we have the following within U:

$$c_{-m-1} < c_{-m-3} < \cdots < p < \cdots < c_{-m-2} < c_{-m}.$$

We will show that $p \notin \omega(c, f)$ and therefore $\omega(c, f) \neq [c_2, c_1]$. Suppose to the contrary that $p \in \omega(c, f)$. Set $\mathcal{W} = \{c_{-m-i} \mid i = 0, 1, 2, \ldots\}$.

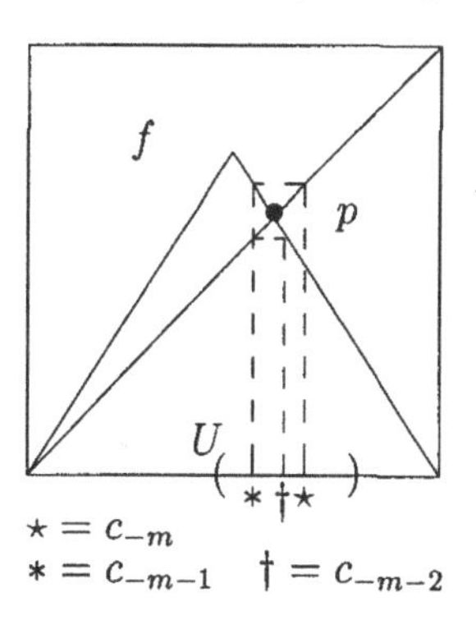

Figure 6.6: Construction of $\{c_{-j}\}_{j \geq m}$

Claim 1: Suppose $D_n \cap \mathcal{W} \neq \emptyset$ with $c_n \in U$ for some n. Then $D_n \cap \mathcal{W}$ has at most one element.

Proof. Suppose to the contrary that D_n contains two adjacent closest preimages, say c_{-w}, c_{-w-2}. Then, $n + w = S_l$ for some l, that is, $n + w$ is a cutting time. Moreover, $f^{-2}(c) \cap f^w(D_n) \neq \emptyset$, and therefore $f^w(D_n)$ contains a component of $(z_1, \hat{z}_1) \setminus \{c\}$ and $c_{n+w} \notin (z_1, \hat{z}_1)$.

Since $c_{n+w} = c_{S_l} \in (z_{Q(l+1)-1}, z_{Q(l+1)}) \cup (\hat{z}_{Q(l+1)}, \hat{z}_{Q(l+1)-1})$ and $c_{n+w} \notin (z_1, \hat{z}_1)$, we obtain $Q(l+1) \leq 1$. On the other hand, $c_i \notin U$ for $0 \leq i \leq S_{k_0}$ and $c_n \in U$ imply that $n > S_{k_0}$, and therefore $S_l = n + w > S_{k_0}$, that is, $l > k_0$. However, $l > k_0$ implies that $Q(l+1) \geq 2$, contradicting $Q(l+1) \leq 1$. This completes the proof of Claim 1.

Since we assume $p \in \omega(c, f)$, we may choose $n = \min \{i \mid c_i \in (c_{-m-3}, c_{-m-2})\}$. It follows from Claim 1 that D_n contains at most one element from $\mathcal{W}$, say $S_{k-1} < n \leq S_k$. Then the definitions of n, $n - S_{k-1} < n$ and D_n containing at most one point from $\mathcal{W}$ imply we cannot have $c_n < c_{-m-4}$ nor $c_n > c_{-m-3}$. Without loss of generality suppose $c_{-m-4} < c_n < c_{-m-2}$. Then, for the same reasons, we have

$$c_{-m-2} < c_{n-S_{k-1}} < c_{-m}; \tag{6.6}$$

see Figure 6.7.

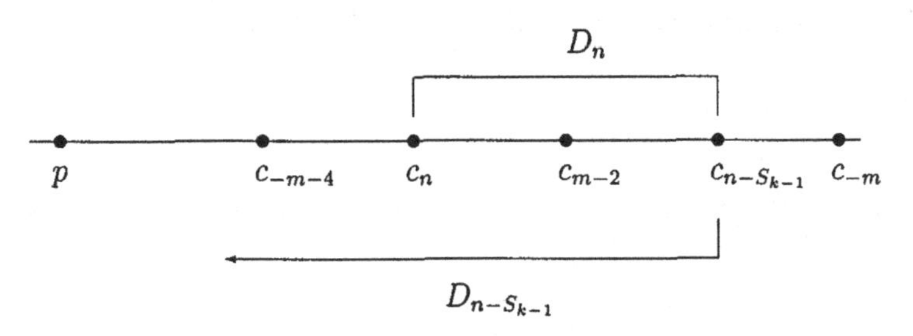

Figure 6.7: View of D_n and $D_{n-S_{k-1}}$

Say $D_{n-S_{k-1}} = \langle c_{n-S_{k-1}}; c_q \rangle$; note $q < n - S_{k-1}$. From $D_n \subset D_{n-S_{k-1}}$, $c_{-m-4} < c_n < c_{-m-2}$, and (6.6) it follows that $D_{n-S_{k-1}} \cap \mathcal{W} \neq \emptyset$. Hence, by Claim 1, $|D_{n-S_{k-1}} \cap \mathcal{W}| = 1$ and therefore $c_q \in (c_{-m-3}, c_{-m-2})$ with $q < n$, a contradiction to the definition of n. Thus $p \notin \omega(c, f)$ and $\omega(c, f) \neq [c_2, c_1]$. □

Remark 6.2.13. Consider the family of symmetric tent maps $\{T_a\}$ for $a \in (\sqrt{2}, 2]$. Any such T_a is not renormalizable and there are no wandering intervals or attracting periodic points. In [35] it is shown that

$$\mathcal{D} = \{a \in (\sqrt{2}, 2] \mid \overline{\{c_i\}}_{i \geq 0} = [c_2, c_1]\} \tag{6.7}$$

has full Lebesgue measure, that is, $\mathcal{D}$ has Lebesgue measure $2 - \sqrt{2}$. Note that for $a \in \mathcal{D}$ we have $\omega(c, T_a) = [c_2, c_1]$. Hence the parameters $a \in (\sqrt{2}, 2]$

such that T_a satisfies the hypothesis of Proposition 6.2.8 or Proposition 6.2.12 has Lebesgue measure 0, that is, the set of a for which $Q_a(k)$ is bounded has Lebesgue measure 0 and the set of a for which $\liminf_{k\geq 0} Q_a(k) \geq 2$ has Lebesgue measure 0. From Proposition 6.2.6 we know that $Q_a(k)$ is bounded if and only if T_a is longbranched. Thus, the set of parameters a for which T_a is longbranched also has Lebesgue measure 0. On the other hand, one can prove that all three of these sets are each dense in $[\sqrt{2}, 2]$.

From Exercise 6.2.2 we have that $Q_a(k) \to \infty$ implies $(\omega(c), T_a)$ is a Cantor set and hence not $[c_2, c_1]$. Thus the set of parameters a for which $Q_a(k) \to \infty$ has Lebesgue measure 0. In Lemma 10.3.13 we prove that the set of a for which $Q_a(k) \to \infty$ is dense in $[\sqrt{2}, 2]$ and is locally uncountable.

6.3 Shadowing

Given a continuous map $f : X \to X$ of a compact metric space (X, d), an orbit for f is a sequence $\{x_i\}_{i\geq 0}$ such that $f(x_i) = x_{i+1}$ for $i \geq 0$. Often, orbits are "calculated" with the aid of a computer and errors can propagate (for example from round-off). Hence, the result may not be an actual orbit, but rather a *pseudo orbit.* Given $\delta > 0$, we say the sequence $\{y_0, y_1, y_2, \dots\}$ is a *δ-pseudo orbit* provided $d(f(y_i), y_{i+1}) \leq \delta$ for all $i \geq 0$. It then becomes important to know when a pseudo orbit can be approximated by an actual orbit.

Definition 6.3.1. Let $f : X \to X$ be a continuous map of a compact metric space. Fix ϵ, $\delta > 0$ and let $\{y_i\}_{i\geq 0}$ be a δ-pseudo orbit. We say the δ-pseudo orbit is *ϵ-shadowed by an actual orbit* provided there exists $x \in X$ such that $d(f^i(x), y_i) \leq \epsilon$ for all $i \geq 0$.

Definition 6.3.2. Let $f : X \to X$ be a continuous map of a compact metric space. We say f has the *shadowing property* provided that, for every $\epsilon > 0$, there is a $\delta > 0$ such that every δ-pseudo orbit can be ϵ-shadowed by an actual orbit.

As before, $\{T_a\}$ is the family of symmetric tent maps, that is, $T_a(x) = ax$ for $0 \leq x \leq \frac{1}{2}$ and $T_a(x) = a(1-x)$ for $\frac{1}{2} \leq x \leq 1$, where $a \in [0, 2]$. The question of which tent maps have the *shadowing* property is answered in [59]. Using the characterization given in [59] (stated here in Theorem 6.3.5), we give an equivalent characterization (Exercise 6.3.6) in terms of kneading maps. Namely, T_a has the shadowing property if and only if either the turning point c is periodic or the kneading map Q_a is not bounded (i.e., T_a is not longbranched). Now to the details.

Definition 6.3.3. [39, page 29] Let f be unimodal with turning point c. We call c_n an *odd return* provided $|f^n(x) - c|$ has a local maximum at c and an *even return* provided $|f^n(x) - c|$ has a local minimum at c. See Figure 6.8.

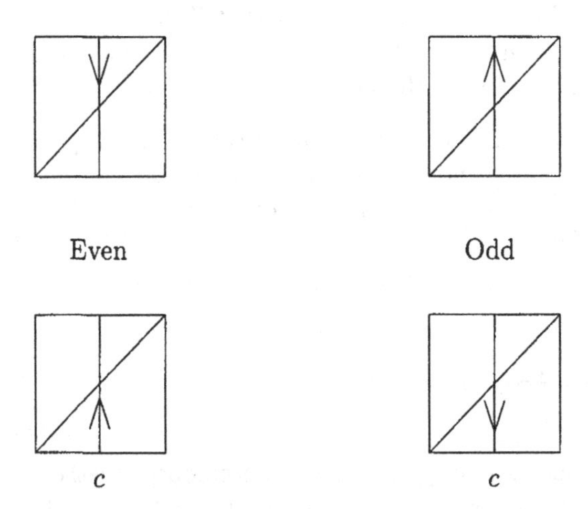

Figure 6.8: Even and odd returns

Exercise 6.3.4. *Let f be unimodal. Prove that c_{S_k} is an odd return for all k.* **HINT:** *Recall Exercise 6.1.4.*

The proof of Theorem 6.3.5 is beyond the scope of the text; see [59] for details. Theorem 6.3.5 provides a characterization of when T_a has the shadowing property. Exercise 6.3.6 provides a way to check for shadowing using the kneading map.

Theorem 6.3.5. *[59] A tent map T_a, with $1 < a < 2$, has the shadowing property if and only if for every $\epsilon > 0$ there exists $n \in \mathbb{N}$ such that $T_a^n(c) = c$* **or** *c_n is an odd return with $|c_n - c| < \epsilon$.*

Exercise 6.3.6. *Prove that a tent map T_a, with $1 < a < 2$, has the shadowing property if and only if one of the following hold:*

- *T_a is* **not** *longbranched.*
- *There exists $m \in \mathbb{N}$ such that $T_a^m(c) = c$.*

Hence for the turning point not periodic we have: T_a has the shadowing property if and only if the kneading map for T_a is **not** *bounded.* **HINT:** *Use Theorem 6.3.5.*

For completeness, Theorem 6.3.7 is the main result in [59].

Theorem 6.3.7. *[59] For almost every $a \in (1,2]$, the associated tent map T_a has the shadowing property. Moreover, the set*

$$\tilde{S} = \{a \in (1,2] \mid T_a \text{ does not have the shadowing property}\}$$

is locally uncountable, *that is, the cardinality of $U \cap \tilde{S}$ is uncountable for every open subinterval $U \subset (1,2]$.*

6.4 Examples of Kneading Maps

In this section we give the construction of [88, Lemmas 2, and 3] in which kneading maps Q are built such that the associated unimodal maps f have either $\omega(c,f) = [c_2, c_1]$ or $\omega(c,f)$ is a Cantor set.

Definition 6.4.1. A sequence

$$\mathcal{F} = (\ 0 = V_0 < U_1 < V_1 < U_2 < V_2 < \cdots\ )$$

of positive integers is called a *frame* provided

$$U_{n+1} \geq n2^{n+V_n} \quad \text{and} \quad V_n \geq n^2 2^{U_n} \quad \text{for} \quad n \geq 1.$$

The *skeleton* $\mathcal{S}(\mathcal{F})$ is defined as the set of all kneading maps $Q : \mathbb{N} \to N \cup \{0\}$ with

$$Q(U_{k+1}) < U_k \quad \text{and} \quad U_k < i \leq V_k \Rightarrow Q(i) = U_k.$$

Definition 6.4.2. Given a frame $\mathcal{F}$, we say a kneading map $Q \in \mathcal{S}(\mathcal{F})$ satisfies property *HK-1* provided

1. $Q(j) = 0$ for $1 \leq j \leq U_1$,
2. $Q(U_n + j) = U_n$ for $1 \leq j \leq V_n - U_n$ and $n \geq 1$,
3. $Q(V_n + j) = U_{n-j}$ for $1 \leq j \leq n-1$ and $n \geq 1$,
4. $Q(V_n + n) = 1$ for $n \geq 1$, and
5. $Q(V_n + n + j) \in \{0, 1\}$ for $1 \leq j \leq U_{n+1} - V_n - n$ for $n \geq 1$.

See Figure 6.9.

Exercise 6.4.3. ◇ *Let $\mathcal{F}$ be a frame and $Q \in \mathcal{S}(\mathcal{F})$ having property HK-1. Let T_a be a symmetric tent map with kneading map Q and kneading sequence $e = e_1, e_2, \ldots$. Prove that*

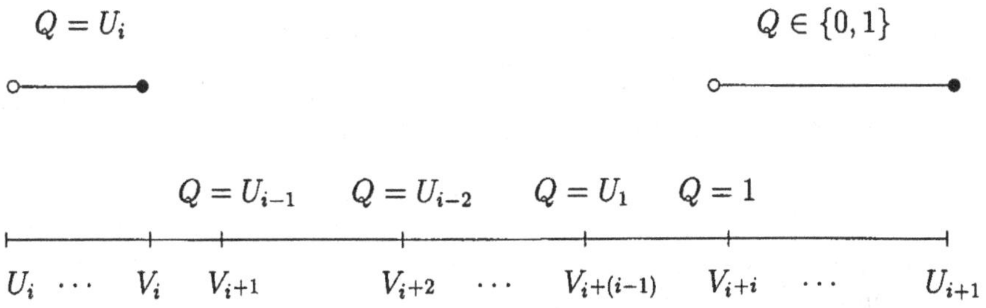

$Q = U_i$ $Q \in \{0,1\}$

$Q = U_{i-1}$ $Q = U_{i-2}$ $Q = U_1$ $Q = 1$

U_i $\cdots$ V_i V_{i+1} V_{i+2} $\cdots$ $V_{i+(i-1)}$ V_{i+i} $\cdots$ U_{i+1}

Figure 6.9: Property HK-1

1. *the string e begins as 100;*
2. *for* $n \geq 1$, *the string* $e_{S_{V_n+n-1}}, \ldots, e_{S_{U_n+1}}$ *consists of blocks* 11 *and* 0 *beginning with* 11*;*
3. *the block* $B = 1010101$ *does not appear in e.*

Conclude that $\omega(c, f)$ *is a Cantor set. See [88, Lemma 2].*

The next exercise is useful in proving that $\omega(c, f) = [c_1, c_2]$ in Example 6.4.5 (see Exercise 6.4.6).

Exercise 6.4.4. ◇ *Let* T_a *be a symmetric tent map with kneading sequence* $e = e_1, e_2, \ldots$; *we are assuming e is an infinite sequence. Let* $x_1, \ldots, x_l \in \{0, 1\}^l$. *Prove there exists* $y \in [0, 1]$ *with* $I(y, T_a)$ *beginning as* $x_1, \ldots, x_l$ *if and only if:*

$$e_1, e_2, \ldots, e_{l-i} \succeq x_{i+1}, x_{i+2}, \ldots, x_l \quad \text{for} \quad 0 \leq i \leq l-1. \tag{6.8}$$

HINT: *See Lemma 10.1.1 and Proposition 10.1.5.*

Example 6.4.5. *[88, Lemma 3] Let* $\mathcal{F}$ *be a frame and* $Q \in \mathcal{S}(\mathcal{F})$ *have property HK-1. We outline a recursive construction that builds a kneading map* $\tilde{Q}$ *such that, if f is unimodal with kneading map* $\tilde{Q}$, *then* $\omega(c, f) = [c_1, c_2]$. *In fact, more strongly,* $\overline{\{c_{\tilde{S}_k}\}} = [c_2, c_1]$.

The recursive construction defines a sequence $\mathcal{B}$ *of blocks (i.e., finite strings of* 0*'s and* 1*'s). Each term of* $\mathcal{B}$ *will represent an interval in the core of the unimodal map f determined by* $\tilde{Q}$. *In order to obtain that* $\overline{\{c_{\tilde{S}_k}\}} = [c_1, c_2]$, *we must make sure that each of these intervals is visited by* $c_{\tilde{S}_k}$ *for some k. This is done in the steps that follow.*

Step One: *Set* $\tilde{Q}(j) = Q(j)$ *for* $1 \leq j \leq U_2$. *Hence,* $\tilde{S}_j$ *for* $0 \leq j \leq U_2$ *and* $e_1, e_2, \ldots, e_{\tilde{S}_{U_2}}$ *are determined. Let* $\mathcal{B}$ *contain all blocks* $x_1, \ldots, x_{\tilde{S}_{U_1}-1}$

such that $e_1 = 1$ *and* $e_1, \ldots, e_{l-i} \succeq x_{i+1}, \ldots, x_l$ *for* $0 \leq i \leq l-1$. *We have now completed Step One.*

Step Two: *Let* $x_1, \ldots, x_m$ *be the first term from* $\mathcal{B}$. *We partition* $x_1, \ldots, x_m$ *into successive strings of maximal lengths that agree with an initial block of* $e_1, e_2, \ldots, e_{\tilde{S}_{U_2}}$. *That is, define positive integers* $s, t_1, t_2, \ldots, t_{s-1} \geq 1$ *such that*

$$\begin{aligned}
x_1, x_2, \ldots, x_{t_1} &= e_1, e_2, \ldots, e_{t_1-1}, \overline{e}_{t_1} \\
x_{t_1+1}, x_{t_1+2}, \ldots, x_{t_1+t_2} &= e_1, e_2, \ldots, e_{t_2-1}, \overline{e}_{t_2} \\
&\cdots \\
&\cdots \\
x_{t_1+\cdots+t_{s-2}+1}, \ldots, x_{t_1+\cdots t_{s-1}} &= e_1, e_2, \ldots, e_{t_{s-1}-1}, \overline{e}_{t_{s-1}} \\
x_{t_1+\cdots+t_{s-1}+1}, \ldots, x_m &= e_1, e_2, \ldots, e_{m-(t_1+\cdots+t_{s-1})}.
\end{aligned}$$

For each $i \in \{1, 2, \ldots, s-1\}$ *there exists* $P(i) \geq 0$ *such that* $t_i = \tilde{S}_{P(i)}$. *Let* u *be minimal such that* $m - (t_1 + \cdots + t_{s-1}) < \tilde{S}_u$. *Since* $m < \tilde{S}_{U_1}$, *we have that* $u \leq U_1$. *Set* $t_s = \tilde{S}_u$. *Then* $m < t_1 + \cdots + t_s$, *and hence we have*

$$x_{m+1}, x_{m+2}, \ldots, x_{t_1+\cdots+t_s} = e_{m-(t_1+\cdots+t_{s-1})+1}, \ldots, e_{t_s-1}, \overline{e}_{t_s}.$$

Thus, we have split $x_1, x_2, \ldots, x_{t_1+\cdots+t_s}$ *into blocks* $e_1, e_2, \ldots, e_{t_i-1}, \overline{e}_{t_i}$ *with* $t_i = \tilde{S}_{P(i)}$ *for* $1 \leq i \leq s$.

Define $\tilde{Q}(j)$ *for* $U_2 < j \leq U_3$ *by*

$$\begin{aligned}
\tilde{Q}(j) = Q(j) \quad & \textit{for} \quad U_2 < j \leq U_3 - s \\
\tilde{Q}(j) = P(j + s - U_3) \quad & \textit{for} \quad U_3 - s + 1 < j \leq U_3.
\end{aligned}$$

Lastly, add to $\mathcal{B}$ *all blocks* $x_1, \ldots x_l$ *such that* $x_1 = 1$, $l = \tilde{S}_{U_2} - 1$, *and (6.8) holds (note that at least one block gets added to* $\mathcal{B}$, *namely an initial segment of* $e_1, e_2, \ldots$). *This completes Step Two.*

Step n**:** *Assume Steps One through* $n-1$ *have been completed. Following the pattern set in the* $n = 2$ *case, let* $x_1, \ldots, x_m$ *be the* $(n-1)$*-th block in* $\mathcal{B}$. *Then* $m \leq \tilde{S}_{U_{n-1}} - 1$. *Define positive integers* $s, t_1, t_2, \ldots, t_{s-1} \geq 1$ *such that*

$$\begin{aligned}
x_1, x_2, \ldots, x_{t_1} &= e_1, e_2, \ldots, e_{t_1-1}, \overline{e}_{t_1} \\
x_{t_1+1}, x_{t_1+2}, \ldots, x_{t_1+t_2} &= e_1, e_2, \ldots, e_{t_2-1}, \overline{e}_{t_2} \\
&\cdots \\
&\cdots \\
x_{t_1+\cdots+t_{s-2}+1}, \ldots, x_{t_1+\cdots t_{s-1}} &= e_1, e_2, \ldots, e_{t_{s-1}-1}, \overline{e}_{t_{s-1}} \\
x_{t_1+\cdots+t_{s-1}+1}, \ldots, x_m &= e_1, e_2, \ldots, e_{m-(t_1+\cdots+t_{s-1})}.
\end{aligned}$$

For each $i \in \{1, 2, \dots, s-1\}$ there exists $P(i) \geq 0$ such that $t_i = \tilde{S}_{P(i)}$. Let u be minimal such that $m - (t_1 + \cdots + t_{s-1}) < \tilde{S}_u$. Since $m < \tilde{S}_{U_{n-1}}$, we have that $u \leq U_{n-1}$. Set $t_s = \tilde{S}_u$. Then $m < t_1 + \cdots + t_s$, and hence we have

$$x_{m+1}, x_{m+2}, \dots, x_{t_1+\cdots+t_s} = e_{m-(t_1+\cdots+t_{s-1})+1}, \dots, e_{t_s-1}, \overline{e}_{t_s}.$$

Thus, we have split $x_1, x_2, \dots, x_{t_1+\cdots+t_s}$ into blocks $e_1, e_2, \dots, e_{t_i-1}, \overline{e}_{t_i}$ with $t_i = \tilde{S}_{P(i)}$ for $1 \leq i \leq s$.

Claim 1: *We have the following:*

- $P(i) \leq U_{n-1}$ *for* $1 \leq i \leq s$.
- $(\tilde{Q}(\tilde{Q}(P(i)) + j))_{1 \leq j \leq s-i} \leq (P(i+j))_{1 \leq j \leq s-i}$ *for* $1 \leq i < s$ *and* $P(i) > 0$.
- $s \leq t_1 + \cdots + t_s \leq m + t_s \leq \tilde{S}_{U_{n-1}} + \tilde{S}_u \leq 2\tilde{S}_{U_{n-1}}$.
- $s \leq t_1 + \cdots + t_s \leq 2^{U_{n-1}+1}$.

This completes Claim 1.

Define $\tilde{Q}(j)$ for $U_n < j \leq U_{n+1}$ by

$$\begin{aligned} \tilde{Q}(j) &= Q(j) && \text{for} \quad U_n < j \leq U_{n+1} - s \\ \tilde{Q}(j) &= P(j + s - U_{n+1}) && \text{for} \quad U_{n+1} - s + 1 < j \leq U_{n+1}. \end{aligned}$$

Lastly, add to $\mathcal{B}$ all blocks $x_1, \dots x_l$ such that $x_1 = 1$, $l = \tilde{S}_{U_n} - 1$, and (6.8) holds This completes Step n.

Claim 2: *The map $\tilde{Q}$ is a valid kneading map, that is, the admissibility condition given in Definition 6.1.12 holds.*

Claim 3: *There exists a symmetric tent map T_a with kneading map $\tilde{Q}$. Moreover, $\overline{\{c_{\tilde{S}_k+1}\}} \supset [c, c_1]$ and therefore $\omega(c, T_a) = [c_2, c_1]$.*

This completes Example 6.4.5.

Exercise 6.4.6. ♣ *[88] Prove Claims 1–3 of Example 6.4.5.*

HINT for Claim 3: *Suppose T_a has kneading map $\tilde{Q}$. Let $U \subset [c, c_1]$ be an open interval and let $x \in U$ be such that $I(x, T_a)$ is infinite. Let $x_1, \dots, x_l$ be an initial segment of $I(x, T_a)$ and $n \in \mathbb{N}$ such that $\tilde{S}_{U_{n-1}} > l$. Using Exercise 6.4.4, establish that $x_1, \dots, x_l$ is an initial subblock of some block added to B in Step n. Conclude that $\overline{\{c_{\tilde{S}_k+1}\}} \supset [c, c_1]$.*

Exercise 6.4.7. *Let T_a have kneading map $\tilde{Q}$ from Example 6.4.5. Prove that $\overline{\{c_{\tilde{S}_k+1}\}} \supset [c, c_1]$ implies $\overline{\{c_{\tilde{S}_k}\}} \supset [c, c_1]$. Conclude that $\overline{\{c_{\tilde{S}_k}\}} = [c_2, c_1]$.*
HINT: *Suppose that $z \in [c, c_1]$ is such that $f(z) = c$. Prove that if $c_{\tilde{S}_k+1} \in [c, z]$, then $\tilde{S}_k + 2$ is a cutting time and $c_{\tilde{S}_k+2} \in [c, c_1]$.*

Question 6.4.8. Construct a kneading map Q for a symmetric tent map T_a such that $\omega(c, T_a) = [c_2, c_1]$ but $\overline{\{c_{S_k}\}} \neq [c_2, c_1]$. Although the "folklore" suggests such a kneading map exists, we are not aware of a publication containing such an example.

Chapter 7

Some Number Theory

We begin this chapter with a discussion of the Farey tree and Farey arithmetic. Next is a brief introduction to continued fractions and lastly a connection between the two (Proposition 7.3.4 and Remark 7.3.6). Both play a role in the combinatorics of one-dimensional dynamics; see [37, 38, 104, 146, 147, 148, 157] and their bibliographies.

In Chapter 8 we discuss circle maps, a first example being rigid rotations. In Proposition 8.2.18 we describe the combinatorics of a rational rigid rotation; for this description we need the Farey tree. In Sections 11.1 to 11.3 we discuss adding machines, irrational rotations of a circle, and unimodal maps; here continued fractions play a role. Irrational rotations are also used in Section 8.3. See [37, 38, 146, 147, 148] for uses of the Farey tree in the study of simple circle maps.

7.1 The Farey Tree

Throughout the text, all fractions are in lowest terms unless otherwise stated. The fractions $\frac{p_1}{q_1}$ and $\frac{p_2}{q_2}$ are said to be *Farey neighbors* if they satisfy the identity $p_1q_2 - p_2q_1 = \pm 1$. An arbitrary pair of fractions $\frac{p_1}{q_1}$ and $\frac{p_2}{q_2}$ can be combined to form their *mediant*, defined as $\frac{p_1+p_2}{q_1+q_2}$, which is not necessarily in lowest terms. In case $\frac{p_1}{q_1}$ and $\frac{p_2}{q_2}$ are Farey neighbors, $\frac{p_1}{q_1} \oplus \frac{p_2}{q_2} \stackrel{\text{def}}{=} \frac{p_1+p_2}{q_1+q_2}$ is necessarily in lowest terms and is called their *Farey sum*.

Exercise 7.1.1. *Prove that if $\frac{p_1}{q_1}$ and $\frac{p_2}{q_2}$ are Farey neighbors, then $\frac{p_1+p_2}{q_1+q_2}$ is in lowest terms.*

Proposition 7.1.2. *[81] Every rational number $\frac{p}{q} \in (0,1)$ can be expressed uniquely as a Farey sum $\frac{p}{q} = \frac{p_1}{q_1} \oplus \frac{p_2}{q_2}$.*

The fractions $\frac{p_1}{q_1}$ and $\frac{p_2}{q_2}$ are the (Farey) parents of $\frac{p}{q} = \frac{p_1}{q_1} \oplus \frac{p_1}{q_1}$. If $\frac{p_1}{q_1} < \frac{p_2}{q_2}$, then $\frac{p_1}{q_1} = L(\frac{p}{q})$ is the *left parent* and $\frac{p_2}{q_2} = R(\frac{p}{q})$ is the *right parent*. The children of $\frac{p}{q} \in (0,1)$ are the *left child* and *right child* defined respectively as $l(\frac{p}{q}) = L(\frac{p}{q}) \oplus \frac{p}{q}$ and $r(\frac{p}{q}) = \frac{p}{q} \oplus R(\frac{p}{q})$. Finally we define one child for each of 0 and 1 by $r(\frac{0}{1}) = l(\frac{1}{1}) = \frac{1}{2}$. Throughout the text, for $\frac{p}{q} \in (0,1)$ we let $\frac{p_1}{q_1} < \frac{p_2}{q_2}$ denote the Farey parents of $\frac{p}{q}$.

The Farey tree **FT** is the infinite tree whose vertices are the set of rational numbers in $[0,1]$ with one edge between every fraction $\frac{p}{q}$ and each of its children; see Figure 7.1. This tree is a subtree of the Stern-Brocot tree [28, 161].

The vertices fall into levels indexed by the nonnegative integers. Vertices $\frac{0}{1}$ and $\frac{1}{1}$ are at level 1, while the vertices at level $n > 1$ are the children of the vertices at level $n-1$. For example, $\frac{2}{5}$ is at level 4, and $\frac{3}{7}$ is at level 5.

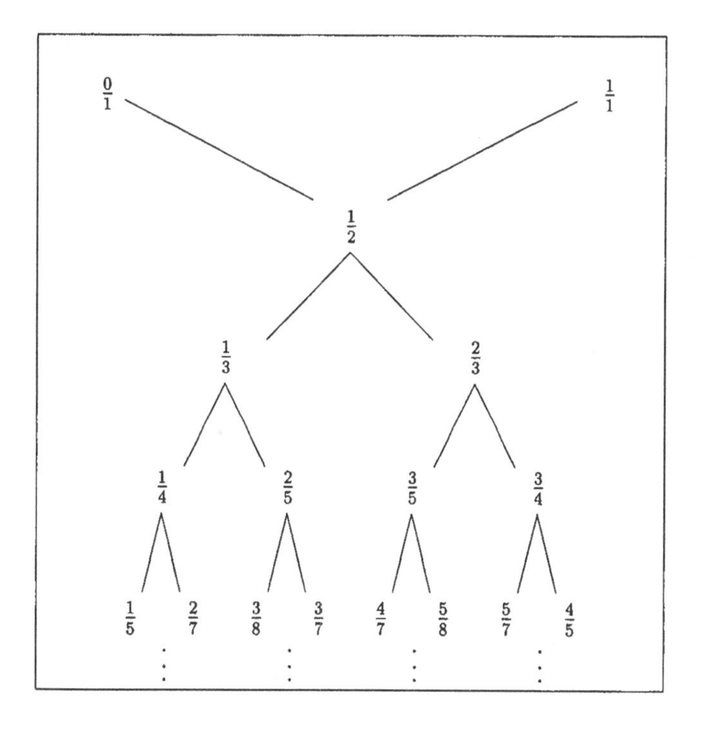

Figure 7.1: Farey tree

Exercise 7.1.3. *Prove , for each rational r in $(0,1)$, there are precisely two infinite paths in* **FT** *that converge to r. For example, $\{\frac{j}{2j+1}\}_{j\geq 1}$ and $\{\frac{j+1}{2j+1}\}_{j\geq 1}$ converge to $\frac{1}{2}$ as $j \to \infty$.*

Exercise 7.1.4. *Prove that, for each irrational r in $(0,1)$, there is exactly one infinite path in* **FT** *that converges to r.*

We next give an equivalent definition of Farey neighbors in terms of a Farey sequence. One can use this equivalence to prove Proposition 7.1.2; see Exercise 7.1.8. (We include this equivalence for the reader interested in proving Proposition 7.1.2; it is not used elsewhere in the text.) Let (m,n) denote the greatest common divisor of m and n.

Definition 7.1.5. [5, 81] For $n \geq 1$ set

$$\mathcal{F}_n = \left\{ \frac{h}{k} \mid 0 \leq h \leq k \leq n \text{ and } (h,k) = 1 \right\}.$$

The ordered set $\mathcal{F}_n$ is called the *Farey sequence of order n.*

Proposition 7.1.6 ([81, Chapter 3]). *Let $n \geq 1$ and $\frac{a}{b}$, $\frac{c}{d}$ be consecutive elements from $\mathcal{F}_n$. Then*

$$bc - ad = 1.$$

Thus $\frac{a}{b}$ and $\frac{c}{d}$ are Farey neighbors.

Proof. We induct on n. The proposition holds for $n = 1$. Let $\frac{a}{b}$ and $\frac{c}{d}$ be consecutive elements from $\mathcal{F}_{n-1}$ and

$$\frac{e}{f} = \min \left\{ \frac{k}{l} \in \mathcal{F}_n \mid \frac{a}{b} < \frac{k}{l} < \frac{c}{d} \right\}.$$

Set

$$be - af = r > 0 \qquad \text{and} \qquad fc - ed = s > 0. \tag{7.1}$$

Solve (7.1) for e, f and use $bc - ad = 1$ to obtain

$$e = sa + rc \qquad \text{and} \qquad f = sb + rd. \tag{7.2}$$

Note that $(r,s) = 1$, since $(e,f) = 1$.

Set

$$S = \left\{ \frac{a'}{b'} = \frac{\mu a + \lambda c}{\mu b + \lambda d} \mid \mu, \lambda \in \mathbb{N} \text{ and } (\lambda, \mu) = 1 \right\}.$$

Then, from (7.2), we have that $\frac{e}{f} \in S$. Each fraction in S lies between $\frac{a}{b}$ and $\frac{c}{d}$ and is in lowest terms; any common divisor of a' and b' would divide

$$\begin{aligned} b(\mu a + \lambda c) - a(\mu b + \lambda d) &= \lambda \\ c(\mu b + \lambda d) - d(\mu a + \lambda c) &= \mu. \end{aligned}$$

Hence each fraction in S appears in $\mathcal{F}_m$ for some m. The first appearance is the smallest, that is, when $\lambda = \mu = 1$. Thus,

$$e = a + c \qquad \text{and} \qquad f = b + d. \tag{7.3}$$

Substituting (7.3) into (7.1), we have $r = s = 1$. □

See [81, Sections 3.1–3.3] for other proofs of Proposition 7.1.6.

Proposition 7.1.7 ([5, Lemma 3.8.4]). *Let $q \geq 2$. Suppose that $\frac{m}{n} < \frac{p}{q} < \frac{r}{s}$ with $\frac{m}{n}, \frac{r}{s} \in \mathcal{F}_{q-1}$ consecutive. Then $p = m + r$ and $q = n + s$, that is, $\frac{p}{q} = \frac{m}{n} \oplus \frac{r}{s}$.*

Proof. Notice that the proposition is easily true for $q = 2$. From Proposition 7.1.6 we have that $rn - ms = 1$. Set $a = pn - mq$ and $b = rq - ps$; thus, $a, b \in \mathbb{N}$. Then,

$$\begin{aligned} q &= qrn - qms = n(ps + b) + s(a - pn) = bn + as \\ p &= prn - pms = r(mq + a) + m(b - rq) = ar + bm. \end{aligned}$$

If $a > 1$, set $p' = (a-1)r + bm$ and $q' = (a-1)s + bn$. Then, $q' < q$ and $\frac{m}{n} < \frac{p'}{q'} < \frac{r}{s}$, contradicting $\frac{m}{n}, \frac{r}{s} \in \mathcal{F}_{q-1}$ consecutive. Thus $a = 1$. Similarly, one shows that $b = 1$. □

Exercise 7.1.8. *Use Proposition 7.1.7 to prove Proposition 7.1.2. (Note that Propositions 7.1.6 and 7.1.7 are independent of Proposition 7.1.2.)*

The following two technical lemmas prove useful in Chapter 8 to describe the combinatorics for rational rigid rotations of a circle; see Proposition 8.2.18.

Lemma 7.1.9. *Let $\frac{u}{v} = \frac{p_1}{q_1} \oplus \frac{p}{q}$, that is, $\frac{u}{v}$ is the left child of $\frac{p}{q}$. Then:*

1. $uq - pv = -1$.
2. $p_1 q - p q_1 = -1$.
3. *If* $i_0 \in \{0, 1, \ldots, q-2\}$, *then* $p_1 q + 1 \leq i_0 + q_1 p \leq p_1 q + (q-1)$.
4. *If* $i_0 \in \{0, 1, \ldots, q-2\}$, *then* $(i_0 + q_1 p) \mod q = i_0 + 1$.

Proof. Item (1) follows from Proposition 7.1.6. From item (1), $v = q_1 + q$, and $u = p_1 + p$, we have that $-1 = uq - pv = (p_1 + p)q - p(q_1 + q)$ and thus (2) holds.

To see (3) notice that $0 \leq i_0 \leq q-2$ gives that $q_1p \leq i_0+q_1p \leq (q-2)+q_1p$. But by item (2), $q_1p = p_1q+1$; therefore, $p_1q+1 \leq i_0+q_1p \leq p_1q+(q-1)$ and (3) holds. For item (4), use that $i_0+q_1p = i_0+p_1q+1$ (by item (2)) and $1 \leq i_0+1 \leq q-1$ to conclude that $[(i_0+q_1p) = (i_0+1+p_1q)] \mod q = i_0+1$. □

Lemma 7.1.10 is proved similarly to Lemma 7.1.9.

Lemma 7.1.10. *Let* $\frac{a}{b} = \frac{p}{q} \oplus \frac{p_2}{q_2}$*, that is,* $\frac{a}{b}$ *is the right child of* $\frac{p}{q}$*. Then:*

1. $aq - pb = 1$.
2. $pq_2 - p_2q = -1$.
3. *If* $1 \leq i_0 \leq q-1$*, then* $p_2q \leq i_0 + q_2p \leq p_2q + (q-2)$.
4. *If* $1 \leq i_0 \leq q-1$*, then* $(i_0 + q_2p) \mod q = i_0 - 1$.

7.2 Continued Fractions

In this section we give a brief overview of continued fractions. See [81, Chapter 10] for more detail. Recall that (m,n) denotes the greatest common divisor of integers m and n.

Definition 7.2.1. Given a nonnegative integer a_0 and positive integers $a_1, \ldots, a_N$, we define the *finite continued fraction* $[a_0, a_1, \ldots, a_N]$ by

$$[a_0, a_1, \ldots, a_N] = a_0 + \cfrac{1}{a_1 + \cfrac{1}{a_2 + \cfrac{1}{\ddots + \cfrac{1}{a_N}}}}.$$

Exercise 7.2.2. *Let* $[a_0, a_1, \ldots, a_N]$ *be given. Show that if* $a_N \geq 2$*, then* $[a_0, a_1, \ldots, a_N] = [a_0, a_1, \ldots, a_N - 1, 1]$*, while if* $a_N = 1$ *we have* $[a_0, a_1, \ldots, a_N] = [a_0, a_1, \ldots, a_{N-1} + 1]$.

Exercise 7.2.3. *Let* $[a_0, a_1, \ldots, a_N]$ *be given and set*

$$\begin{array}{llll} q_0 = 1 & q_1 = a_1, & q_n = a_nq_{n-1} + q_{n-2} & \text{for } 2 \leq n \leq N, \\ p_0 = a_0 & p_1 = a_1a_0 + 1, & p_n = a_np_{n-1} + p_{n-2} & \text{for } 2 \leq n \leq N. \end{array}$$

Prove

$$[a_0, a_1, \ldots, a_n] = \frac{p_n}{q_n} \qquad \text{for } 0 \leq n \leq N.$$

We call each $\frac{p_n}{q_n}$ a convergent *of* $[a_0, a_1, \ldots, a_N]$. **HINT:** *Induct on* n.

Exercise 7.2.4. *Let $[a_0, a_1, \dots, a_N]$ be given. Prove each of the following.*

1. $p_n q_{n-1} - p_{n-1} q_n = (-1)^{n-1}$ *for* $\leq n \leq N$.
2. $(p_n, q_n) = 1$ *for each* n.

Exercise 7.2.5. ◇ *Let $[a_0, a_1, \dots, a_N]$ be given and let $\frac{p_n}{q_n}$, $0 \leq n \leq N$, be the convergents. Prove:*

$$\frac{p_0}{q_0} < \frac{p_2}{q_2} < \cdots < \frac{p_{N-2}}{q_{N-2}} < \frac{p_N}{q_N} = [a_0, a_1, \dots, a_N] < \frac{p_{N-1}}{q_{N-1}} < \cdots < \frac{p_3}{q_3} < \frac{p_1}{q_1}$$

if N is even and

$$\frac{p_0}{q_0} < \frac{p_2}{q_2} < \cdots < \frac{p_{N-1}}{q_{N-1}} < \frac{p_N}{q_N} = [a_0, a_1, \dots, a_N] < \frac{p_{N-2}}{q_{N-2}} < \cdots < \frac{p_3}{q_3} < \frac{p_1}{q_1}$$

if N is odd.

Lemma 7.2.6. *Let $[a_0, a_1, \dots, a_N]$ be given and let $\frac{p_n}{q_n}$, $0 \leq n \leq N$, be the convergents. Then*

$$q_n \geq n,$$

for $n \geq 0$, with inequality for $n > 3$.

Proof. For $n = 0, 1$ we have $q_0 = 1$ and $q_1 = a_1 \geq 1$. If $n > 2$, then $q_n = a_n q_{n-1} + q_{n-2} \geq q_{n-1} + 1$. Hence, $q_n > q_{n-1}$ and $q_n \geq n$. For $n > 3$, $q_n \geq q_{n-1} + q_{n-2} > q_{n-1} + 1 \geq n$; thus $q_n > n$. □

Lemma 7.2.7. *Let a_0 be a nonnegative integer; $a_1, a_2, a_3, \dots$ be positive integers; and $\frac{p_n}{q_n} = [a_0, a_1, \dots, a_n]$ for $n \geq 0$. Then $\lim_{n\to\infty} \frac{p_n}{q_n}$ exists.*

Proof. By Exercise 7.2.5 (note: a monotone bounded sequence of real numbers converges) we have

$$\lim_{n\to\infty} \frac{p_{2n}}{q_{2n}} = L \leq U = \lim_{n\to\infty} \frac{p_{2n+1}}{q_{2n+1}}.$$

It remains to show $L = U$. From Exercise 7.2.4(1) and Lemma 7.2.6 we have

$$\left| \frac{p_{2n}}{q_{2n}} - \frac{p_{2n-1}}{q_{2n-1}} \right| = \frac{1}{q_{2n} q_{2n-1}} \leq \frac{1}{2n(2n-1)}.$$

Since $\lim_{n\to\infty} \frac{1}{2n(2n-1)} = 0$, we obtain $L = U$. □

Definition 7.2.8. Let a_0 be a nonnegative integer; $a_1, a_2, a_3, \dots$ be positive integers; and $\frac{p_n}{q_n} = [a_0, a_1, \dots, a_n]$ for $n \geq 0$. Set $x = \lim_{n\to\infty} \frac{p_n}{q_n}$. We call $[a_0, a_1, \dots]$ an *infinite continued fraction* and refer to $[a_0, a_1, \dots]$ as the *continued fraction expansion of x.*

Given $x \in (0,1)$, we are interested in finding a continued fraction expansion for x. If x is irrational, then there is precisely one infinite continued fraction expansion for x; whereas if x is rational, then there are two finite continued expansions for x (one with an even number of convergents and the other with an odd number of convergents) [81, Chapter 10]. We next show how to find the continued fraction expansion for x. For $z \in \mathbb{R}$, let $\lfloor z \rfloor$ denote the integer part of z, for example, $\lfloor 2.3 \rfloor = 2$, $\lfloor 0.67 \rfloor = 0$, and $\lfloor -3.1 \rfloor = -3$. The algorithm given in the next definition can be written for $x \in \mathbb{R}$; however, we focus on $x \in (0,1)$.

Definition 7.2.9. Fix $x \in (0,1)$. Set $a_0 = 0$ and $x_0 = x$. For $k \geq 0$ set

$$x_{k+1} = \begin{cases} \dfrac{1}{x_k - a_k} & \text{if } x_k \neq a_k \\ 0 & \text{if } x_k = a_k \end{cases} \qquad \text{and} \qquad a_{k+1} = \lfloor x_{k+1} \rfloor. \tag{7.4}$$

The algorithm given by (7.4) is called the *continued fraction algorithm.* If $x_k = a_k$ for some k, then there exists a smallest $N \in \mathbb{N}$ such that $a_n = 0$ for all $n > N$. In this case, we obtain the finite continued fraction $[a_0, a_1, \ldots, a_N]$. Otherwise we obtain an infinite continued fraction.

Exercise 7.2.10. ◊ *Fix $x \in (0,1)$. Let $[a_0, a_1, \ldots]$ be obtained from the continued fraction algorithm. Prove*

1. *If x is rational, then there exists $N \in \mathbb{N}$ such that $n \geq N$ implies $a_n = 0$.*

2. *If x is irrational, then $a_i > 0$ for all $i > 0$.*

3. *$[a_0, a_1, \ldots]$ is a continued expansion for x, that is,*

$$\lim_{n\to\infty} \left[\frac{p_n}{q_n} = [a_0, a_1, \ldots, a_n] \right] = x.$$

7.3 Continued Fractions and the Farey Tree

In this section we relate the Farey tree to the continued fraction expansions of numbers from $[0,1]$. As we are working in $[0,1]$, we assume $a_0 = 0$ (recall Definition 7.2.1). Hence we write

$$[a_1, a_2, \ldots, a_N] = \cfrac{1}{a_1 + \cfrac{1}{a_2 + \cfrac{1}{\ddots + \cfrac{1}{a_N}}}}.$$

We use the notation of Sections 7.1 and 7.2.

Proposition 7.3.1. *Let $\frac{p}{q} = [a_1, a_2, \ldots, a_t] \in (0,1)$, where $t = 2n$ or $t = 2n+1$ for some n. The left and right children in the Farey tree of $\frac{p}{q}$ are given by:*

$$\begin{aligned} l([a_1, a_2, \ldots, a_{2n}]) &= [a_1, a_2, \ldots, a_{2n} - 1, 2] \\ r([a_1, a_2, \ldots, a_{2n}]) &= [a_1, a_2, \ldots, a_{2n} + 1] \\ l([a_1, a_2, \ldots, a_{2n+1}]) &= [a_1, a_2, \ldots, a_{2n+1} + 1] \\ r([a_1, a_2, \ldots, a_{2n+1}]) &= [a_1, a_2, \ldots, a_{2n+1} - 1, 2]. \end{aligned}$$

Proof. We induct on n, the levels in the Farey tree. Recall that vertices $\frac{0}{1}$ and $\frac{1}{1}$ are at level 1, while the vertices at level n are the children of the vertices at level $n-1$.

Assume the result holds up to level m. We show the result holds at level $m+1$. One argues by cases. We do one case and leave the remaining cases to the reader.

Suppose that $\frac{y}{z} = [a_1, \ldots, a_{2n}]$ is at level m. Then, by the induction hypotheses, $r(\frac{y}{z}) = [a_1, \ldots, a_{2n} + 1] := \frac{u}{v}$. We need to show that $l(\frac{u}{v}) = [a_1, \ldots, a_{2n}, 2]$.

For $1 \leq k \leq 2n$, set $\frac{y_k}{z_k} = [a_1, \ldots, a_k]$. Using Exercise 7.2.3, we have:

$$\begin{aligned} \frac{y}{z} &= [a_1, \ldots, a_{2n}] = \frac{a_{2n} y_{2n-1} + y_{2n-2}}{a_{2n} z_{2n-1} + z_{2n-2}} \\ \frac{u}{v} &= [a_1, \ldots, a_{2n} + 1] = \frac{(a_{2n} + 1) y_{2n-1} + y_{2n-2}}{(a_{2n} + 1) z_{2n-1} + z_{2n-2}}. \end{aligned}$$

Thus (again using Exercise 7.2.3),

$$\begin{aligned} l(\frac{u}{v}) &= \frac{y+u}{z+v} \\ &= \frac{a_{2n} y_{2n-1} + y_{2n-2} + (a_{2n}+1) y_{2n-1} + y_{2n-2}}{a_{2n} z_{2n-1} + z_{2n-2} + (a_{2n}+1) z_{2n-1} + z_{2n-2}} \\ &= \frac{2\left[a_{2n} y_{2n-1} + y_{2n-2}\right] + y_{2n-1}}{2\left[a_{2n} z_{2n-1} + z_{2n-2}\right] + z_{2n-1}} \\ &= \frac{2 y_{2n} + y_{2n-1}}{2 z_{2n} + z_{2n-1}} \\ &= [a_1, \ldots, a_{2n}, 2]. \end{aligned}$$

Hence, $l(\frac{u}{v}) = [a_1, \ldots, a_{2n}, 2]$. □

Exercise 7.3.2. *Complete the details for the other cases in the proof of Proposition 7.3.1.*

In Section 7.1, we saw that each rational number $x \in [0,1]$ appears exactly once in the Farey tree, we defined left/right children/parents, and we discussed paths in the Farey tree. For example, there is a path in the Farey tree from $\frac{1}{2}$ to $\frac{11}{17}$, namely $\frac{11}{17} = r(r(r(r(l(r(\frac{1}{2}))))))$. Recall that $l(x)$ denotes the left child of x and $r(x)$ denotes the right child of x.

Lemma 7.3.3. *Let $x = [a_1, a_2, \ldots, a_t]$ with $a_i \in \mathbb{N}$ (thus, $a_i > 0$ for all i) and let $z \in N$. Then there exists a unique path in the Farey tree joining x to $[a_1, a_2, \ldots, a_t, z]$.*

Proof. We need only prove there exists some path as the uniqueness of a path follows from Proposition 7.1.2 and the fact that the Farey tree is a binary tree (thus, each entry in the tree has exactly one path leading to it).

We break the proof into two cases, depending on whether t is even or odd. We do the case for t even, as the other case is done similarly. Assume $t = 2n$. We induct on z. Set $y = [a_1, a_2, \ldots, a_{2n}, z]$.

Suppose that $z = 1$. Then $y = [a_1, a_2, \ldots, a_{2n}, 1] = [a_1, a_2, \ldots, a_{2n} + 1] = r(x)$, and thus there is a path from x to y. Assume that there is a path from x to $[a_1, a_2, \ldots, a_{2n}, k-1]$. Since $l([a_1, a_2, \ldots, a_{2n}, k-1]) = [a_1, a_2, \ldots, a_{2n}, k]$, we have a path from x to $[a_1, a_2, \ldots, a_{2n}, k]$. □

Proposition 7.3.4. *Let $r = [a_1, a_2, \ldots] \in (0,1)$ be irrational and let $\beta(r)$ be the unique path in the Farey tree that converges to r (recall Exercise 7.1.4). Then each convergent $\frac{p_n}{q_n} = [a_1, a_2, \ldots, a_n]$ belongs to $\beta(r)$.*

Proof. The proposition is immediate from Exercise 7.1.4 and Lemma 7.3.3 □

Exercise 7.3.5. *Let $r = [a_1, a_2, \ldots] \in (0,1)$ be irrational, and let $\frac{p_n}{q_n} = [a_1, \ldots, a_n]$, $\frac{p_{n+1}}{q_{n+1}} = [a_1, \ldots, a_{n+1}]$ be consecutive convergents of r. Show there exists $m \geq 0$ such that*

- *if $\frac{p_n}{q_n} < r$, then $\frac{p_{n+1}}{q_{n+1}} = l^m \circ r(\frac{p_n}{q_n})$ and*
- *if $\frac{p_n}{q_n} > r$, then $\frac{p_{n+1}}{q_{n+1}} = r^m \circ l(\frac{p_n}{q_n})$.*

HINT: *Recall Exercise 7.2.4, that is, consecutive convergents of $r = [a_1, a_2, \ldots]$ are Farey neighbors.*

Remark 7.3.6. Let $r = [a_1, \ldots] \in (0,1)$ be irrational and let $\beta(r)$ be the unique path in the Farey tree that converges to r (recall Exercise 7.1.4). It follows from Exercise 7.3.5 that the convergents $\frac{p_n}{q_n}$'s of r occur at the "turns" in $\beta(r)$. More precisely, convergents immediately precede turns in the path $\beta(r)$ and each turn in $\beta(r)$ is preceded by a convergent.

Chapter 8

Circle Maps

The chapter begins with an introduction to orientation preserving homeomorphisms of a circle; the collection of such homeomorphisms is denoted by $\mathcal{H}$. We discuss rotation numbers, rigid rotations, and characterize maps in $\mathcal{H}$ that are topologically conjugate to an irrational rigid rotation (Theorem 8.1.11). Next we investigate degree one circle maps, the combinatorics of a rational rigid rotation (Proposition 8.2.18 and Example 8.2.19), and first return maps for irrational rigid rotations. In the last section we discuss two results. First, we consider the possible ω-limit sets for homeomorphisms of the circle with an irrational rotation number (Theorem 8.4.1). Second, given your favorite Cantor set $K \subset [0,1)$ and irrational $\rho \in (0,1)$, one can construct a homeomorphism of the circle h with rotation number ρ and such that $\omega(x,h) = K$ for all x (Theorem 8.4.2).

Sections 3.1 through 3.4 along with Section 7.2 contain background material for this chapter. Sections 8.1 and 8.2 are needed for Chapter 12, and Section 8.3 for Section 11.3.

8.1 Circle Homeomorphisms

In this section we discuss orientation preserving homeomorphisms of the circle group $\mathbb{R}/\mathbb{Z}$ (given by $x, y \in \mathbb{R}$ are *equivalent* if and only if $x - y \in \mathbb{Z}$). Let $\mathcal{H}$ denote the collection of such homeomorphisms. We can *lift* any $f \in \mathcal{H}$ to $\mathbb{R}$ by considering continuous increasing functions $F : \mathbb{R} \to \mathbb{R}$ that satisfy the following two conditions.

1. $F(x+1) = F(x) + 1$ for all $x \in \mathbb{R}$.

2. $\pi \circ F = f \circ \pi$, where $\pi : \mathbb{R} \to \mathbb{R}/\mathbb{Z}$ is given by $\pi(x) = x \mod 1$.

Condition (1) is not artificial. Wrapping $[0,1]$ into a circle and identifying endpoints allows us to view $[0,1)$ as a circle. Thus, looking at self-maps of $[0,1)$ is simply looking at circle maps. More generally, replace $[0,1)$ by $\mathbb{R}/\mathbb{Z}$. Hence, condition (1) is needed to guarantee we are dealing with circle maps.

A lift of $f \in \mathcal{H}$ is determined up to an additive integer (more precisely, if F is a lift for some $f \in \mathcal{H}$, then so also is $F+k$ for any $k \in Z$). However, for ease, we choose coordinates such that $0 \leq F(0) < 1$. Thus, given $f \in \mathcal{H}$, there is one associated (well-defined) lift $F : \mathbb{R} \to \mathbb{R}$.

Proposition 8.1.1. *For $f \in \mathcal{H}$ we have $\rho(f) := \lim_{i\to\infty} \frac{1}{i}F^i(0)$ exists and for all $x \in \mathbb{R}$*

$$\rho(f) = \lim_{i\to\infty} \frac{1}{i}\left(F^i(x) - x\right) \text{ uniformly.}$$

Proof. [92] Let $f \in \mathcal{H}$ and its lift F be fixed. For $0 \leq x \leq 1$ and $j \geq 0$ we have $F^j(0) \leq F^j(x) \leq F^j(0)+1$, and hence for $x, y \in [0,1]$ we have

$$\left|\left(F^j(x) - x\right) - \left(F^j(y) - y\right)\right| \leq 1. \tag{8.1}$$

As $F(x+1) = F(x)+1$ for all $x \in \mathbb{R}$, (8.1) holds for all $x \in \mathbb{R}$. Setting $M = \max_{x\in\mathbb{R}} |F(x) - x|$ and using $F^j(x) - x = \sum_{l=0}^{j-1} \left(F(F^l(x)) - F^l(x)\right)$, we obtain:

$$\max_{x\in\mathbb{R}} \left|F^j(x) - x\right| \leq jM. \tag{8.2}$$

For $k > i$ with $k = Li + r$ and $0 \leq r < i$, write

$$F^k(x) - x = F^k(x) - F^{Li}(x) + \sum_{l=0}^{L-1} \left[F^i(F^{il}(x)) - F^{il}(x)\right]. \tag{8.3}$$

By (8.1), each term in the sum from (8.3) differs from $F^i(0)$ by at most 1 and there are L terms in that sum. From (8.2) we have that $|F^k(x) - F^{Li}(x)| = |F^r(F^{Li}(x)) - F^{Li}(x)| \leq rM$. Hence, (8.1), (8.2), and (8.3) give

$$\left|F^k(x) - x - LF^i(0)\right| < rM + L < iM + L.$$

Assuming $k > i^2$, we have $\frac{L}{k} \approx \frac{1}{i}$, and therefore there exists a constant Γ such that

$$\left|\frac{1}{k}\left(F^k(x) - x\right) - \frac{1}{i}F^i(0)\right| < \frac{\Gamma}{i}. \tag{8.4}$$

Setting $x = 0$ in (8.4), we have that $\{\frac{1}{i}F^i(0)\}$ is Cauchy and therefore convergent. The result now follows from (8.4). $\square$

We call $\rho(f)$ the *rotation number* of f. Given our assumption $0 \leq F(0) < 1$, we obtain $0 \leq \rho(f) < 1$.

Definition 8.1.2. For each $\alpha \in (0,1)$ define $R_\alpha : \mathbb{R} \to \mathbb{R}$ by $R_\alpha(x) = x+\alpha$. Often we view R_α as a map of the circle via $R_\alpha(x) = x + \alpha \mod 1$. To study R_α it suffices to study R_α resticted to $[0,1)$; see Figure 8.1. We call R_α a *rational* or *irrational rigid rotation of the circle* depending on α.

Rigid rotations are first examples of maps in $\mathcal{H}$. Yet maps $f \in \mathcal{H}$ with $\rho(f) = \alpha$ irrational and whose orbits are dense in $\mathbb{R}/\mathbb{Z}$ are topologically conjugate to the rigid rotation R_α (see Theorem 8.1.11).

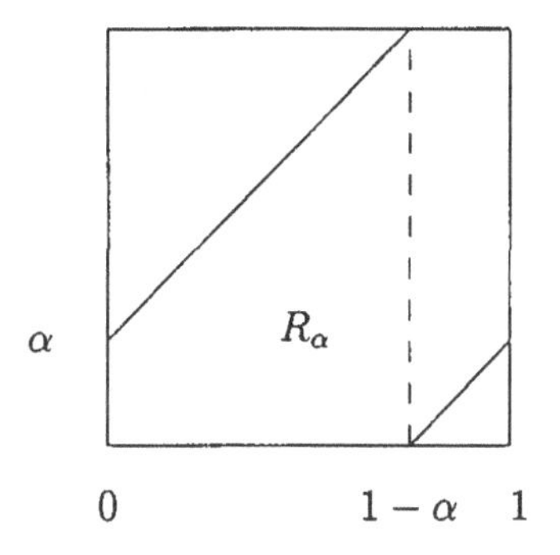

Figure 8.1: Rigid rotation R_α

Exercise 8.1.3. *Consider the map:* $R_\alpha(x) = x + \alpha \mod 1$. *Prove:*

1. *If* $\alpha = \frac{p}{q}$ *is rational, then each* $x \in [0,1)$ *is periodic with period* q *(recall that* $\frac{p}{q}$ *is in lowest terms).*

2. *If* α *is irrational, then* $\overline{\{R_\alpha^j(x)\}}_{j\geq 0} = [0,1]$ *for each* $x \in [0,1)$. **HINT:** *Fix* $x \in [0,1)$ *and* $\epsilon > 0$. *Notice that* $R_\alpha^j(x) \neq R_\alpha^k(x)$ *for* $j \neq k$. *Thus the set* $\{R_\alpha^i(x) \mid i \geq 0\}$ *has an accumulation point. Hence there exist* $m < n$ *such that* $|R_\alpha^n(x) - R_\alpha^m(x)| < \epsilon$ *and thus setting* $k = n - m$ *we have* $|R_\alpha^k(x) - x| < \epsilon$. *As* R_α *preserves lengths, we can partition* $[0,1)$ *(using* $x, R_\alpha^k(x), R_\alpha^{2k}(x), \ldots$*) into arcs of length less than* ϵ.

Exercise 8.1.4. *Consider the lift* $R_\alpha(x) = x + \alpha$. *Show that* $\rho(R_\alpha) = \{\alpha\}$.

Exercise 8.1.5. *Prove each of the following for* $f \in \mathcal{H}$.

1. $\rho(f^k) = k\rho(f)$ *for all* k.

2. $\rho(h \circ f \circ h^{-1}) = \rho(f)$ *for all* $h \in \mathcal{H}$.

3. $\min_{x\in\mathbb{R}}(F(x)-x) \le \rho(f) \le \max_{x\in\mathbb{R}}(F(x)-x)$.

4. *$\rho(f)$ is rational if and only if f has periodic points.*

The map $f \mapsto \rho(f)$ is continuous on $\mathcal{H}$. See Proposition 8.2.7.

Lemma 8.1.6. *[92] Let $f \in \mathcal{H}$ and $s, k \in \mathbb{Z}$ such that $s < k\rho(f) < s+1$. Then, $s < F^k(x) - x < s+1$ for all $x \in \mathbb{R}$.*

Proof. Suppose $\min_{x\in\mathbb{R}}\left(F^k(x)-x\right) \le s$. Then (from Exercise 8.1.5 part 4) $F^k(x') = x' + s$ for some $x' \in \mathbb{R}$ and therefore $\rho(f) = \frac{s}{k}$. Similarly for max. □

Exercise 8.1.7. *Let $f \in \mathcal{H}$, $\alpha = \rho(f)$ be irrational and $n \in \mathbb{Z}$. Prove*

$$\lfloor F^n(x) - x \rfloor = \lfloor n\alpha \rfloor.$$

Recall that $\lfloor x \rfloor$ denotes the integer part of x. **HINT:** *Use Lemma 8.1.6.*

Exercise 8.1.8. *Let $\alpha \in (0,1)$ be irrational and $k < s$. Prove*

$$k\alpha \mod 1 < s\alpha \mod 1 < (k+1)\alpha \mod 1$$

iff

$$\lfloor s\alpha - (k+1)\alpha \rfloor < \lfloor s\alpha - k\alpha \rfloor.$$

Exercise 8.1.9. *Let $f \in \mathcal{H}$, $\rho(f) = \alpha$ be irrational, $x \in \mathbb{R}$, $t = x \mod 1$, and $k < s \in \mathbb{Z}$. Prove*

$$f^k(t) < f^s(t)$$

iff

$$\lfloor F^{s-k-1}(x) - x \rfloor < \lfloor F^{s-k}(x) - x \rfloor.$$

Theorem 8.1.10 (Poincaré [92]). *Let $f \in \mathcal{H}$ with $\alpha = \rho(f)$ irrational. Fix $t \in \mathbb{R}/\mathbb{Z}$. Then the points $\{t_j := f^j(t)\}_{j\in\mathbb{Z}}$ lie on $\mathbb{R}/\mathbb{Z}$ in the same order as the points $\{j\alpha \mod 1\}_{j\in\mathbb{Z}}$.*

Proof. Let $k, m, s \in \mathbb{Z}$. We want to prove that

$$k\alpha \mod 1 < s\alpha \mod 1 < m\alpha \mod 1 \tag{8.5}$$

$$\Longleftrightarrow$$

$$t_k < t_s < t_m. \tag{8.6}$$

Without loss of generality assume that $k, m < s$ and $k\alpha \mod 1 < m\alpha \mod 1 < (k+1)\alpha \mod 1$. Using Exercise 8.1.8, we have formula (8.5) equivalent to

$$\lfloor (s-k-1)\alpha \rfloor < \lfloor (s-k)\alpha \rfloor \text{ and } \lfloor (s-m)\alpha \rfloor < \lfloor (s-m+1)\alpha \rfloor.$$

Similarly, using Exericse 8.1.9, we have formula (8.6) equivalent to

$$\lfloor F^{s-k-1}(x) - x \rfloor < \lfloor F^{s-k}(x) - x \rfloor \text{ and } \lfloor F^{s-m}(x) - x \rfloor < \lfloor F^{s-m+1}(x) - x \rfloor.$$

The result now follows from Exercise 8.1.7. □

Theorem 8.1.11. *Let $f \in \mathcal{H}$ and $\alpha = \rho(f)$ be irrational. Then f is topologically conjugate in $\mathcal{H}$ (i.e., the conjugacy is in $\mathcal{H}$) to R_α if and only if one (and hence all) f-orbits are dense in $\mathbb{R}/\mathbb{Z}$.*

Proof. Fix $t \in \mathbb{R}/\mathbb{Z}$ and set $t_j = f^j(t)$ for $j \in \mathbb{Z}$. Define

$$h(t_j) := j\alpha \mod 1. \tag{8.7}$$

By Theorem 8.1.10, h is an order preserving mapping of $\{t_j\}_{j\in\mathbb{Z}}$ onto a dense (recall Exercise 8.1.3(2)) subset of $\mathbb{R}/\mathbb{Z}$. The map h admits a unique continuous extension of $\mathbb{R}/\mathbb{Z}$ into itself. Call this extension h. If $\{t_j\}_{j\in\mathbb{Z}}$ is dense in $\mathbb{R}/\mathbb{Z}$, then h is a homeomorphism. From (8.7) we immediately have

$$h \circ f = R_\alpha \circ h.$$

Hence, f is semiconjugate to R_α. If $\{t_j\}_{j\in\mathbb{Z}}$ is dense in $\mathbb{R}/\mathbb{Z}$, then $h \in \mathcal{H}$ and f is topologically conjugate to R_α. Lastly, if f is conjugate to R_α, then all orbits of f are dense. □

Let $f, g \in \mathcal{H}$ with $\rho(f) = \alpha$ and $\rho(g) = \beta$ irrational. Suppose that f-orbits and g-orbits are dense in $\mathbb{R}/\mathbb{Z}$. It follows from Theorem 8.1.11 that f and g are topologically conjugate if and only if $\alpha = \beta$. On the other hand, if $f, g \in \mathcal{H}$ are such that neither map has a dense orbit and the maps are topologically conjugate, then $\rho(f) = \rho(g)$.

8.2 Degree One Circle Maps

In this subsection we give a brief introduction to degree 1 circle maps; for a detailed presentation and history see [5].

Our first examples of degree one circle maps were the *rigid rotations* defined in the Section 8.1. For rational rotations every point is periodic of the same period and hence the dynamics are not so interesting. However, the combinatorics of the periodic orbit are described via the Farey tree. We present this fact in Proposition 8.2.18; for more detail on the combinatorics of periodic orbits see [5]. In Sections 11.1 to 11.3 we present results of [48] showing that any irrational rotation can be obtained as a factor of a unimodal system.

Definition 8.2.1. Let $F : \mathbb{R} \to \mathbb{R}$ be continuous. We call F a *degree one lift* provided $F(x+1) = F(x)+1$ for all $x \in \mathbb{R}$. To each degree one lift F is associated a *degree one* map f of the circle group $\mathbb{R}/\mathbb{Z}$ such that $\pi \circ F = f \circ \pi$, where $\pi : \mathbb{R} \to \mathbb{R}/\mathbb{Z}$ is given by $\pi(x) = x \mod 1$.

Note that the lift F of any $f \in \mathcal{H}$ (from Section 8.1) is a degree one lift.

Exercise 8.2.2. *Show that the rigid rotation R_α is a degree one lift.*

Exercise 8.2.3. *[5, Proposition 3.1.7] Let F be a degree one lift. Prove that $(F+k)^n(x) = F^n(x) + kn$ for all $x \in \mathbb{R}$, $k \in \mathbb{Z}$, and $n \in \mathbb{N}$.*

Definition 8.2.4. Let F be a degree one lift. For each $x \in \mathbb{R}$ define:

$$\underline{\rho}_F(x) = \liminf_{m\to\infty} \frac{F^m(x) - x}{m} \quad \text{and} \quad \overline{\rho}_F(x) = \limsup_{m\to\infty} \frac{F^m(x) - x}{m}.$$

Set:

$$\begin{aligned} \underline{Rot}(F) &= \{\underline{\rho}_F(x) \mid x \in \mathbb{R}\} \\ \overline{Rot}(F) &= \{\overline{\rho}_F(x) \mid x \in \mathbb{R}\} \\ Rot(F) &= \{\rho_F(x) \mid x \in \mathbb{R} \text{ and } \rho_F(x) = \underline{\rho}_F(x) = \overline{\rho}_F(x)\}. \end{aligned}$$

It is known that $\underline{Rot}(F) = \overline{Rot}(F) = Rot(F)$ and that this set is either a single point or a nondegenerate closed interval [5, Theorem 3.7.20]. The interested reader can obtain proofs for Lemmas 8.2.5 and 8.2.15 and Theorem 8.2.12 from [5].

Lemma 8.2.5 ([5, Proposition 3.7.11]). *Let F be a nondecreasing degree one lift. Then, for every $x \in \mathbb{R}$, we have $\underline{\rho}_F(x) = \overline{\rho}_F$, and this number is independent of x. Thus $Rot(F)$ is a single point.*

Notice that if F is the lift of some $f \in \mathcal{H}$, then $\rho(f)$ defined in Section 8.1 is precisely $Rot(f)$.

Exercise 8.2.6. *For F, G degree one lifts, set*

$$||F - G||_\infty = \sup_{x\in\mathbb{R}}\{|F(x) - G(x)|\}.$$

Prove that $||\cdot||_\infty$ is a metric on the space of degree one lifts.

Proposition 8.2.7 ([5, Lemma 3.7.12]). *The function $F \mapsto \rho(F)$ defined on the collection of all nondecreasing degree one lifts with metric $||\cdot||_\infty$ is continuous.*

Proof. Let F be a nondecreasing degree one lift. Suppose $p \in \mathbb{Z}$, $q \in \mathbb{N}$, $\rho(F) \neq \frac{p}{q}$, and define $G : \mathbb{R} \to \mathbb{R}$ by $G(x) = F^q(x) - p - x$. By continuity, either $G(x) > 0$ for all x or $G(x) < x$ for all x. The first is equivalent to $\rho(F) > \frac{p}{q}$ and the second to $\rho(F) < \frac{p}{q}$.

The condition $G(x) > 0$ for all $x \in \mathbb{R}$ is open, that is, there exists $\epsilon > 0$ such that, if F' is a degree one lift and $||G - F'||_\infty < \epsilon$, then $F'(x) > 0$ for all $x \in \mathbb{R}$. Similarly for $G(x) < 0$. Also, the definition of G gives that G depends continuously on F. Hence, if $\rho(F) < \frac{p}{q}$ ($\rho(F) > \frac{p}{q}$), then $\rho(F') < \frac{p}{q}$ ($\rho(F') > \frac{p}{q}$) for all nondecreasing degree one lifts F' from some neighborhood of F. As this holds for all rationals $\frac{p}{q}$, we obtain the result. □

Definition 8.2.8. Let F be a degree one lift. Define maps F_l and F_u as follows:

$$\begin{aligned} F_l(x) &= \inf\{F(y) \mid y \geq x\} \\ F_u(x) &= \sup\{F(y) \mid y \leq x\}. \end{aligned}$$

We call F_l and F_u the *lower* and *upper* maps , respectively. See Figure 8.2.

One can obtain F_u by taking the graph of F and pouring water into it until the water pours out over the edges [5, page 143]. See Figure 8.2. The map F_l is obtained similarly, except that the water is poured by folks on the Mauritius Islands.

Exercise 8.2.9. ◇ *Let F be a degree one lift. Prove:*

1. *$F_l(x) \leq F(x) \leq F_u(x)$ for all $x \in \mathbb{R}$.*
2. *If F is nondecreasing, then $F_l = F = F_u$.*
3. *The maps F_l and F_u are degree one lifts and are nondecreasing.*
4. *The functions $F \mapsto F_l$ and $F \mapsto F_u$ are continuous on the space of degree one lifts with metric $||\cdot||_\infty$.* **HINT:** *$F_l(x) = \inf\{F(y) \mid y \geq x\} = \inf\{F(y) \mid x + 1 \geq y \geq x\}$ and $||F_l - G_l||_\infty = \sup\{|F_l(x) - G_l(x)| \mid x \in [0, 1]\}$.*

Remark 8.2.10. It follows from Exercise 8.2.9(4) and Proposition 8.2.7 that the functions $F \mapsto \rho(F_l)$ and $F \mapsto \rho(F_u)$ are continuous on the space of degree one lifts with metric $||\cdot||_\infty$.

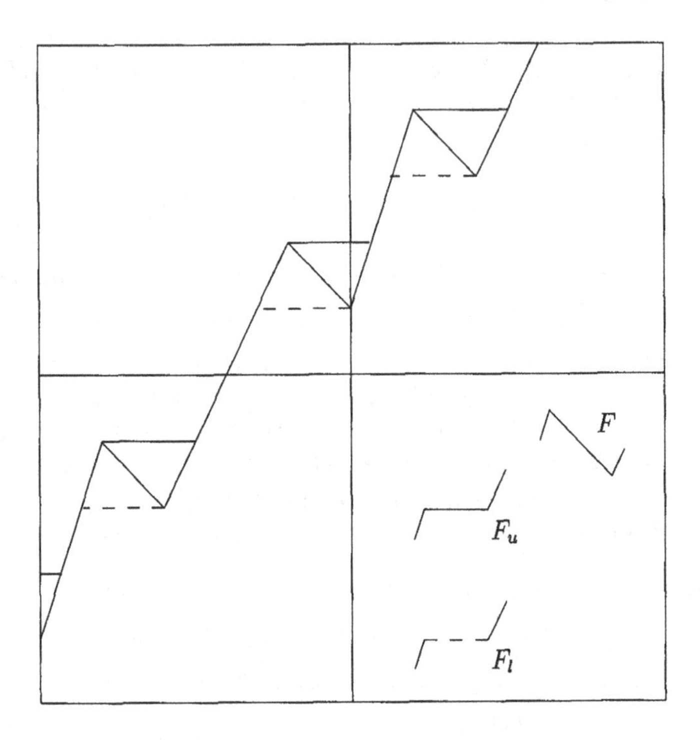

Figure 8.2: Maps F, F_l, and F_u

Definition 8.2.11. [5, Sections 3.2, 3.3, and 3.7] Let F be a degree one lift and let $x \in \mathbb{R}$. The *lift-orbit of* x is defined to be $orb(x, F) + \mathbb{Z}$, where $orb(x, F) = \{x, F(x), F^2(x), \dots\}$. We say that the lift-orbit of x is a *twisted lift-orbit* provided that F restricted to $orb(x, F) + \mathbb{Z}$ is nondecreasing. We say that $orb(x, F) + \mathbb{Z}$ is a *lifted cycle* provided there is some $n \in \mathbb{N}$ such that $F^n(x) - x \in \mathbb{Z}$. The lift-orbit $orb(x, F) + \mathbb{Z}$ is called a *lifted m-cycle* if $orb(x, F) + \mathbb{Z}$ is a lifted cycle and if $m = \min\{n \in \mathbb{N} \mid F^n(x) - x \in \mathbb{Z}\}$. Lastly, $orb(x, F) + \mathbb{Z}$ is a *twist lifted m-cycle* if $orb(x, F) + \mathbb{Z}$ is a lifted m-cycle and a twisted lift-orbit (in this case we say that x or F has a twist lifted m-cycle).

Let $F : \mathbb{R} \to \mathbb{R}$ be continuous. Set

$$\text{Const}(F) = \{x \in \mathbb{R} \mid F|U \ \ is\ constant\ for\ some\ open\ interval\ \ U \ni x\}.$$

Thus, Const(F) is the maximal open set on which F is locally constant.

Theorem 8.2.12 ([5, Theorem 3.7.20]). *Let F be a degree one lift. Then:*

1. $Rot(F) = \underline{Rot}(F) = \overline{Rot}(F) = [\rho(F_l), \rho(F_u)]$.

2. *For every $a \in Rot(F)$ there exists a twist lifted orbit of F with rotation number a and disjoint from $Const(F)$.*

3. *For every $a \in \mathbb{Q} \cap Rot(F)$ there exists a twist lifted cycle of F with rotation number a and disjoint from $Const(F)$.*

Remark 8.2.13. From Remark 8.2.10 and Theorem 8.2.12(1) we obtain that the endpoints of the rotation interval of a degree one lift F depend continuously on F.

The remainder of this section discusses the combinatorics of twist lifted cycles.

Exercise 8.2.14. ♣ *[5, Proposition 3.2.2] Let F be a degree one lift and suppose that $orb(x, F) + \mathbb{Z}$ is a lifted m-cycle. Then, the cardinality of $(orb(x, F) + \mathbb{Z}) \cap [y, y + 1) = m$ for every $y \in \mathbb{R}$.*

Lemma 8.2.15 ([5, Lemma 3.7.2 and Corollary 3.7.6]). *If F is a degree one lift, $orb(x, F) + \mathbb{Z}$ a twist lifted m-cycle, and $l \in \mathbb{Z}$ is such that $F^m(x) - x = l$, then m, l are coprime and for every $y \in orb(x, F) + \mathbb{Z}$ we have $\rho_F(y) = \frac{l}{m}$*

Exercise 8.2.16. *[5, Lemma 3.7.4] Let F be a degree one lift and let $x \in \mathbb{R}$ have a twist lifted q-cycle with $F^q(x) - x = p$. Set $orb(x, F) + \mathbb{Z} = \{\dots, y_{-2}, y_{-1}, y_0, y_1, y_2, \dots\}$ with $\dots < y_{-2} < y_{-1} < y_0 < y_1 < y_2 < \cdots$. Then $F(y_i) = y_{i+p}$ for all $i \in \mathbb{Z}$.*

Lemma 8.2.17. *Let F, x, q, p, and $\{y_i\}_{-\infty}^{\infty}$ be as in Exercise 8.2.16. Let $(orb(x, F) + \mathbb{Z}) \cap [0, 1) = \{y_0, y_1, \dots, y_{q-1}\}$; recall Exercise 8.2.14. Then, $y_{i+p} \mod 1 = y_{(i+p) \mod q}$ for each $i \in \mathbb{Z}$.*

Proof. Fix $i \in \mathbb{Z}$. Choose the unique $m \in \mathbb{Z}$ such that $i + p \in \{mq, mq + 1, \dots, (m + 1)q - 1\}$; we can this do since $\{mq + j \mid m \in \mathbb{Z} \text{ and } 0 \leq j \leq q - 1\} = \mathbb{Z}$, say $i + p = mq + t$ with $0 \leq t \leq q - 1$. Then, $(i + p) \mod q = t$. Hence, if we show that $y_{mq+t} \mod 1 = y_t$, then we are done. But $orb(x, F) + \mathbb{Z} = (orb(x, F) + \mathbb{Z}) + \mathbb{Z}$ and Exercise 8.2.14 imply that $y_i + mq = y_{mq+i}$ for $0 \leq i \leq q - 1$. Thus, $y_{mq+t} \mod 1 = (y_t + mq) \mod 1 = y_t$. This completes the proof. □

Proposition 8.2.18 gives the combinatorics (or permutation) of a twist orbit and hence describes the forward orbit of any $x \in \mathbb{R}$ under a rational rigid rotation $R_{\frac{p}{q}}$.

Proposition 8.2.18. *Let F be a degree one lift and let $x \in \mathbb{R}$ have a twist lifted q-cycle with $F^q(x)-x = p$. Set $orb(x,F)+\mathbb{Z} = \{\dots, y_{-2}, y_{-1}, y_0, y_1, y_2, \dots\}$ with $\dots < y_{-2} < y_{-1} < y_0 < y_1 < y_2 < \dots$ and with $(orb(x,F)+\mathbb{Z}) \cap [0,1) = \{y_0, y_1, \dots y_{q-1}\}$. For emphasis recall that $\frac{p}{q} = \frac{p_1}{q_1} \oplus \frac{p_2}{q_2}$. Lastly, let $i_0 \in \{0, 1, \dots, q-2\}$ and $i_1 \in \{1, 2, \dots q-1\}$. Then:*

1. $F^{q_1}(y_{i_0}) \mod 1 = y_{i_0+1}$,
2. $F^{q_1}(y_{i_0}) \in [p_1, p_1+1)$,
3. $F^{q_2}(y_{i_1}) \mod 1 = y_{i_1-1}$, *and*
4. $F^{q_2}(y_{i_1}) \in [p_2, p_2+1)$.

Proof. We first prove items 1 and 2. By Exercise 8.2.16, $F^{q_1}(y_{i_0}) = y_{i_0+q_1p}$ and therefore $(F^{q_1}(y_{i_0}) = y_{i_0+q_1p}) \mod 1 = y_{(i_0+q_1p) \mod q}$, by Lemma 8.2.17. However, $(i_0+q_1p) \mod q = i_0+1$ by Lemma 7.1.9(2–4) and hence item 1 holds. To see item 2, first note that $p_1q+1 \leq i_0+q_1p \leq p_1q+(q-1)$ (by Lemma 7.1.9(3) and then use Exercise 8.2.14 along with $F^{q_1}(y_{i_0}) = y_{i_0+q_1p}$ (Exercise 8.2.16).

Items 3 and 4 are proven in a similar manner using Lemma 7.1.10 in place of Lemma 7.1.9. □

Example 8.2.19. *Let $F(x) = x+\frac{3}{5}$ and set $y_0 = \frac{0}{1}$, $y_1 = \frac{1}{5}$, $y_2 = \frac{2}{5}$, $y_3 = \frac{3}{5}$, $y_4 = \frac{3}{5}$, $y_5 = \frac{4}{5}$. We have*

$$\frac{3}{5} = \frac{1}{2} \oplus \frac{2}{3}.$$

Then, for $i_0 \in \{0,1,2,3\}$ and $i_1 \in \{1,2,3,4\}$, we have

$$\begin{array}{rclcl} F^2(y_{i_0}) \mod 1 & = & y_{i_0+1} & \text{and} & F^2(y_{i_0}) \in [1,2) \\ F^3(y_{i_1}) \mod 1 & = & y_{i_1-1} & \text{and} & F^3(y_{i_1}) \in [2,3). \end{array}$$

Thus, setting $\frac{p}{q} = \frac{3}{5}$, $\frac{p_1}{q_1} = \frac{1}{2}$, and $\frac{p_2}{q_2} = \frac{2}{3}$ we see items 1–4 of Proposition 8.2.18 holding. Stated informally, we have:

1. *Under q_1 iterations of the map F, y_i maps to $y_{i+1} \mod (1)$ and the interger part of $F^{q_1}(y_i)$ is precisely p_1.*
2. *Under q_2 iterations of the map F, y_i maps to $y_{i-1} \mod (1)$ and the integer part of $F^{q_2}(y_i)$ is precisely p_2.*

8.3 Irrational Rotations and Return Maps

In this section we discuss irrational rotations, first return maps, and continued fraction expansions. Given an irrational rotation R_ρ, with $\rho \in (0,1)$, we show that the first return map is again an irrational rotation $R_{\rho'}$. Moreover, $\rho \in (0, \frac{1}{2}) \iff \rho' \in (\frac{1}{2}, 1)$. We establish a relation between the first return times and the continued fraction expansion of ρ; see Exercise 8.3.7. This material is used in Sections 11.2 and 11.3. For further information see [115, Chapter I.1].

Example 8.3.1. *Fix $\rho \in (0, \frac{1}{2})$ and set $R_\rho(x) = x + \rho \mod 1$. Define:*

$$\begin{aligned} J' &= (1-\rho, 1) \\ J'' &= (0, 1-\rho) \\ a(R_\rho) &= \max\{j \mid j\rho \in (0,1)\} - 1 \\ J(R_\rho) &= \overline{R_\rho^{a(R_\rho)+1}(J') \cup J'}. \end{aligned}$$

Let $\mathcal{R}(R_\rho)$ be the first return map of R_ρ to $J(R_\rho)$, that is, $\mathcal{R}(R_\rho)(x) = R_\rho^{k(x)}(x)$, where $k(x) = \min\{j \geq 1 \mid R_\rho^j(x) \in J(R_\rho)\}$. For ease of notation set $n = a(R_\rho) + 1$. Then,

$$J(R_\rho) = [(n-1)\rho, 1]$$

and

$$\mathcal{R}(R_\rho) = \begin{cases} x + n\rho \mod 1 & \text{if } x \in [1-\rho, 1] \\ x + \rho & \text{if } x \in [(n-1)\rho, 1-\rho). \end{cases}$$

Notice that $0 < (n-1)\rho < 1-\rho$ and that $\mathcal{R}(R_\rho)$ is defined on $[(n-1)\rho, 1]$. We do an affine change of coordinates to rescale $\mathcal{R}(R_\rho)$ to $[0,1]$. More precisely, set:

$$\begin{aligned} h_\rho(x) &= (1-(n-1)\rho)x + (n-1)\rho \\ \rho' &= \frac{\rho}{1-(n-1)\rho}. \end{aligned}$$

Then,

$$h_\rho^{-1} \circ \mathcal{R}(R_\rho) \circ h(x) = R_{\rho'}(x) = x + \rho' \mod 1,$$

with $\frac{1}{2} < \rho' < 1$. See Figure 8.3, where $* = (n-1)\rho$.

Example 8.3.2. *Fix $\rho \in (\frac{1}{2}, 1)$ and set $R_\rho(x) = x + \rho \mod 1$. Define:*

$$\begin{aligned} J' &= (0, 1-\rho) \\ J'' &= (1-\rho, 1) \\ a(R_\rho) &= \max\{j \mid j(1-\rho) \in (0,1)\} - 1 \\ J(R_\rho) &= \overline{R_\rho^{a(R_\rho)+1}(J') \cup J'}. \end{aligned}$$

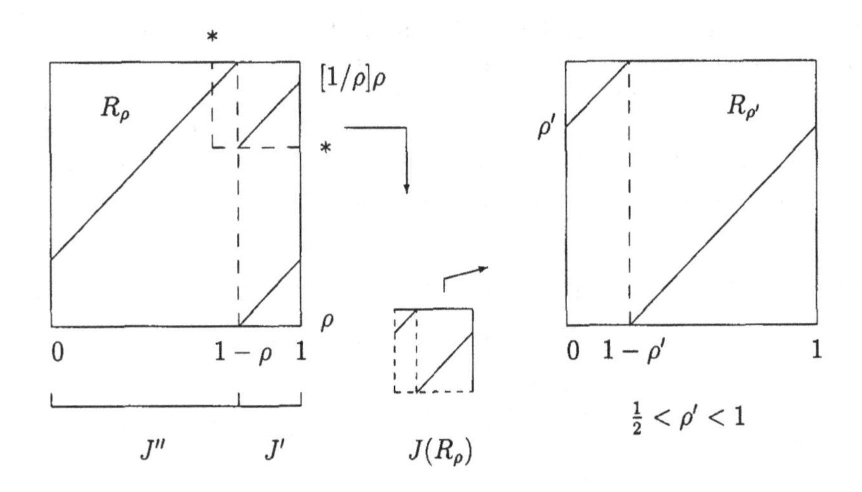

Figure 8.3: First return map $\mathcal{R}(R_\rho)$ rescaled to $R_{\rho'}$ for $0<\rho<\frac{1}{2}$

Let $\mathcal{R}(R_\rho)$ be the first return map of R_ρ to $J(R_\rho)$, that is, $\mathcal{R}(R_\rho)(x) = R_\rho^{k(x)}(x)$, where $k(x) = \min\{j \geq 1 \mid R_\rho^j(x) \in J(R_\rho)\}$. For ease of notation set $n = a(R_\rho)+1$. Then,

$$J(R_\rho) = [0, 1-(n-1)(1-\rho)]$$

and

$$\mathcal{R}(R_\rho) = \begin{cases} x+n\rho \mod 1 & \text{if } x \in [0,1-\rho] \\ x+\rho & \text{if } x \in (1-\rho, 1-(n-1)(1-\rho)]. \end{cases}$$

Notice that $1-\rho < 1-(n-1)(1-\rho) < 1$ and that $\mathcal{R}(R_\rho)$ is defined on $[0,1-(n-1)(1-\rho)]$. Again, we do an affine change of coordinates to rescale $\mathcal{R}(R_\rho)$ to $[0,1]$. More precisely, set:

$$\begin{aligned} h_\rho(x) &= (1-(n-1)(1-\rho))x \\ \rho' &= \frac{1-n(1-\rho)}{(1-\rho)+1-n(1-\rho)}. \end{aligned}$$

Then,

$$h_\rho^{-1} \circ \mathcal{R}(R_\rho) \circ h(x) = R_{\rho'}(x) = x+\rho' \mod 1,$$

with $\frac{1}{2}<\rho'<1$. See Figure 8.4, where $* = 1-(n-1)(1-\rho)$.

Exercise 8.3.3. ♣ *Verify the details of Examples 8.3.1 and 8.3.2.* **HINT:** *See [115, Section 1.1].*

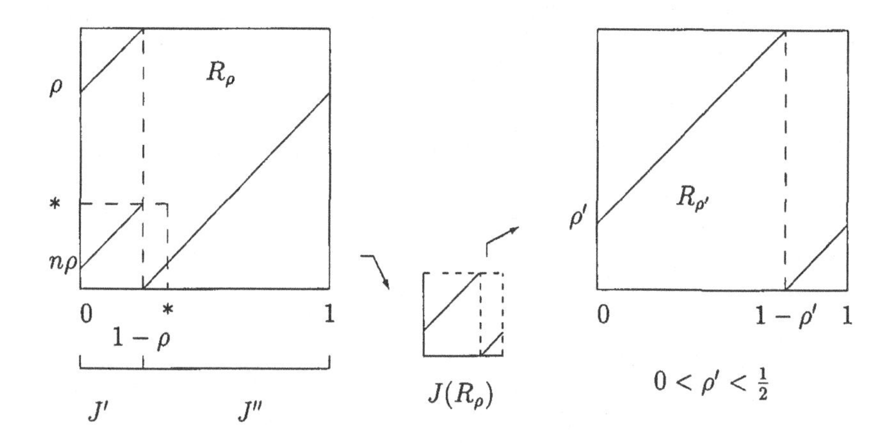

Figure 8.4: First return map $\mathcal{R}(R_\rho)$ rescaled to $R_{\rho'}$ for $\frac{1}{2}<\rho<1$

Definition 8.3.4. For $\alpha \in (0,1)\setminus\mathbb{Q}$ and $R_\alpha(x) = x+\alpha \mod 1$, set $c = c_\alpha = 1-\alpha$.

Definition 8.3.5. Let $\rho \in (0,\frac{1}{2})\setminus\mathbb{Q}$ and $R_\rho(x) = x+\rho \mod 1$. Set $a_1 = a(R_\rho)+1$, $\varphi_1 = \mathcal{R}(R_\rho)$, and $J_1 = J(R_\rho)$. For $n \geq 2$ set

$$\begin{aligned} a_n &= a(\mathcal{R}^{n-1}(R_\rho)) \\ \varphi_n &= \mathcal{R}^n(R_\rho). \end{aligned}$$

For $n \geq 1$ set

$$J'_n = \begin{cases} \text{interior of left component of } J_n\setminus\{c\} & \text{if } n \text{ is odd} \\ \text{interior of right component of } J_n\setminus\{c\} & \text{if } n \text{ is even.} \end{cases}$$

Lastly, let J''_n be the interior of the remaining component of $J_n\setminus\{c\}$. (Notice: throughout the inductive process, $c = 1-\rho$.)

Definition 8.3.6. Let $\rho \in (\frac{1}{2},1)\setminus\mathbb{Q}$ and $R_\rho(x) = x+\rho \mod 1$. Set $a_1 = 1$, $\varphi_1 = R_\rho$, and $J_1 = [0,1]$. For $n \geq 2$ set

$$\begin{aligned} a_n &= a(\mathcal{R}^{n-2}(R_\rho)) \\ \varphi_n &= \mathcal{R}^{n-1}(R_\rho). \end{aligned}$$

For $n \geq 1$ set

$$J'_n = \begin{cases} \text{interior of left component of } J_n\setminus\{c\} & \text{if } n \text{ is odd} \\ \text{interior of right component of } J_n\setminus\{c\} & \text{if } n \text{ is even.} \end{cases}$$

Lastly, let J''_n be the interior of the remaining component of $J_n \backslash \{c\}$. (Notice: throughout the inductive process, $c = 1 - \rho$.)

Exercise 8.3.7. ♣ *Let* $\rho \in (0,1) \backslash \mathbb{Q}$ *and* $R_\rho(x) = x + \rho \mod 1$. *Let* $\{a_n\}$, $\{J'_n\}$, $\{J''_n\}$, *and* $\{\varphi_n\}$ *be as in Definitions 8.3.5 and 8.3.6. Set* $q_0 = 1$, $q_1 = a_1$, $p_0 = 0$, $p_1 = 1$, *and for* $n \geq 1$

$$\begin{aligned} q_{n+1} &= a_{n+1}q_n + q_{n-1} \\ p_{n+1} &= a_{n+1}p_n + p_{n-1}. \end{aligned}$$

Prove:

1. $\varphi_n | J'_n = R_\rho^{q_{n-1}}$ *and* $\varphi_n | J''_n = R_\rho^{q_n}$.

2. $J_n = (R_\rho^{q_{n-1}}(1-\rho), R_\rho^{q_n}(1-\rho))$.

3. *The continued fraction expansion of* ρ *is precisely* $[0, a_1, a_2, \ldots]$ *with convergents* $\frac{p_i}{q_i}$ *(i.e.,* $\frac{p_i}{q_i} = [0, a_1, a_2, \ldots, a_i]$*). See [81].*

HINT: *See [115, Section 1.1].*

8.4 Cantor Thread

In this section we consider homeomorphisms h of the circle group $\mathbb{R}/\mathbb{Z}$ with an irrational rotation number and ask, what can an ω-limit set be? From here on out, we interchangeably use $S^1 = \{(x,y) \in \mathbb{R}^2 \mid x^2 + y^2 = 1\}$, $\mathbb{R}/\mathbb{Z}$, and $[0,1)$ with endpoints identified, as all are simply a circle. We prove two theorems.

Theorem 8.4.1. *If* h *is a homeomorphism of the circle* S^1 *and* $\rho(h) \notin \mathbb{Q}$, *then* $\omega(x,h)$ *is a minimal set and* $\omega(x,h)$ *is the same for all* x. *Furthermore,* $\omega(x,h) = S^1$ *unless* h *has wandering intervals. If there are wandering intervals, then* $\omega(x,h)$ *is a Cantor set.*

Theorem 8.4.2. *Given any Cantor set* $K \subset [0,1)$ *and any irrational number* $\gamma \in (0,1)$, *there is a homeomorphism* h *of the circle* S^1 *such that*

- $\rho(h) = \gamma$ *and*
- $\omega(x,h) = K$ *for all* x.

Remark 8.4.3. Circle homeomorphisms with wandering intervals were first studied by Denjoy [63]. If h is sufficiently smooth with an irrational rotation number $\rho(h)$, (i.e., h is C^2 or, more precisely, f is a C^1 diffeomorphism and $\log h'$ has bounded variation; see [115]), then h cannot have wandering intervals. In this case one can prove that h is topologically conjugate to a circle rotation over an angle $\rho(h)$. It follows from Theorem 8.1.11 that all h orbits are dense and hence indeed $\omega(x, h) = S^1$ for all x.

Proof of Theorem 8.4.1: Let x and y be two different points on S^1. By Exercise 8.1.5(5) and the assumption $\rho(h) \notin Q$, we know that x and y have infinite orbits. We will show that any point $z \in \omega(x, h)$ also belongs to $\omega(y, h)$. By changing the role of x and y, we obtain $\omega(x, h) = \omega(y, h)$.

We can assume that x and y have different orbits, because otherwise the statement is immediate. Let $z \in \omega(x, h)$ and $\epsilon > 0$ be arbitrary. Find integers $n, k > 0$ such that $|h^n(x) - z| < \epsilon$ and $|h^{n+k}(x) - z| < \epsilon$. Look at the orbit of y under h^k. There exists i such that $h^{ik}(y)$ lies between x and $h^k(x)$. Indeed, choose some orientation $<$ on the circle and find i such that $h^{ik}(y) \leq h^k(x) < h^{(i+1)k}(y)$. Then $h^{ik}(y)$ must lie between x and $h^k(x)$, because otherwise h^k does not preserve orientation.

It follows that $h^{n+ik}(y)$ lies between $h^n(x)$ and $h^{n+k}(x)$, and therefore $|h^{n+ik}(y) - z| < \epsilon$. Because ϵ was arbitrary, we have $z \in \omega(y, h)$ as claimed.

We have proved that $\omega(x, h) = \omega(y, h)$ for all $x, y \in S^1$. It follows immediately that $\omega(x, h)$ is a minimal set.

Now, if $\omega(x, h) \neq S^1$, then there exists an open interval $J \supset S^1 \setminus \omega(x, h)$. We claim that $h^i(J) \cap J = \emptyset$ for all $i > 0$. Indeed, if $h^i(J) \cap J \neq \emptyset$, then neither $J \supset h^i(J)$ nor $h^i(J) \supset J$. This would contradict that h and hence h^i preserves orientation. It follows that $J, h^i(J), h^{2i}(J), \dots$ is a "monotone" sequence of intervals, and the next interval in the sequence always intersects the previous. As $\cup_k h^{ik}(J)$ is disjoint from $\omega(x, h)$, we have $\cup_k h^{ik}(J) \neq S^1$. Hence $\cup_k h^{ik}(J)$ is an interval with an endpoint that is fixed by h^i. But this would imply that the rotation number of h is rational, contrary to our assumption that $\rho(h) \notin \mathbb{Q}$. Hence J is indeed a wandering interval.

For the remaining statement, namely that $\omega(x, h)$ is a Cantor set if there are wandering intervals, note that $\omega(x, h)$ is a closed set. If $\omega(x, h)$ contains an isolated point z, then z is periodic, contrary to our assumption. It is also easy to see that if $\omega(x, h)$ contains an interval, then $\omega(x, h) = S^1$. Hence a Cantor set is the only possibility. This completes the proof of Theorem 8.4.1.

Proof of Theorem 8.4.2: [92] Enumerate the components $I_{n \in \mathbb{Z}}$ of $S^1 \setminus K$ such that their order on S^1 is the same as that of $\{n\gamma\}_{n \in \mathbb{Z}}$ on S^1. See Figure 8.5.

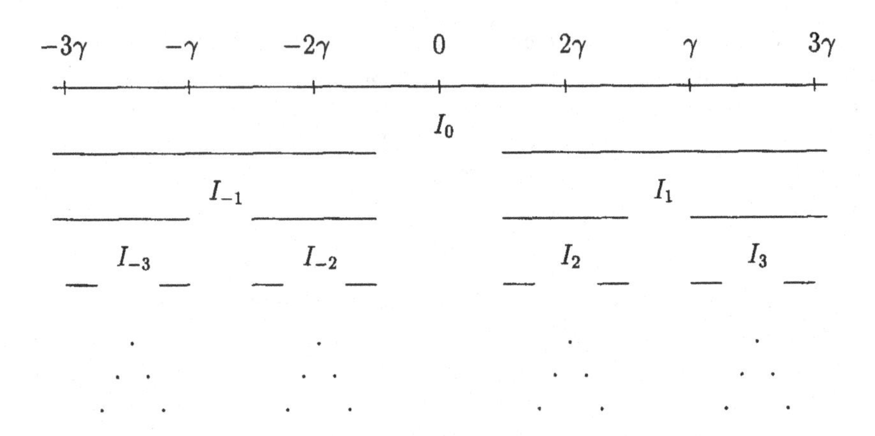

Figure 8.5: Construction of a homeomorphism h for given γ

For each $n \in \mathbb{Z}$ define $h : I_n \to I_{n+1}$ to be linear, onto, and increasing. Extend h to all of S^1 by continuity.

If $t \in S^1 \setminus K$, the orbit of t enters each I_n exactly once, and hence $\omega(t, h) \subset K$; recall that each I_n is an open interval. As any point in K is a accumulation point of a subsequence of $\{I_n\}$, we have that $\omega(t, h) = K$. Similarly, $\omega(t, h) = K$ for $t \in K$. Theorem 8.1.10 gives that $\rho(h) = \gamma$. This completes the proof of Theorem 8.4.2.

The construction in the proof of Theorem 8.4.2 is very close to the original idea of Denjoy for constructing circle maps h with wandering intervals. In fact, by cleverly choosing the intervals I_n, he managed to make h differentiable. However, as previously remarked (see Remark 8.4.3), C^2 is impossible.

Chapter 9

Topological Entropy

Given a compact metric space (X, d) and a map $f : X \to X$ (not necessarily continuous) one asks: How many orbits does the dynamical system $f : X \to X$ have? Of course, there are as many orbits as there are points, because every point has its own orbit. However, in many cases, these orbits behave in the same way. For example, take $f : S^1 \to S^1$ a rigid circle rotation (recall Definition 8.1.2) and $x, y \in S^1$. If x and y are ε apart, then $f(x)$ and $f(y)$ are also ε apart, and for every $n \in \mathbb{Z}$, $|f^n(x) - f^n(y)| = \varepsilon$. So the orbits of x and y behave in essentially the same way.

How many essentially different orbits does a dynamical system have? This depends on what we think is "essentially different"; however, it is reasonable to say that $\text{orb}(x)$ and $\text{orb}(y)$ are at least ε apart if there exists i such that $d(f^i(x), f^i(y)) > \varepsilon$. If "essentially different" means "at least ε apart," then a rigid circle rotation has "$1/\varepsilon$ essentially different orbits." A dynamical system with *sensitive dependence*, on the other hand, has infinitely many essentially different orbits: In every neighborhood of every x, there is a y that eventually gets ε apart from x, so x and y have essentially different orbits.

If x and y are very close together, it could take many iterations before (or if) we find $d(f^i(x), f^i(y)) > \epsilon$. It is therefore more useful to compute how many essentially different orbits there are up to the nth iteration. Hence, define a new metric on X by:

$$d_n(x, y) = \max\{d(f^i(x), f^i(y)) \mid i = 0, \ldots, n\}.$$

Exercise 9.0.4. *Verify that d_n is a metric for each n.*

The number of ε-different orbits in this metric, in general, still tends to infinity as $n \to \infty$; but the rate at which this happens is an important measure of the complexity of a dynamical system. It is called the *(topological) entropy* of the system, and it is an intrinsic quantity of the system: If two dynamical systems are conjugate, then they have the same entropy.

In the next section, we give a precise definition of entropy and derive its basic properties. The remainder of the chapter investigates entropy in various settings. The reader may find [168, Chapter 7] helpful while working through this chapter. Recall that $\#S$ denotes the cardinality of a set S.

Chapter 3, in particular Section 3.6, and Section 6.1 contain background material for this chapter. Results from Section 9.4 are used in Chapter 10 and Section 12.4 uses the notion of topological entropy.

9.1 Basic Properties of Topological Entropy

Let (X, d) be a compact metric space and $f : X \to X$ a map (not necessarily continuous). A set $S \subset X$ is called an (n, ε)-*spanning set* if, for every $x \in X$, there exists $y \in S$ such that $d_n(x, y) \leq \varepsilon$. In other words, for every $x \in X$, there is a $y \in S$ that ε-shadows the orbit of x up to iterate n. Observe that the compactness of X implies the existence of finite (n, ε)-spanning sets. Indeed, let $\{U_1, \dots, U_m\}$ be an open cover of X with $\operatorname{diam}(U_i) < \varepsilon$ for all i. Choose a point in every nonempty set of the form

$$\cap_{j=0}^{n} f^{-j}(U_{i_j}), \tag{9.1}$$

where $1 \leq i_j \leq m$. We have selected at most m^n points, and these points determine an (n, ε)-spanning set. For example, let $x \in X$ and for each $0 \leq j \leq n$ take i_j such that $f^j(x) \in U_{i_j}$. Then x belongs to the set in formula (9.1), and if w was the point selected from that set, then $d(f^j(x), f^j(w)) \leq \varepsilon$ for $0 \leq j \leq n$ as both $f^j(x)$ and $f^j(w)$ belong to U_{i_j} with diameter $\leq \varepsilon$. Let $s_{n,\varepsilon}(X, f)$ be the smallest cardinality of an (n, ε)-spanning set:

$$s_{n,\varepsilon}(X, f) = \inf\{\#S \mid S \text{ is } (n, \varepsilon)\text{-spanning}\}.$$

For ease of notation we write $s_{n,\varepsilon}(X, f) = s_{n,\varepsilon}(X) = s_{n,\varepsilon}(f) = s_{n,\varepsilon}$ when X and/or f are understood. We have that $s_{n,\varepsilon}(X, f) \leq m^n$ and hence

$$\limsup_{n\to\infty} \frac{1}{n} \log(s_{n,\varepsilon}(X, f)) \tag{9.2}$$

is finite. One cannot replace "lim sup" in formula (9.2) with "lim" [112, Chapter IV-7].

Exercise 9.1.1. *Show that S is (n, ε)-spanning if and only if*

$$X \subset \cup_{y\in S} \cap_{i=0}^{n} f^{-i}(\overline{B_d(f^i(y), \varepsilon)}).$$

Exercise 9.1.2. *Prove that for $n \in \mathbb{N}$ and $\varepsilon_1 > \varepsilon_2$ we have $s_{n,\varepsilon_1}(X, f) \leq s_{n,\varepsilon_2}(X, f)$.*

Definition 9.1.3. Let (X, d) be a compact metric space and $f : X \to X$ a map (not necessarily continuous). The *topological entropy* of f is the limit (as $\varepsilon \to 0$) of the exponential growth rate of $s_{n,\varepsilon}(X, f)$:

$$h_{top}(f) = \lim_{\varepsilon \to 0} \limsup_{n} \frac{1}{n} \log s_{n,\varepsilon}(X, f).$$

Exercise 9.1.4. *Show that the limit in Definition 9.1.3 exists and is non-negative. (This limit may be infinite.)* **HINT:** *From Exercise 9.1.2 we have that*

$$\varepsilon_1 < \varepsilon_2 \Rightarrow \limsup_{n} \frac{1}{n} \log s_{n,\varepsilon_1}(X, f) \geq \limsup_{n} \frac{1}{n} \log s_{n,\varepsilon_2}(X, f).$$

Topological entropy was first defined by Adler, Konheim, and McAndrew [2], using a definition based on open covers of X. Their definition requires that X be a compact metric space and that f be continuous. We would like to discuss entropy for piecewise continuous maps on compact spaces, and hence we use Definition 9.1.3, developed by Bowen [25] and Dinaburg [66]. In the setting of compact metric spaces and continuous maps, these two definitions agree. See [91, 168] for an extensive treatment of topological entropy. In general, topological entropy is difficult to compute explicitly.

Exercise 9.1.5. *Show that $h_{top}(f) = 0$ for $f : S^1 \to S^1$ a rigid circle rotation. Show, more generally, that the entropy of every isometry is 0. (Let (X_1, d_1), (X_2, d_2) be compact metric spaces and f a map from X_1 to X_2. We call f an* isometry *provided $d_1(x, y) = d_2(f(x), f(y))$ for all $x, y \in X_1$. Note that an isometry is continuous.)*

Exercise 9.1.6. ♣ *[168, Theorem 7.14] Show that every homeomorphism $h : S^1 \to S^1$ has entropy 0.*

Exercise 9.1.7. ♣ *[168, Corollary 7.14.1] Prove that any homeomorphism of $[0, 1]$ has zero topological entropy. This result is immediate from Theorem 9.4.1 of Section 9.4. Can you provide a proof without using Theorem 9.4.1?*

In the next section, we see examples of homeomorphisms on compact spaces with positive topological entropy; see Exercise 9.2.2. Thus, Exercise 9.1.7 does not generalize off the interval.

Exercise 9.1.8. *Let $T_2 : S^1 \to S^1$ be the angle doubling map $T_2(x) = 2x$* mod 1.

1. *Take $\varepsilon = 1/m$. Find an (n, ε)-spanning set and show that $s_{n,\varepsilon} \approx m2^n$.*

2. *Show that $h_{top}(T_2) = \log 2$.*

3. *Repeat the exercise for the map $T_r(x) = rx \mod 1$ for $r \in \mathbb{N}$.*

Theorem 9.1.9 tells us that, for compact spaces and continuous maps, topological entropy (Definition 9.1.3) is conjugacy invariant. This is not the case if we allow bounded but not compact spaces; see Exercise 9.1.11.

Theorem 9.1.9. *Let (X, d) and $(\tilde{X}, \tilde{d})$ be compact metric spaces and $f : X \to X$, $\tilde{f} : \tilde{X} \to \tilde{X}$ be continuous maps. If (X, d, f) is topologically conjugate to $(\tilde{X}, \tilde{d}, \tilde{f})$, then $h_{top}(f) = h_{top}(\tilde{f})$.*

Proof. Let $h : X \to \tilde{X}$ be the conjugacy. Since X is compact, h is uniformly continuous (Exercise 1.1.32), and hence for every $\varepsilon > 0$ there exists $\delta > 0$ such that for every $x, y \in X$ with $d(x, y) < \delta$ we have $d(h(x), h(y)) < \varepsilon$. If S is an (n, δ)-spanning set of (X, f), then $h(S)$ is an (n, ε)-spanning set. Indeed, if $\tilde{x} = h(x) \in \tilde{X}$ is an arbitrary point and $y \in S$ is such that $d_n(x, y) < \delta$, then $\tilde{d}_n(h(x), h(y)) < \varepsilon$. It follows that $\limsup_n \frac{1}{n} \log s_{n,\delta}(f) \geq \limsup_n \frac{1}{n} \log s_{n,\varepsilon}(\tilde{f})$. In the limit $\varepsilon \to 0$ we obtain $h_{top}(f) \geq h_{top}(\tilde{f})$. The reverse inequality follows from changing the role of f and $\tilde{f}$. □

Exercise 9.1.10. *Show that T_7 and T_3 (from Exercise 9.1.8) are not topologically conjugate.* **HINT:** *Use Theorem 9.1.9.*

Exercise 9.1.11. *This exercise shows that Theorem 9.1.9 does not hold if the spaces are bounded but not compact.*

1. *Let X be a spiral in the unit disk, defined in polar coordinates by $\{(r, \varphi) \mid \varphi \in [0, \infty), r = r(\varphi) = \frac{2}{\pi} \arctan \varphi\}$. Let $f : X \to X$ be defined by $(r(\varphi), \varphi) \mapsto (r(2\varphi), 2\varphi)$. Show (by comparing to Exercise 9.1.8) that $h_{top}(f) \geq \log 2$.*

2. *Show that f is topologically conjugate to the map $g : [0, 1) \to [0, 1)$ defined by $g(t) = \frac{2}{\pi} \arctan(2 \tan \frac{\pi}{2} t)$.*

3. *Prove that $h_{top}(g) = 0$.*

Corollary 9.1.12. *In the setting of Theorem 9.1.9, if f and $\tilde{f}$ are semi-conjugate, that is, $\tilde{f} \circ h = h \circ f$ for a continuous map $h : X \to \tilde{X}$, then $h_{top}(f) \geq h_{top}(\tilde{f})$. (Thus, the entropy of a factor cannot be greater than the entropy of the system in which it sits.)*

Proof. Use the first part of the proof of Theorem 9.1.9. □

The next proposition, Proposition 9.1.13, is a first observation regarding topological entropy. However, if the condition that the map f be continuous is dropped, Proposition 9.1.13 can fail. Can you provide an example?

Proposition 9.1.13. *Let (X, d) be a compact metric space and $f : X \to X$ continuous. Then $h_{top}(f^N) = Nh_{top}(f)$ for each $N \geq 0$.*

Proof. Fix N and let $\varepsilon > 0$ be arbitrary. Since f is continuous and X is compact, the map f, as well as each of its iterates, is uniformly continuous. Hence, there exists $0 < \delta < \varepsilon$ such that, if $d(x, y) < \delta$, then $\max_{0 \leq i < N} d(f^i(x), f^i(y)) < \varepsilon$. Now, if S is an (n, δ)-spanning set for f^N, then it is also an (nN, ε)-spanning set for f. On the other hand, an (nN, ϵ)-spanning set for f is also an (n, ϵ)-spanning set for f^N. Therefore we have

$$s_{nN,\delta}(f) \geq s_{n,\delta}(f^N) \geq s_{nN,\varepsilon}(f), \tag{9.3}$$

and hence

$$\begin{aligned} Nh_{top}(f) &= N \lim_{\varepsilon \to 0} \limsup_n \frac{1}{nN} \log s_{nN,\varepsilon}(f) \\ &\geq \lim_{\varepsilon \to 0} \limsup_n \frac{1}{n} \log s_{n,\varepsilon}(f^N) = h_{top}(f^N). \end{aligned}$$

The reverse inequality follows from the second inequality in (9.3). □

We have defined topological entropy for maps (not necessarily continuous) on compact metric spaces. It follows from Exercise 9.1.11 that Definition 9.1.3 applied to noncompact spaces is no longer a conjugacy invariant. One might ask for a definition of entropy when dealing with noncompact spaces; see [52]. Indeed, for arbitrary metric spaces and uniformly continuous maps we have Definition 9.1.14 [168]. Again, Exercise 9.1.11 shows that Definition 9.1.14 is not conjugacy invariant for noncompact spaces. The two definitions (Definition 9.1.3 and Definition 9.1.14) agree when the space is compact and the map is continuous, that is, if (X, d) is a compact metric space and $f : X \to X$ is continuous, then $h_{top}(f) = \overline{h}_{top}(f)$; moreover, in this setting, topological entropy is invariant under conjugacy. (Recall that a continuous map on a compact space is uniformly continuous; Exercise 1.1.32.) Exercise 9.1.15 provides an example, on a noncompact space, where $h_{top}(f) \neq \overline{h}_{top}(f)$.

Definition 9.1.14. Let (X, d) be a metric space and $f : X \to X$ be uniformly continuous. Let $\varepsilon > 0$, $n \in \mathbb{N}$, and $K \subset X$ be compact. A set $H \subset X$ is said to (n, ε)-span K with respect to the map f provided that, for every $x \in K$, there exists $y \in H$ with $d_n(x, y) \leq \varepsilon$. Let $s_{n,\varepsilon}(K, f)$ denote the smallest cardinality of any (n, ε)-spanning set for K with respect to f. Set

$$h(K, f) = \lim_{\varepsilon \to 0} \limsup_{n \to \infty} \frac{1}{n} \log s_{n,\varepsilon}(K, f).$$

Lastly, set

$$\overline{h}_{top}(f) = \sup\{h(K, f) \mid K \subset X \text{ compact}\}.$$

Exercise 9.1.15. *Let $f(x) = 2x$, for $x \in \mathbb{R}$. Show that $s_{n,\varepsilon}(\mathbb{R}, f) = \infty$ for all $\varepsilon < \infty$. Show that $\overline{h}_{top}(f) = \log 2$. Conclude that $h_{top}(f)$ need not equal $\overline{h}_{top}(f)$.*

Proposition 9.1.20 provides (yet) another way to calculate topological entropy for continuous maps on compact spaces. In Proposition 9.1.20 one uses "lim" to calculate h_{top}, whereas in Definition 9.1.3 one uses "lim sup" to calculate h_{top}. For Proposition 9.1.20 we consider covers of X by open sets U of d_n-diameter $\leq \varepsilon$, where d_n-diameter$(U) = \sup\{d_n(x, y) \mid x, y \in U\}$ [91, Section 3.1]. Let $\tilde{s}_{n,\varepsilon}(X, f)$ be the smallest cardinality of a cover of X with open sets of d_n-diameter $\leq \varepsilon$. The trick to being able to use "lim" in place of "lim sup" is that this new sequence $\{\tilde{s}_{n,\varepsilon}(X, f)\}_n$ is *submultiplicative* (Exercise 9.1.18), whereas the sequence $\{s_{n,\varepsilon}(X, f)\}_n$ may not be submultiplicative.

Definition 9.1.16. A sequence (a_n), $a_n \geq 0$, is called *subadditive* if $a_n + a_m \geq a_{n+m}$ for all $n, m \in \mathbb{N}$.

Proposition 9.1.17. *If (a_n) is subadditive, then $\lim_n \frac{1}{n} a_n$ exists and*

$$\lim_n \frac{1}{n} a_n = \liminf_n \frac{1}{n} a_n = \inf_n \frac{1}{n} a_n.$$

Proof. Fix m and write $n = qm + r$ for $0 \leq r < m$. Put for completeness $a_0 = 0$. Then

$$\frac{a_n}{n} \leq \frac{q a_m + a_r}{qm + r} \leq \frac{a_m}{m} + \max_{0 \leq i < m-1} \frac{a_i}{n}.$$

Then $\limsup_n \frac{a_n}{n} \leq \frac{a_m}{m}$. As this is true for all n, we have

$$\limsup_n \frac{a_n}{n} \leq \inf_m \frac{a_m}{m} \leq \liminf_n \frac{a_n}{n}.$$

□

Exercise 9.1.18. *A sequence (a_n) is called* submultiplicative *if $a_n a_m \geq a_{n+m}$ for all $n, m \in \mathbb{N}$. Verify that if (a_n) is submultiplicative, then $(\log a_n)$ is subadditive; hence*

$$\lim_{n \to \infty} \frac{1}{n} \log a_n = \liminf_{n \to \infty} \frac{1}{n} \log a_n = \inf_n \frac{1}{n} \log a_n. \tag{9.4}$$

Exercise 9.1.19. *Show that every cover $\mathcal{U}$ with open sets U of d_n-diameter $\leq \varepsilon$ gives rise to an (n, ε)-spanning set of the same cardinality.*

On the other hand, every (n, ε)-spanning set gives rise to a cover $\mathcal{U}'$ with open sets of d_n-diameter $\leq 2\varepsilon$. Conclude that

$$\tilde{s}_{n,2\varepsilon}(X, f) \leq s_{n,\varepsilon}(X, f) \leq \tilde{s}_{n,\varepsilon}(X, f). \tag{9.5}$$

Proposition 9.1.20. *Let (X, d) be a compact metric space and $f : X \to X$ be continuous. Then,*

$$h_{top}(f) = \lim_{\varepsilon \to 0} \lim_{n \to \infty} \frac{1}{n} \log \tilde{s}_{n,\varepsilon}(X, f). \tag{9.6}$$

Proof. In view of equation (9.4), it suffices to show that $\tilde{s}_{n,\varepsilon}(X, f)$ is a submultiplicative sequence. Let A be a set of d_n-diameter $\leq \varepsilon$ and B a set of d_m-diameter $\leq \varepsilon$. Then $A \cap f^{-n}(B)$ is a set of d_{n+m}-diameter $\leq \varepsilon$. Therefore, if $\mathcal{U}_n$ and $\mathcal{U}_m$ are covers with d_n-diameter resp. d_m-diameter $\leq \varepsilon$, then $\{A \cap f^{-n}(B) \mid A \in \mathcal{U}_n,\, B \in \mathcal{U}_m\}$ is a cover (of d_{n+m}-diameter $\leq \varepsilon$) of cardinality $\leq \#\mathcal{U}_n \cdot \#\mathcal{U}_m$. It follows that $\tilde{s}_{n,\varepsilon}(X, f)$ is indeed submultiplicative.

Lastly, equation (9.6) follows immediately from (9.5). This completes the proof. □

9.2 Entropy of Subshifts

In Section 3.6 we discussed shift spaces in general and the particular class of shift spaces given by the shifts of finite type. Recall that for each $n \in \mathbb{N}$ we have the full n-shift $X_n = \{0, 1, \dots, n-1\}^{\mathbb{N}}$, the shift map $\sigma : X_n \to X_n$, and the following metric on X_n:

$$d(s, t) = \sum_{i \geq 1} 2^{-i} \delta(s_i, t_i),$$

where $\delta(a, b) = 1$ if $a \neq b$ and $\delta(a, b) = 0$ otherwise. We saw that (X_n, d) is a compact metric space, X_n is a Cantor set, and σ is continuous. A set $S \subset X_n$ is called a shift space provided it is closed and is invariant under the shift map σ; we refer to such S's as *subshifts*. We saw that every shift space S can be expressed as $X_{\mathcal{F}}$ for some collection of forbidden words $\mathcal{F}$. When the collection of forbidden words is finite, we have a shift of finite type.

In this section, we first define the *complexity function* for a shift space and then use this function to calculate the topological entropy of the shift map on the shift space. Using this result, we prove that the topological entropy of the shift map on an irreducible shift of finite type is precisely the log of the spectral radius of its transition matrix (recall Exercise 3.6.9).

Fix $n \in \mathbb{N}$ and let $S \subset X_n$ be a shift space. A *cylinder set* $C_m(S)$ of length m is given by

$$C_m(S) = \{s \in S \mid s_1 s_2 \ldots s_m = c_1 c_2 \ldots c_m\}$$

for some choice $c_1 \ldots c_m \in \{0, 1, \ldots n-1\}^m$. The cylinder $C_m(S)$ is thus determined by the string $c_1 \ldots c_m$. Let $\mathcal{C}_m(S)$ be the collection of all nonempty m-cylinders (i.e., cylinder sets of length m). The *complexity* function of the subshift S is given by

$$p(m) := \#\mathcal{C}_m(S)$$

for $m \in \mathbb{N}$.

The system (S, d, σ) is a dynamical system. Indeed, (S, d) is a compact metric space and $\sigma : S \to S$ is a continuous map. Hence, we may compute the topological entropy of σ.

Theorem 9.2.1. *Let $S \subset X_l$ be a shift space. Then,*

$$h_{top}(\sigma, S) = \lim_{n\to\infty} \frac{1}{n} \log p(n).$$

Proof. First, we show that the limit exists. It is easy to verify

$$\#\mathcal{C}_n \leq \#\mathcal{C}_{n+m} \leq \#\mathcal{C}_n \cdot \#\mathcal{C}_m. \tag{9.7}$$

Therefore $p(n)$ is submultiplicative, and, by Exercise 9.1.18, $\lim_n \frac{1}{n} \log p(n)$ exists.

To prove the theorem, let us take $\varepsilon > 0$ arbitrary and N such that $2^{-N-1} < \varepsilon \leq 2^{-N}$. If C_{n+N+2} is an $n + N + 2$ cylinder, then $\sigma^n(C_{n+N+2})$ is still an $N+2$ cylinder. Therefore, if $s, t \in C_{n+N+2}$, then $d_n(s,t) \leq 2^{-N-1} < \varepsilon$. Therefore, by taking one sequence from each $n + N + 2$ cylinder, we obtain an (n, ε)-spanning set. It follows from (9.7) that

$$\begin{aligned}
\lim_{\varepsilon\to 0} \limsup_{n\to\infty} \frac{1}{n} \log s_{n,\varepsilon}(S, \sigma) &\leq \lim_{N=N(\varepsilon)\to\infty} \limsup_{n\to\infty} \frac{1}{n} \log p(n + N + 2) \\
&\leq \lim_{N\to\infty} \lim_{n\to\infty} \frac{1}{n}(\log p(n) + \log p(N + 2)) \\
&= \lim_{n\to\infty} \frac{1}{n} \log p(n).
\end{aligned}$$

On the other hand, if $s \in C_{n+N}$ and $t \in \tilde{C}_{n+N} \neq C_{n+N}$, then there exists $0 \leq i \leq n$ such that $d(\sigma^i(s), \sigma^i(t)) \geq 2^{-N+1} > \varepsilon$, so $d_n(s,t) > \varepsilon$. Therefore, any (n, ε)-spanning set contains at least one point from each $n+N$ cylinder. Because $p(n) \leq p(n+N)$, we have

$$\begin{aligned} \lim_{\varepsilon \to 0} \limsup_{n \to \infty} \frac{1}{n} \log s_{n,\varepsilon}(S, \sigma) &\geq \lim_{N=N(\varepsilon) \to \infty} \limsup_{n \to \infty} \frac{1}{n} \log p(n+N) \\ &\geq \lim_{N \to \infty} \lim_{n \to \infty} \frac{1}{n} \log p(n) \\ &= \lim_{n \to \infty} \frac{1}{n} \log p(n). \end{aligned}$$

This proves the theorem. □

Exercise 9.2.2. *Show that the topological entropy of the shift map on the full n-shift is precisely* $\log n$. *In addition, if* $S \subset X_n$ *is a shift space, prove the entropy of* σ *on* S *is no more than* $\log n$.

It follows from Exercise 9.2.2 and Theorem 9.1.9 that (X_n, σ) is not topologically conjugate to (X_m, σ) for $n \neq m$.

Exercise 9.2.3. *Let* $t = t_1 t_2 \ldots t_p t_1 t_2 \ldots t_p t_1 \ldots$ *be a periodic string from* X_2 *and set* $S = \overline{orb_\sigma(t)} \subset X_2$. *What is* $h_{top}(\sigma, S)$*?*

Exercise 9.2.4. *Let t be the sequence of formula (5.1):*

$$t = 1011101010111011101110101011101 0\ldots,$$

and set $S = \overline{orb_\sigma(t)} \subset X_2$. *What is* $h_{top}(\sigma, S)$*?*

We now consider the case where the shift space S is an irreducible shift of finite type. We prove that, in this setting, the topological entropy of the shift map is precisely the log of the spectral radius of the associated transition matrix. First, a few results from matrix analysis [89].

For $n \in \mathbb{N}$, let M_n denote the collection of square $n \times n$ matrices with real entries. Each M_n is a vector space over the reals with the usual addition operation. Given $A \in M_n$, the *spectral radius* of A, denoted $\rho(A)$, is defined to be

$$\rho(A) = \max\{|\lambda| \mid \lambda \text{ is an eigenvalue of } A\}.$$

A function $|||\cdot||| : M_n \to \mathbb{R}$ is a *matrix norm* provided the following hold for all $A, B \in M_n$:

1. $|||A||| \geq 0$.

2. $|||A||| = 0$ if and only if $A = 0$.

3. $|||cA||| = |c||||A|||$ for all $c \in \mathbb{R}$.

4. $|||A + B||| \leq |||A||| + |||B|||$.

5. $|||AB||| \leq |||A||| \, |||B|||$.

Exercise 9.2.5. *[89, page 291] Fix $n \in \mathbb{N}$ and define a function $|| \cdot ||_1 : M_n \to \mathbb{R}$ by*

$$||A||_1 = \sum_{i,j=0}^{n-1} |a_{ij}|,$$

where a_{ij} is the entry of A in row i and column j. Prove $|| \cdot ||_1$ is a matrix norm.

The proof of the next proposition can be found in [89, Corollary 5.6.14].

Proposition 9.2.6. *Let $||| \cdot |||$ be a matrix norm on M_n. Then*

$$\rho(A) = \lim_{k \to \infty} |||A^k|||^{\frac{1}{k}}$$

for all $A \in M_n$.

We are now set to determine the entropy of the shift map on an irreducible shift of finite type.

Theorem 9.2.7. *Let $A = (a_{ij})_{0 \leq i,j \leq n-1}$ be an irreducible $\{0, 1\}$ matrix. Then*

$$h_{top}(\sigma, X_A) = \log \rho(A).$$

Proof. As A is irreducible, we have

$$\begin{aligned} p(m) &= \#\{[i_0, \ldots, i_{m-1}] \in \{0, 1, \ldots, n-1\}^m \mid a_{i_0 i_1}, a_{i_1 i_2}, \ldots, a_{i_{m-2} i_{m-1}} = 1\} \\ &= \sum_{i,j} (A^m)_{ij} \\ &= ||A^m||_1. \end{aligned}$$

The theorem now follows from Theorem 9.2.1 and Proposition 9.2.6. □

Hence one way to explicitly determine the topological entropy of the shift map on an irreducible shift of finite type X_A is to determine the spectral radius of A. For $n \in \mathbb{N}$ large and $A \in M_n$, one can use Proposition 9.2.6 to estimate the spectral radius of A. If in addition the matrix A is primitive, then $\rho(A) = \lim_{m \to \infty} [a_{ij}^{(m)}]^{\frac{1}{m}}$ [89, Problem 2, page 521], which is perhaps an easier calculation.

Exercise 9.2.8. *Let (S, σ) be the golden mean shift; recall Example 3.6.5. Prove that $h_{top}(\sigma) = \log(\frac{1+\sqrt{5}}{2})$.* **HINT:** *Use $S = X_A$, where*

$$A = \begin{bmatrix} 1 & 1 \\ 1 & 0 \end{bmatrix}.$$

Exercise 9.2.9. ♣ *[89, Corollary 8.4.10] What is $h_{top}(\sigma, X_A)$ if A is not irreducible?*

Exercise 9.2.10. ◇ *Let X_A be a shift of finite type. Assume that somewhere on the diagonal of A^m there is the 2×2 block $\begin{pmatrix} 1 & 1 \\ 1 & 1 \end{pmatrix}$. Show that the entropy $h_{top}(\sigma) \geq \frac{1}{m} \log 2$.*

Example 9.2.11. *Let T denote the unique symmetric tent map such that $\frac{1}{2}$ is periodic of period 3. From Exercise 3.1.4 we have that the slope of T is $(1+\sqrt{5})/2$, the golden mean. We will see in Exercise 9.4.8 that the topological entropy of T is precisely $\log((1+\sqrt{5})/2)$. Here, we define a shift of finite type X_A using the graph of T^3 such that the following hold:*

1. *$h_{top}(\sigma, X_A) = log((1+\sqrt{5})/2) = h_{top}(T, [c_2, c_1])$.*

2. *There is a continuous and one-to-one map $h : X_A \to [c_2, c_1]$ such that the following diagram commutes.*

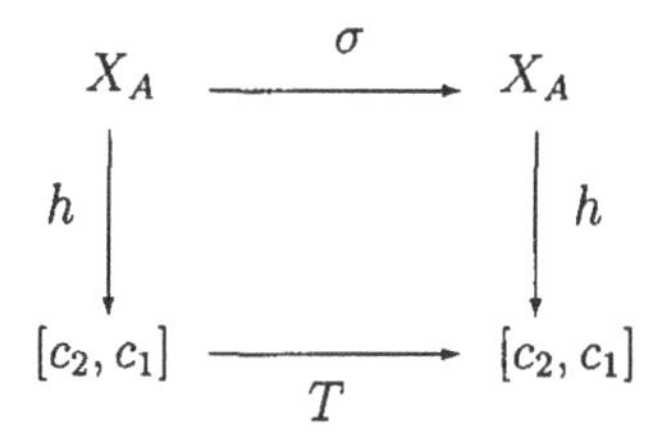

The map h is not onto. However, the points in $[c_2, c_1]$ with no preimage under h are precisely the preimages under T of $c = \frac{1}{2}$ and therefore is a countable set.

From Figure 9.1 we see that $T^3|[c_2, c_1]$ has exactly five maximal (in length) open subintervals of monotonicity. If J is one such interval of monotonicity, the first three entries in the itinerary of any $z \in J$ agree.

Let $J_0, J_1, \ldots, J_4$ denote these open intervals moving from left to right. The itinerary of any $z \in J_0$ begins as 011, of any $z \in J_1$ begins as 010, of

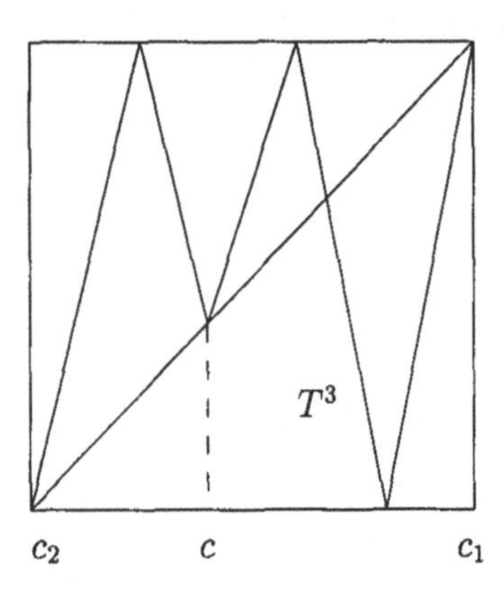

Figure 9.1: Graph of T^3 on the core $[c_2, c_1]$

any $z \in J_2$ begins as 110, *of any $z \in J_3$ begins as* 111, *and of any $z \in J_4$ begins as* 101. *We have that T maps J_0 into $J_2 \cup J_3$, maps J_1 into J_4, maps J_2 into J_4, maps J_3 into $J_3 \cup J_2$, and maps J_4 into $J_0 \cup J_1$. Using this information, form a graph as in Figure 9.2 with associated transition matrix*

$$A = \begin{bmatrix} 0 & 0 & 1 & 1 & 0 \\ 0 & 0 & 0 & 0 & 1 \\ 0 & 0 & 0 & 0 & 1 \\ 0 & 0 & 1 & 1 & 0 \\ 1 & 1 & 0 & 0 & 0 \end{bmatrix}.$$

One can check that the spectral radius of A is precisely the golden mean. For each $x \in X_A$ there exists a unique $z_x \in [c_2, c_1]$ with $I(z_x, T) = x$. Set $h(x) = z_x$.

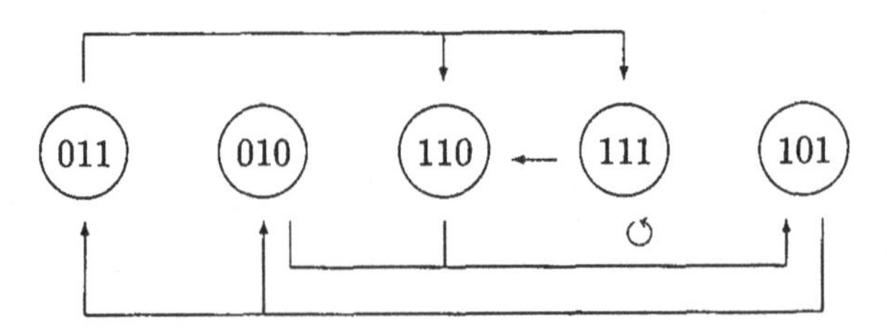

Figure 9.2: Graph G_A

Exercise 9.2.12. *Let S be a two-sided shift, that is, a subshift of $\mathcal{A}^{\mathbb{Z}}$ for some finite alphabet $\mathcal{A}$. Now the shift map σ is given by*

$$\sigma(s) = \sigma(\dots s_{-2}s_{-1}s_0.s_1s_2s_3\dots) = \dots s_{-1}s_0s_1.s_2s_3s_4\dots,$$

and it has an inverse given by

$$\sigma^{-1}(s) = \sigma^{-1}(\dots s_{-2}s_{-1}s_0.s_1s_2s_3\dots) = \dots s_{-3}s_{-2}s_{-1}.s_0s_1s_2\dots.$$

Let C_n be the collection of all $2n+1$ cylinders

$$C = \{s \in S \mid s_{-n}s_{-n+1}\cdots s_{n-1}s_n = c_{-n}c_{-n+1}\cdots c_{n-1}c_n\}$$

for some choice of $c_{-n}c_{-n+1}\cdots c_{n-1}c_n$. Show that $h_{top}(\sigma, S) = \lim_n \frac{1}{n}\log \#C_n$.

9.3 Lapnumbers and Markov Extensions

In this section, we concentrate on *piecewise monotone* maps f of an interval and introduce two tools: *lapnumbers* and *Markov extensions*. The exponential growth rate of the lapnumbers turns out to be the topological entropy of f; we make this statement precise in formula (9.8) and Theorem 9.4.1. Markov extensions are used to prove this result. First, we fix the class of maps under consideration.

Definition 9.3.1. An interval map $f : [0,1] \to [0,1]$ is called *piecewise monotone* if there exist points $0 = a_0 < a_1 < \cdots < a_N = 1$ such that $f|(a_{i-1}, a_i)$ is continuous and monotone. We call $C = \{a_0, \dots, a_N\}$ the *critical set.* We will always take the intervals (a_{i-1}, a_i) maximal such that $f|(a_{i-1}, a_i)$ is continuous and monotone. The map need not be continuous at the a_i's; in fact, $f(a_i)$ need not even be defined. We use the notation

$$f([a_{i-1}, a_i]) = [\lim_{a \searrow a_{i-1}} f(a), \lim_{a \nearrow a_i} f(a)]$$

for both continuous and discontinuous maps.

Examples of piecewise monotone maps include unimodal maps and also the β-transformations: $T_\beta(x) = \beta x \pmod 1$ (see Chapter 12).

Definition 9.3.2. Let f be a piecewise monotone map. If J is a maximal interval on which $f|J$ is continuous and monotone, then $f : J \to f(J)$ is called a *branch* or *lap* of f. The *lapnumber*, $l(f)$, is the number of laps of f.

A unimodal map has two laps. A piecewise monotone map f from Definition 9.3.1 has N laps. We are interested in counting the laps of f^n for large n because the exponential growth rate of the lapnumber equals the topological entropy:

$$h_{top}(f) = \lim_{n\to\infty} \frac{1}{n} \log l(f^n). \tag{9.8}$$

The next lemma along with Exercise 9.1.18 establish that the limit in formula (9.8) exists. In the next section we prove (Theorem 9.4) that the limit is indeed the entropy. Much of the work in this section is to prove Proposition 9.3.15, which plays a major role in the proof of Theorem 9.4.

Lemma 9.3.3. *Let f be a piecewise monotone map. Then, the sequence of lapnumbers$\{l(f^n)\}_{n\geq 1}$ is submultiplicative, that is, $l(f^{n+m}) \leq l(f^n) \cdot l(f^m)$. Hence, the limit $\lim_n \frac{1}{n} \log l(f^n)$ exists.*

Proof. If $f^n : J \to f^n(J)$ is a branch of f^n, then $f^n(J) \subset [0,1]$, and, in m more iterates, $f^n(J)$ divides into at most $l(f^m)$ branches. This shows that $\{l(f^n)\}$ is submultiplicative. The existence of the limit follows from formula (9.4). □

The second tool of this section is the *Markov extension* associated to a piecewise monotone map f. In Section 3.6 we saw transition graphs and matrices, and by taking powers of the transition matrix we were able to compute the number of n-paths in the transition graph. The use of a Markov extension is similar, except that the extension is an infinite graph, and an n-path corresponds to a lap of f^n. These ideas go back to Hofbauer and Hofbauer and Keller, for example, [83, 85, 86]. The graphs were originally called *Markov graphs*, and in the unimodal case they became known as *Hofbauer towers.* We will see the precise relation between Markov extensions and Hofbauer towers later in this section.

Let us now turn to the construction of the Markov extension associated to a piecewise monotone map f with critical set $C = \{a_0, \dots, a_N\}$. The Markov extension G is an infinite graph whose vertices are subintervals of $[0,1]$. Each such subinterval D comes with an *upper endpoint* $x_u = f^u(a_i)$ and a *lower endpoint* $x_l = f^l(a_j)$ for some $a_i, a_j \in C$ and $l, u \geq 0$; thus $D = [x_u, x_l]$ or $[x_l, x_u]$. We use the words upper and lower because we will always assume $u \geq l$. We refer to u as the *upper level* of D and l as the *lower level* of D. Two intervals D and D' represent the same vertex in G if and only if (1) $x_u = x_{u'}$, $x_l = x_{l'}$, and $u = u'$, $l = l'$ or (2) $x_u = x_{l'}$, $x_l = x_{u'}$ and $u = u' = l = l'$.

The construction starts with the *base* of the graph $B = [0,1]$, with $x_u = 0$, $u = 0$, and $x_l = 1$, $l = 0$. We continue inductively. Thus, assume $D = [x_u, x_l]$ is a vertex. For each i such that $D \cap (a_{i-1}, a_i) \neq \emptyset$, we draw an arrow $D \to D'$, where $D' = f(D \cap [a_{i-1}, a_i])$.[1] The upper and lower endpoints of D' are found as follows:

[1]We might need limits if f is discontinuous or undefined at critical points. We will use limits in the sequel if necessary, without further notice.

1. If $x_u \in [a_{i-1}, a_i]$, then $x'_{u'} = f(x_u)$ and $u' = u + 1$.

2. If $x_l \in [a_{i-1}, a_i]$, then $f(x_l)$ is an endpoint of D'. It is the upper endpoint $x'_{u'}$ (and $u' = l + 1$) if case 1 does not apply. Otherwise, $f(x_l) = x'_{l'}$ and $l' = l + 1$.

3. If D' still has endpoints for which 1 and 2 don't apply, then such endpoint is $f(a_{i-1})$ or $f(a_i)$. The *level* of such an endpoint is 1, and, depending on the other endpoint, you can decide if it is a lower or upper endpoint. For example, if $[a_{i-1}, a_i] \subset D^\circ$, then $D' = [f(a_{i-1}), f(a_i)]$ and you can choose $x'_{u'} = f(a_{i-1})$, $u' = 1$ and $x'_{l'} = f(a_i)$, $l' = 1$.

In the event that D' already exists in the graph (i.e., $D' = D''$ as described above for some vertex D''), we do not make a new vertex, but rather draw the arrow from D to the existing D'. We write G_K for the subgraph of G consisting of all vertices with upper level $\leq K$ and let $\#G_K$ denote the number of vertices in G_K.

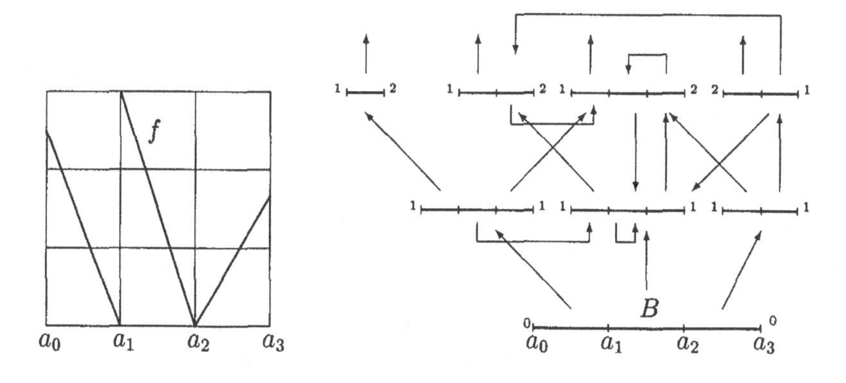

Figure 9.3: An interval map and some vertices of its Markov extension. The numbers indicate the upper and lower levels of a vertex.

An example is provided in Figure 9.3. The base B has $N = 3$ outgoing arrows, and the middle arrow goes to the interval $D = [0, 1]$. Yet D is different from B because the upper and lower levels are different (both 0 for B and both 1 for D). However, the three vertices $f([a_{i-1}, a_i])$ all have an outgoing arrow to the same vertex in the third row.

The three vertices $f([a_{i-1}, a_i])$ in the second row, and their outgoing arrows, can be read off directly from the graph of f. For the vertices on higher rows and their outgoing arrows, one needs to compute iterates of f.

By an *n-path in G* we mean a path in G with $n+1$ vertices and n arrows. The next exercise establishes that every n-path in the graph G starting from B corresponds to a lap of f^n, justifying:

$$l(f^n) = \#\{n\text{-paths of } G \text{ starting in } B\}. \tag{9.9}$$

Exercise 9.3.4. *Prove each of the following.*

1. *Each lap of f^n corresponds to precisely one n-path in G starting from B.*
2. *The image of a lap $f^n : J \to f^n(J)$ is a vertex in G with upper level less than or equal to n.*

Conclude that (9.9) holds.

Exercise 9.3.5. *Verify that, from each vertex $D \in G$, there are at most N outgoing arrows. Conclude that $\lim_n \frac{1}{n} \log l(f^n) \leq \log N$.*

Exercise 9.3.6. *Given K, show that G contains at most $2N$ vertices with upper level K. Conclude that $\#G_K \leq 2KN + 1$.*

Exercise 9.3.7. *Show that G has the* Markov property, *that is, for vertices D, D' in G we have:*

$$D \to D' \Rightarrow f(D) \supset D'. \tag{9.10}$$

This property is the reason why G is called a Markov extension or Markov graph.

The Markov extension resembles the Hofbauer tower for unimodal maps defined in Section 6.1. The Hofbauer tower was introduced as a collection of intervals not yet connected by arrows. However, this connection can easily be done, resulting in the Hofbauer tower appearing as a subgraph of the Markov extension. The Markov extension for the unimodal map with the Fibonacci combinatorics (recall Exercise 6.1.2) is given in Figure 9.4. The intervals from the Hofbauer tower are indicated by thick lines.

Exercise 9.3.8. *In Figure 9.4, consider the Hofbauer tower $\{D_i\}_{i\geq 2}$ as a subgraph of G. Recall from Exercise 6.1.6 that $D_i = \langle c_i, c_{\beta(i)}\rangle$ with $\beta(i) = i - \max\{S_k \mid S_k < i\}$. Prove the upper and lower levels of D_i in G are precisely i and $\beta(i)$, respectively. Moreover, G contains precisely two intervals with upper level i, one of which belongs to the Hofbauer tower.*

In G prove that $D_i \to D_{i+1}$ for $i \geq 2$, $D_{S_k} \to D_{S_{Q(k)}+1}$ for all cutting times $S_k \geq 2$, and there are no other arrows in G involving intervals from the Hofbauer tower. Compare to Exercise 6.1.16.

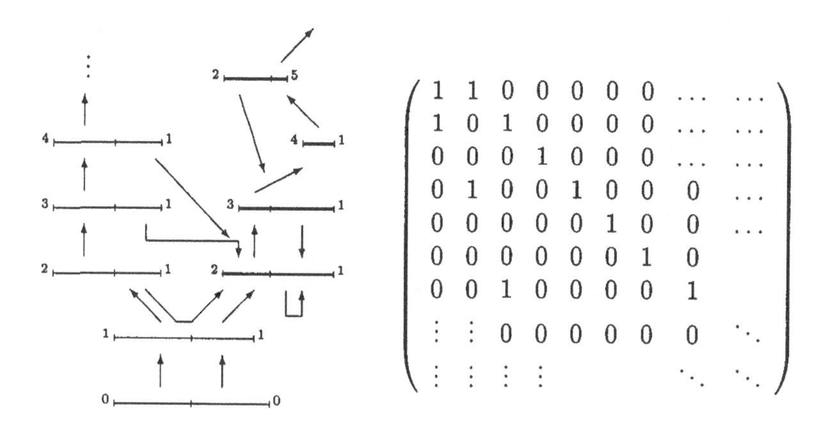

Figure 9.4: Left: The Markov extension for the Fibonacci map, with the Hofbauer tower in thick lines. (The numbers indicate the levels of the endpoints). Right: The transition matrix of the Hofbauer tower.

Exercise 9.3.9. *Let $f:[0,1]\to[0,1]$ be unimodal with the Fibonacci combinatorics. Show that, for $n\geq 2$, we have the following:*

$$l(f^n|[0,1]) = 2 + 2\sum_{k=0}^{n-2} l(f^k|[c_2,c_1]). \tag{9.11}$$

Conclude that $\lim_n \frac{1}{n}\log l(f^n|[0,1]) = \lim_n \frac{1}{n}\log l(f^n|[c_2,c_1])$. Does formula (9.11) hold for arbitrary unimodal maps?

The Hofbauer tower in Figure 9.4 has an associated infinite transition graph M, also shown in Figure 9.4. We have $M=(m_{i,j})_{i,j=2}^{\infty}$, with $m_{i,j}=1$ if $D_i\to D_j$ and $m_{i,j}=0$ otherwise.

Let M_N be the upper left $N-1\times N-1$ corner of M. Then the sum of the first row of $(M_N)^n$ is the number of n-paths in G starting from D_2 and passing only through vertices D_k with $k\leq N$. This number of n-paths gives a good approximation of $l(f^n|[c_2,c_1])$ and is in turn well approximated by $(\rho_N)^n$, where ρ_N is the spectral radius of M_N. (This follows from Proposition 9.3.15 below.) By formula (9.8) we conclude that

$$\log\rho_N \to h_{top}(f) \text{ as } N\to\infty,$$

providing a practical algorithm to compute entropy; see [41, 44, 84].

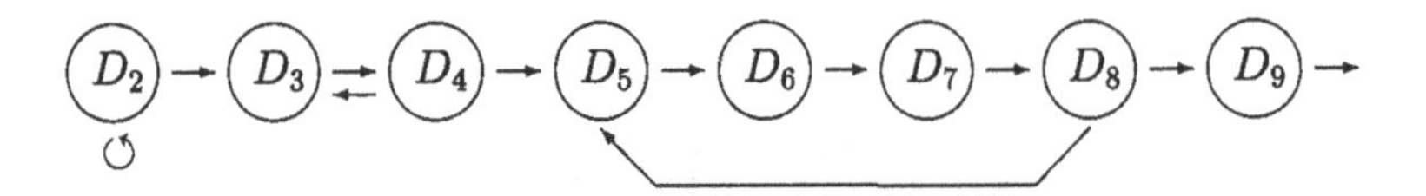

Figure 9.5: Transition graph for the Feigenbaum map

Exercise 9.3.10. *Figure 9.5 gives a schematic picture of the Hofbauer tower for the Feigenbaum map (recall Chapter 5). Verify that it is correct. This exercise aims at counting the lapnumbers of $f^n|[c_2, c_1]$. For brevity, write $a_n = l(f^n|[c_2, c_1])$ and $b_n = l(f^n|[c_3, c_1])$. Show:*

(a) $a_n = a_{n-1} + b_{n-1}$.

(b) $b_{2n} = b_{2n+1}$. *(Use the fact that vertices D_k have only one outgoing arrow for k odd.)*

(c) $b_{2n} = a_n$. *(Use the "self-similarity" of the graph.)*

(d) $a_0 = 1$, $a_1 = 2$, $a_n = a_{n-1} + a_{\lfloor (n-1)/2 \rfloor}$.

(e) Conclude that a_n grows subexponentially ($\lim_n \frac{1}{n} \log a_n = 0$) but faster than any polynomial. See [123, Table 1.5] for a discussion of this growth rate.

Let us come back to the Markov extension, G, of a piecewise monotone map f with critical set $C = \{a_0, \dots, a_N\}$. Recall that G_K is the subgraph of G consisting of all vertices with upper level $\leq K$. Let M_K denote the finite transition matrix associated to G_K and ρ_K its spectral radius. Let

$$l_K(f^n) = \#\{\ n\text{-paths in } G_K \text{ starting in } B\}.$$

Exercise 9.3.11. *Prove that $\{l_K(f^n)\}_n$ is submultiplicative.* **HINT:** *Adapt the argument given in Lemma 9.3.3 for $l(f^n)$.*

It follows from Theorem 9.2.7 and Exercise 9.3.4 that $\log \rho_K = \lim_n \frac{1}{n} \log l_K(f^n)$. Notice that $l_K(f^n)$ is nondecreasing with K and hence so also is ρ_K. We would like to conclude that $\log \rho_K \to \lim_n \frac{1}{n} \log l(f^n)$ as $K \to \infty$; this is precisely Proposition 9.3.15 below. In [143, Lemma 6], this is done somewhat implicitly, using results from infinite matrix theory [96, 154]. We provide a different proof.

Before moving to the proof of Proposition 9.3.15 we recall a tool to compute the characteristic polynomial (see [89]) of a matrix. It was developed by Block et al. [20], as was the following terminology.

Definition 9.3.12. Let G be a graph. A subgraph $\mathcal{R} \subset G$ is called a *rome* if $G \setminus \mathcal{R}$ contains no loops. In other words, all paths starting outside $\mathcal{R}$ eventually lead to $\mathcal{R}$.

Let G be a finite graph, and let M be the associated transition matrix. If $\mathcal{R}$ is a rome, then we say that $D_0 \to \cdots \to D_n$ is a *simple path* if D_0 and D_n belong to $\mathcal{R}$ but $D_i \notin \mathcal{R}$ for $0 < i < n$. Simple paths contained in $\mathcal{R}$ therefore must have length 1. Let $\#G$ and $\#\mathcal{R}$ denote the number of vertices in G and $\mathcal{R}$, respectively. Build the matrix $A = (a_{i,j})_{i,j=1}^{\#\mathcal{R}}$ by putting

$$a_{i,j} = \sum_p x^{1-l(p)},$$

where the sum is over all simple paths from vertex $i \in \mathcal{R}$ to vertex $j \in \mathcal{R}$ and $l(p)$ stands for the length of the path. We refer to A as the *rome matrix for* R. (Here, we allow ourselves a slightly different notation from that in [20].)

Theorem 9.3.13. *[20] The characteristic polynomial of M is equal to*

$$(-x)^{\#G-\#\mathcal{R}} \det(A - xI),$$

where I is the $\#\mathcal{R} \times \#\mathcal{R}$ identity matrix.

See [20] for a proof. Note that if $\mathcal{R} = G$, then $A = M$ and we retrieve the usual definition of the characteristic polynomial $\det(M - xI)$.

Exercise 9.3.14. *Assume that G consists of two loops of length $l_1 \leq l_2$ with one vertex in common. Use Theorem 9.3.13 to show that the growth rate $\lim_n \frac{1}{n} \log \#\{n\text{-path in } G\}$ is the largest root of the equation $x^{l_2} = x^{l_2-l_1} + 1$. In particular, this root is greater than* 1.

We now have things in place to state and prove the main result of this section.

Proposition 9.3.15. *Let G be the Markov extension for some piecewise monotone map f with critical set $C = \{a_0, \dots, a_N\}$. Then, for every $\delta > 0$, there exists $\mathcal{K}$ such that $K \geq \mathcal{K}$ implies*

$$\lim_n \frac{1}{n} \log l_K(f^n) \geq \lim_n \frac{1}{n} \log l(f^n) \; - \delta. \tag{9.12}$$

Note that Proposition 9.3.15 fails for arbitrary graphs. For example, consider the *binary graph* where every vertex D has one incoming (none if $D = B$) and two outgoing arrows. For each n there are 2^n n-paths; yet, $l_K(f^n)$ would be undefined for $n > K$, since G_K would contain no paths longer than K. The binary graph, however, does *not* serve as the Markov extension for any piecewise monotone map.

Proof. Set $\rho = \exp(\ \lim_n \frac{1}{n} \log l(f^n)\ )$ and fix $\delta > 0$. Recall that G_K is the subgraph of G consisting of all vertices with upper level $\leq K$, M_K is the associated transition matrix, and ρ_K is the spectral radius of M_K. We construct a graph $\tilde{G}_K$ with transition matrix $\tilde{M}_K$ and spectral radius $\tilde{\rho}_K$ such that $\rho_K \leq \rho \leq \tilde{\rho}_K$. Using Theorem 9.3.13, we show that $\tilde{\rho}_K - \rho_K \to 0$ as $K \to \infty$, proving the assertion.

Choose $t \in \mathbb{N}$ such that $b > a + 2b^{1-t}$ for all $1 \leq a < a + \delta \leq b$. Take $\mathcal{K} = 2t + 1$, $K \geq \mathcal{K}$ arbitrary, and let $D \in G_K$ be a vertex with upper level $K - t$. There are at most $2N$ such vertices; see Exercise 9.3.6. Clearly, $f^t|D$ can have many laps, but only the outer two, say $f^t|J_0$ and $f^t|J_1$, correspond to paths from D that have not "fallen" in the graph G to some vertex in $G_t \subset G_{K-t}$. Hence, there are at most two paths from D that lead out of G_K before first entering G_{K-t} again. The images $f^t(J_0)$ and $f^t(J_1) \subset [0,1] = B$, so if we add to G_K extra vertices D_i that make two t-paths $D \to D_1 \to \cdots \to D_{t-1} \to B$, then the new graph has at least as many paths from D as exist in the Markov extension G. We do this for every D with upper level $K - t$. Call the resulting graph $\tilde{G}_K$. Since the number of n-paths in $\tilde{G}_K$ grows at least as fast as $l(f^n)$, we obtain $\tilde{\rho}_K \geq \rho$. As G_K is a subgraph of G, the inequality $\rho_K \leq \rho$ is immediate. Thus, we have $\rho_K \leq \rho \leq \tilde{\rho}_K$.

Let $e \leq 2Nt$ be the total number of extra vertices in $\tilde{G}_K$ as compared to G_K. Now G_K is a rome of $\tilde{G}_K$. Therefore Theorem 9.3.13 applies, and the rome matrix for G_K can be written as

$$A(x) = M_K + x^{1-t}\Delta,$$

where Δ is an $\#G_K \times \#G_K$ matrix that consists of zeroes except for the first column. Namely, for each row corresponding to a vertex $D \in G_K$ with upper level $K - t$, there is a 2 in the first column. Now Theorem 9.3.13 tells us that the characteristic polynomial of $\tilde{M}_K$ is $(-x)^e \det(A(x) - xI)$. In particular, the largest root of this equation, that is, the spectral radius $\tilde{\rho}_K$, is the largest root of the equation $\det(A(x) - xI) = 0$. This root is then the spectral radius of $A(\tilde{\rho}_K)$. Let v be the corresponding eigenvector of norm $\|v\| = \sum_i |v_i| = 1$. Then we find

$$\tilde{\rho}_K = \|Av\| \leq \|M_K v\| + \tilde{\rho}_K^{1-t}\ \|\Delta v\| \leq \rho_K + 2(\tilde{\rho}_K)^{1-t}.$$

Recalling the choice of t gives that $\tilde{\rho}_K < \rho_K + \delta$, proving the proposition. □

9.4 Lapnumbers and Entropy

The previous section provided the preliminary work needed to relate entropy to lapnumbers. In this section, we prove the main result in this direction.

Theorem 9.4.1. *Let $f : [0,1] \to [0,1]$ be a piecewise monotone map with critical set $C = \{a_0, \dots, a_N\}$. Then $h_{top}(f) = \lim_{n\to\infty} \frac{1}{n} \log l(f^n)$.*

This theorem is due independently to Rothschild [149] and Misiurewicz and Szlenk [125]. A major difference between these proofs and the proof provided here is that we use the definition of entropy based on (n,ε)-separated sets, whereas earlier proofs build on the (original) definition based on open covers [2]. Hence, we provide a different proof.

Proof. We use Definition 9.1.3 to calculate entropy; here the compact metric space X is simply $[0,1]$.

First we establish an upper bound; namely, we prove the following:

$$h_{top} \le \lim_n \frac{1}{n} \log l(f^n). \tag{9.13}$$

Choose $\varepsilon > 0$ arbitrary. For each branch J of f^n, one can choose a finite set S such that $f^n(S) \supset \partial f^n(J)$ and such that every two neighboring points in $f^n(S)$ lie less than ε apart. Clearly one may choose S such that $\#S < 2/\varepsilon$. For $x \in J$, find $y', y'' \in S$ such that $y' \le x < y''$. Then both $|f^n(y') - f^n(x)| < \varepsilon$ and $|f^n(y'') - f^n(x)| \le \varepsilon$. Now, if $|f^{n-1}(y) - f^{n-1}(x)| > \varepsilon$ for both $y = y'$ and $y = y''$, insert new points $y \in (y', y'')$ such that $|f^{n-1}(y) - f^{n-1}(x)| < \varepsilon$. No more than $2/\varepsilon$ points need to be inserted this way. Doing this for the iterates $n-2$ down to 0, we obtain an (n,ε)-spanning set of cardinality $\le 2(n+1)/\varepsilon$. Hence,

$$\begin{aligned} h_{top}(f) &\le \lim_{\varepsilon\to 0} \limsup_{n\to\infty} \frac{1}{n} \log\left[2(n+1)l(f^n)/\varepsilon\right] \\ &= \lim_{\varepsilon\to 0} \limsup_{n\to\infty} \frac{1}{n} \log l(f^n) + \frac{1}{n} \log 2(n+1)/\varepsilon \\ &= \limsup_{n\to\infty} \frac{1}{n} \log l(f^n) = \lim_{n\to\infty} \frac{1}{n} \log l(f^n). \end{aligned}$$

We now have the upper bound, that is, we have formula (9.13).

Consider the Markov extension G of f and its finite subgraphs G_K. Recall that $l_K(f^n)$ is the number of n-paths in G starting from B that stay within G_K. We prove that, given K, ε sufficiently small and $n > K$, we have:

$$\text{an } (n,\varepsilon)\text{-spanning set has at least } \frac{l_K(f^n)}{N^K} \text{ elements.} \tag{9.14}$$

Once we establish (9.14) we have that $h_{top}(f) \ge \liminf_n \frac{1}{n} \log(l_K(f^n)/N^K)$. Using the fact from Exercise 9.3.11 that $l_K(f^n)$ is submultiplicative together with Proposition 9.1.17, we obtain

$$h_{top}(f) \ge \lim_n \frac{1}{n} \log l_K(f^n). \tag{9.15}$$

From the upper bound (9.13), we now have

$$\lim_{n\to\infty}\frac{1}{n}\log l_K(f^n)\le h_{top}(f)\le\lim_{n\to\infty}\frac{1}{n}\log l(f^n),$$

with $\lim_n \frac{1}{n}\log l_K(f^n)$ converging to $\lim_n \frac{1}{n}\log l(f^n)$ as $K\to\infty$ by Proposition 9.3.15. The theorem follows.

It remains to prove (9.14). Fix K, take $n>K$, and set $C_{K+1}=\{x\in[0,1]\mid f^i(x)\in C \text{ for some } 0\le i<K+1\}$. This is the set of critical and discontinuity points of f^{K+1}. As it is a finite set, we may set $\varepsilon_0:=\min\{d(x,y)\mid x\ne y\in C_{K+1}\}$ as positive. Choose $\varepsilon\in(0,\varepsilon_0)$ arbitrary.

We select a collection $\mathcal{J}$ of branch domains of f^{n-K} such that every (n,ε)-spanning set S intersects every $J\in\mathcal{J}$. The collection $\mathcal{J}$ will have at least $l_K(f^n)/N^K$ elements. We would then have claim (9.14), as desired.

We now construct the collection $\mathcal{J}$. Every n-path $B\to D_1\to\cdots\to D_n$ within G_K corresponds to a branch $f^n: J_0\to f^n(J_0)$. The domain J_0 is contained in exactly one branch domain J of f^{n-K}, and there are at most N^K domains J_0 corresponding to the same J; this follows easily from Exercise 9.3.5. Let $\mathcal{J}$ be the collection of all the branch domains J obtained in this way. Then, $\#\mathcal{J}\ge l_K(f^n)/N^K$.

Let S be an (n,ε)-spanning set. We show that S intersects each $J\in\mathcal{J}$. Indeed, assume by contradiction that $J\cap S=\emptyset$ for some $J\in\mathcal{J}$. Take $y\in J_0\subset J$ arbitrary, where, as before, J_0 is the domain of a branch of f^n such that its n-path $B\to D_1\to\cdots\to D_n$ stays in G_K. As S is an (n,ε)-spanning set, we may choose $x\in S$ such that $d_n(x,y)\le\varepsilon$. Notice that $x\notin J$ since $J\cap S=\emptyset$. Let U be the smallest interval containing J_0 and x, and let $k\ge 0$ be minimal such that $f^k(U)$ contains a point in C, say a_i. Then $f^k(J_0)$ intersects an ε-neighborhood of a_i. Because both endpoints of J map into C in the first $n-K$ iterates and $x\notin J$, we have $k<n-K$. Lastly, the choice of ε, $f^k(J_0)\subset D_k$, and $d(f^k(x),f^k(y))<\varepsilon<\varepsilon_0$ give that the upper level of D_{k+K+1} is at least $K+1$; thus, $D_{k+K+1}\notin G_K$ with $k+K+1\le n$. This contradicts the entire path $B\to D_i\to\cdots\to D_n$ staying inside G_K. Therefore $J\cap S\ne\emptyset$, as desired, proving (9.14). □

Exercise 9.4.2. *Use Theorem 9.4.1 to show that every homeomorphism $h:[0,1]\to[0,1]$ has entropy 0. Recall Exercise 9.1.7.*

Exercise 9.4.3. *Show that the Feigenbaum map has entropy 0.* **HINT:** *Use Exercises 9.3.10 and 9.3.9*

Corollary 9.4.4. *Let f be a piecewise monotone map and for $\gamma>0$ set $l_\gamma(f^n):=\#\{J \text{ is a branch of } f^n\mid |f^n(J)|\ge\gamma\}$. Then, for every $\delta>0$, there exists $\gamma>0$ such that*

$$h_{top}(f)\ge\liminf_n\frac{1}{n}\log l_\gamma(f^n)\ge h_{top}(f)-\delta. \tag{9.16}$$

Proof. The upper bound follows directly from Theorem 9.4.1. Given $\delta > 0$, choose K as in (9.12). Next take $\gamma := \min\{|D| \mid D \in G_K\}$. Then each branch J of f^n whose path stays within G_K has $|f^n(J)| \geq \gamma$. The corollary now follows from (9.12). □

We close this section with two more ways to compute the entropy of a piecewise monotone map; both are well known (see, e.g., [5]).

Definition 9.4.5. Let $f : [0,1] \to \mathbb{R}$. The *variation* of f is

$$\mathrm{Var}_{[0,1]}(f) = \sup\{|f(x_i) - f(x_{i-1})| \mid 0 = x_0 < x_1 < \cdots < x_N = 1\},$$

so the supremum is taken over all finite partitions of $[0,1]$.

Exercise 9.4.6. *Let $f : [0,1] \to \mathbb{R}$ be continuous and monotone. Prove each of the following:*

1. $Var_{[0,1]}(f) = |f(1) - f(0)|$.
2. *For every $a \in [0,1]$, we have $Var_{[0,1]}(f) = Var_{[0,a]}(f) + Var_{[a,1]}(f)$.*

Proposition 9.4.7. *If $f : [0,1] \to [0,1]$ is a piecewise monotone map, then*

$$h_{top}(f) = \max\{0, \lim_{n\to\infty} \frac{1}{n} \log Var_{[0,1]}(f^n)\}.$$

Proof. Using Exercise 9.4.6, we have $\mathrm{Var}_{[0,1]}(f) = \sum_J |f^n(J)| \leq l(f^n)$, where the sum is over all branches of f^n. For any $\gamma > 0$ we have that

$$\mathrm{Var}_{[0,1]}(f^n) \geq \sum_{J, |f^n(J)| \geq \gamma} |f^n(J)| \geq \gamma \#\{J \mid |f^n(J)| \geq \gamma\}.$$

Choose γ from Corollary 9.4.4. Then

$$\begin{aligned} h_{top}(f) - \delta &\leq \liminf_{n\to\infty} \frac{1}{n} \log \gamma \#\{J; |f^n(J)| \geq \gamma\} \\ &\leq \lim_{n\to\infty} \frac{1}{n} \log \mathrm{Var}_{[0,1]}(f^n) \leq h_{top}(f). \end{aligned}$$

Because $\delta > 0$ is arbitrary, the proposition follows. □

Exercise 9.4.8. *Let T_a be a symmetric tent map with $a \in [1,2]$ (recall Definition 3.1.3). Show that $h_{top}(T_a) = \log a$.* **HINT:** *Use Proposition 9.4.7.*

Proposition 9.4.9. *Let $f : [0,1] \to [0,1]$ be a piecewise monotone map, which is leo onto $[0,1]$. Let per_n be the number of periodic points with period n (not necessarily prime period). Then*

$$h_{top}(f) = \max\{0, \lim_{n\to\infty} \frac{1}{n} \log per_n\}.$$

Proof. In view of Theorem 9.4.1, we need to show that $\lim_n \frac{1}{n} \log l(f^n) = \lim_n \frac{1}{n} \log \mathrm{per}_n$. It is clear that $\mathrm{per}_n \le l(f^n)$ because, if $f^n : J \to f^n(J)$ is a lap of f^n, then J contains at most one point of period n. (Note that we use leo here, to ensure that J contains no nontrivial interval J' such that $f^n(J') = J'$.)

For the other inequality, let $\delta > 0$ be arbitrary and choose $\gamma > 0$ such that inequality (9.16) holds. Since f is leo, there exists $N = N(\gamma)$ such that, for any interval U of length $|U| \ge \gamma$, we have $f^N(U) = [0,1]$.

This implies that if $f^n : J \to f^n(J)$ is a branch with $|f^n(J)| \ge \gamma$, then $f^{n+N}(J) \supset J$; so J contains a point of period $n + N$. Hence

$$\liminf_n \frac{1}{n} \log \mathrm{per}_{n+N} \ge \lim \frac{1}{n} \log l_\gamma(f^n) \ge \lim_n \frac{1}{n} \log l(f^n) - \delta.$$

Since $\delta > 0$ is arbitrary and N does not depend on n, $\liminf_n \frac{1}{n} \log \mathrm{per}_n \ge \lim_n \frac{1}{n} \log l(f^n)$. This concludes the proof. □

Exercise 9.4.10. ◇ *Proposition 9.4.9 dealt with periodic points whose period n is not necessarily the prime period. Can you strengthen the proof to: $h_{top}(f) = \max\{0, \lim_n \frac{1}{n} \log \#\{\text{points of prime period } n\}\}$?*

9.5 Semiconjugacy to a Piecewise Linear Map

The purpose of this section is to prove Theorem 9.5.1. This result goes back to Parry [134, 135]; he dealt mainly with interval maps that can be described by a Markov chain (i.e., by a subshift of finite type). The general case for continuous maps was done in [123], among much deeper results on continuity and monotonicity of the entropy (as function of the map). We will say a little more about this in the next section (also see the exposition in [115]). In [5] another proof of Theorem 9.5.1 is given, which also applies to maps of the circle.

Theorem 9.5.1. *Let $f : [0,1] \to [0,1]$ be a piecewise monotone map with critical set $C = \{a_i\}$ and entropy $h_{top}(f) = \log \lambda > 0$. Then there exists a piecewise monotone, piecewise linear map $g : [0,1] \to [0,1]$ with slope $\pm\lambda$ on each lap such that $g \circ h = h \circ f$ for some semiconjugacy h.*

Exercise 9.5.2. *Conclude that $h_{top}(f) = h_{top}(g)$ from Theorem 9.5.1 and hence h preserves topological entropy.*

In Theorem 9.5.1, f need not be defined at the points a_i. The same lack of definiteness should be expected of g. Theorem 9.5.1 is not valid for the case $h_{top}(f) = 0$. For example, if $f(x) = x(1-x)$, then $h_{top}(f) = 0$ (verify!) and therefore $\lambda = 1$. The candidate piecewise linear map would be $g(x) = T_1 = \max(x, 1-x)$. But g has a whole interval of fixed points, so the semiconjugacy cannot be onto. Buzzi and Hubert [51] study the entropy 0 case in detail.

Exercise 9.5.3. *Show that Theorem 9.5.1 applies to circle maps as well.*

Let us now prove Theorem 9.5.1.

Proof. We use lapnumbers $l(f^n)$ to build a *formal power series* $p(t) = \sum_{n=0}^{\infty} l(f^n)t^n$. Recall from calculus that such a power series has a *radius of convergence* R, which can be computed as $1/R = \lim_n \sqrt[n]{l(f^n)}$. Since $h_{top}(f) = \log\lambda > 0$, Theorem 9.4.1 tells us that $\lambda = \lim_n \sqrt[n]{l(f^n)}$. Hence, $R = 1/\lambda$. Therefore $p(t)$ converges for $|t| < 1/\lambda$ and diverges for $|t| > 1/\lambda$. For $t = 1/\lambda$, convergence is in general more difficult to decide. In our case, however, we know that $\{l(f^n)\}$ is submultiplicative and therefore (from the proof of Proposition 9.1.17)

$$\frac{1}{n}\log l(f^n) \geq \inf_{k\geq 1}\frac{1}{k}\log l(f^k) = \lim_{k\to\infty}\frac{1}{k}\log l(f^k) = \log\lambda$$

holds for all $n \geq 1$. Thus, $l(f^n) \geq \lambda^n$, and hence all the terms in $p(1/\lambda)$ are ≥ 1. It follows that $p(1/\lambda) = \infty$, and that for all $N \in \mathbb{N}$ we have

$$\lim_{t\nearrow 1/\lambda} p(t) \geq \lim_{t\nearrow 1/\lambda}\sum_{n=0}^{N} l(f^n)t^n = \sum_{n=0}^{N} l(f^n)\lambda^{-n} \geq N.$$

Hence, $\lim_{t\nearrow 1/\lambda} p(t) = \infty$.

For $0 \leq a \leq b \leq 1$, consider the quantity

$$\rho(a,b) = \lim_{t\nearrow 1/\lambda}\frac{\sum_{n=0}^{\infty} l(f^n|[a,b])t^n}{\sum_{n=0}^{\infty} l(f^n)t^n}. \tag{9.17}$$

Since $l(f^n|[a,b]) \leq l(f^n)$, the quotient in this expression is ≤ 1 for all $t \in (0, 1/\lambda)$. Therefore we can take the limit $t \nearrow 1/\lambda$, even though the limits of the numerator and denominator can be ∞.

Somewhat imprecisely said, $\rho(a,b)$ measures, for large n, the relative amount of laps of f^n on the interval $[a,b]$ compared to the total lapnumber.

For example, if f is unimodal, $a = 0$, and b is the critical point, then $[a, b]$ contains exactly half of the laps of f^n. You can verify that indeed $\rho(a, b) = 1/2$ in this case. Also, $\rho(0, 1) = 1$; this is true for every map f.

In fact, ρ is a semimetric on $[0, 1]$. The conditions of Definition 1.1.1 are easy to check. It is clear that $\rho(a, b) \geq 0$. With the convention that $[a, b]$ and $[b, a]$ indicate the same interval, we have $\rho(a, b) = \rho(b, a)$. The triangle inequality follows from $l(f^n|[a, b]) \leq l(f^n|[a, c]) + l(f^n|[c, b])$. However, $\rho(a, b) = 0$ is possible for $a \neq b$. For example, if f^n is homeomorphic on $[a, b]$ for all n, then $l(f^n|[a, b]) \equiv 1$. The denominator in (9.17) tends to ∞, so $\rho(a, b) = 0$.

First we define the semiconjugacy h and then, using h, we determine the map g.

Define $h : [0, 1] \to [0, 1]$ by $h(s) = \rho(0, s)$. Let's check that h is continuous and onto. It is clear that $h(0) = 0$, $h(1) = 1$, and h is nondecreasing (the above example shows that h need not be strictly increasing; namely, if $\rho(a, b) = 0$, then $h(a) = h(b)$). Hence, once we have that h is continuous, it follows that h is onto.

For $s \in [0, 1]$ and $\epsilon > 0$ we have

$$h(s + \epsilon) - h(s) = \rho(0, s + \epsilon) - \rho(0, s) \leq \rho(s, s + \epsilon).$$

Thus, to prove that h is continuous at s it suffices to show that for every $\delta > 0$ there exists $\epsilon > 0$ such that $\rho(s, s + \epsilon) < \delta$.

Fix $s \in [0, 1]$ and $\delta > 0$. Choose $N \in \mathbb{N}$ such that $2\lambda^{-N} < \delta$ (notice that it is here that we are using $\lambda > 1$). Fix $\epsilon > 0$ such that $f^N|[s, s + \epsilon]$ has at most two laps. Then

$$l(f^n|[s, s + \epsilon]) \leq 2l(f^{n-N}) \text{ for all } n \geq N. \tag{9.18}$$

Thus,

$$\begin{aligned}
\rho(s, s + \epsilon) &= \lim_{t \to 1/\lambda} \frac{\sum_{n=0}^{\infty} l(f^n|[s, s + \epsilon])t^n}{\sum_{n=0}^{\infty} l(f^n)t^n} \\
&\leq \lim_{t \to 1/\lambda} \frac{\sum_{n=N}^{\infty} l(f^n|[s, s + \epsilon])t^n + \sum_{n=0}^{N-1} t^n}{\sum_{n=0}^{\infty} l(f^n)t^n} \\
&\leq \lim_{t \to 1/\lambda} \frac{2t^N \sum_{m=0}^{\infty} l(f^m)t^m + 1/(1 - t)}{\sum_{n=0}^{\infty} l(f^n)t^n} \quad \text{use (9.18)} \\
&\leq 2\lambda^{-N} < \delta.
\end{aligned}$$

Now we come to the definition of g. It must satisfy $g \circ h = h \circ f$. Thus, for any $y \in [0, 1]$, take x such that $h(x) = y$ and define $g(y) = h \circ f(x)$. This cannot be done if $x = a_i$ is a discontinuity point of f. After all, $f(a_i)$ need not be defined without ambiguity. However, this does not matter, as g need only be piecewise linear. Hence, let us leave $g(y)$ undefined if $y = h(a_i)$.

Next we check that g is well defined, that is, $g(y)$ does not depend on the point x such that $h(x) = y$. To show this, suppose that $x < x'$ are such that $h(x) = h(x') = y$. Then $l(f^n|[x, x'])$ grows so slowly that $\rho(x, x') = 0$. If $[x, x']$ contains a discontinuity point a_i of f, then g is not defined on $h([x, x']) = h(a_i)$. Thus, we need only consider the case that $f|[x, x']$ is continuous. Let $a_j < \cdots < a_k$ be the critical (turning) points between x and x'. Then $l(f^n|[x, x']) = l(f^{n-1}|f([x, a_j])) + l(f^{n-1}|f([a_j, a_{j+1}])) + \cdots + l(f^{n-1}|f([a_k, x']))$, so the lapnumbers $l(f^{n-1}|\cdot)$ on each of the images $f([x, a_j]) \dots f([a_k, x'])$ do not grow faster than $l(f^n|[x, x'])$. By continuity, $[f(x), f(x')] \subset f([x, a_j]) \cup \cdots \cup f([a_k, x'])$, and hence $\rho(f(x), f(x')) = 0$ and $h(f(x)) = h(f(x'))$. Therefore $g(y) = h \circ f(x) = h \circ f(x')$ is well defined.

We now have a map g, with a critical set contained in the set $\{h(a_i)\}$. Let us show that g has slope $\pm\lambda$ between these points. Take x, x' in the same interval (a_{i-1}, a_i). Then $f|[x, x']$ is monotone and continuous, and hence

$$\sum_{n=0}^{\infty} l(f^n|[f(x), f(x')])t^n = \sum_{n=0}^{\infty} l(f^{n+1}|[x, x'])t^n = \frac{1}{t}\left(\sum_{n=0}^{\infty} l(f^n|[x, x'])t^n - 1\right).$$

Thus

$$\rho(f(x), f(x')) = \lim_{t \to 1/\lambda} \frac{1}{t} \frac{\sum_{n=0}^{\infty} l(f^n|[x, x'])t^n - 1}{\sum_{n=0}^{\infty} l(f^n)t^n} = \lambda\rho(x, x'), \qquad (9.19)$$

because the denominator tends to ∞ as $t \nearrow 1/\lambda$. But this means that $|g \circ h(x) - g \circ h(x')| = \lambda|h(x) - h(x')|$. Therefore g is indeed piecewise continuous, piecewise linear with constant slope $\pm\lambda$. □

The off-shot of this theorem is that the semiconjugacy h squeezes an interval $[a, b]$ to a point if the lapnumber $l(f^n|[a, b])$ grows "considerably slower" than the total lapnumber $l(f^n)$. "Considerably slower" means: For every ε there exists n_0 such that $l(f^n|[a, b]) \leq \varepsilon l(f^n)$ for all $n \geq n_0$. Such intervals are for instance, intervals that are attracted to an attracting periodic point, but restrictive intervals (recall Definition 3.4.21) also can meet with this fate.

The solution g need not be unique, as Figure 9.6 shows. In this case, f is already piecewise linear, with entropy $\log 2$, and hence $g = f$ is a solution. This is also the solution from the proof of Theorem 9.5.1. Other solutions g_1 and g_2 emerge from h squeezing the intervals $[0, \frac{1}{2}]$ resp. $[\frac{1}{2}, 1]$ to points.

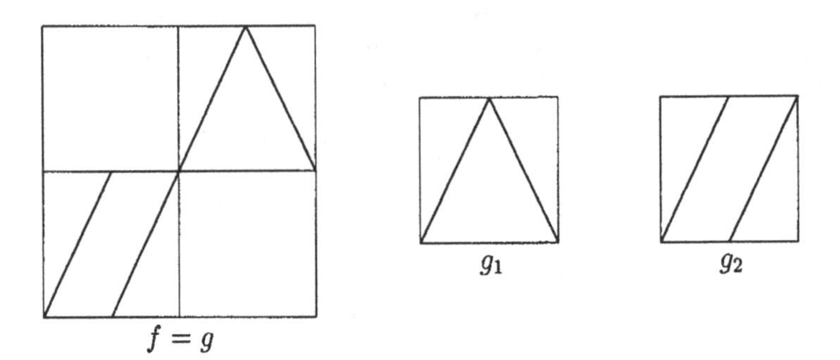

Figure 9.6: Piecewise linear maps semiconjugate to a fixed map

9.6 The Monotonicity Problem

If T_a is the tent family, then the map $a \mapsto h_{top}(T_a)$ is monotone (recall Exercise 9.4.8). It had been an open problem for a long time whether $a \mapsto h_{top}(g_a)$ is monotone for the quadratic family $g_a(x) = ax(1-x)$ with $a \in [0, 4]$. A first partial result was by Hofbauer [84]; among other things, he showed that the map $\nu \mapsto h_{top}(\nu)$ is monotone, where ν is a kneading invariant and $h_{top}(\nu)$ the topological entropy of a unimodal map with kneading invariant ν.

A full proof was given in [123, 163]. Different, and very nice, proofs can be found in [67, 165, 166]. In all of these proofs, the quadratic map is extended to the complex plane, and the structure of complex analytic maps is used. Therefore people remain curious about what "real" (i.e., not moving into the complex plane) properties are needed for the monotonicity of entropy within families of unimodal maps.

Monotonicity results for broader classes of families are proven by Tsujii, [166] and Brucks et al. [36]. But there are also counterexamples: The first is due to Zdunik [171] (piecewise linear, not approximable by maps with negative Schwarzian derivative). Another stems from Nusse and Yorke [131], but there the map is not convex. Other examples are given by Kolyada [99], Bruin [41], and Bier and Bountis [14]. Multimodal versions have been studied by Dawson et al. [62] and Milnor and Tresser [124]. In this case "monotonicity" means: The regions in parameter space of constant entropy are connected.

In searching for a proof that $a \mapsto h_{top}(g_a)$ is nondecreasing, which involves only "real techniques," the collection of *barn maps* was investigated. An example of a barn map is given in Figure 9.7. Using f_A from Figure 9.7 we form a one-parameter family of maps by scaling: $\{af_A \mid a \in [0, 1]\}$. We then ask whether the map $a \mapsto h_{top}(af_A)$ is nondecreasing.

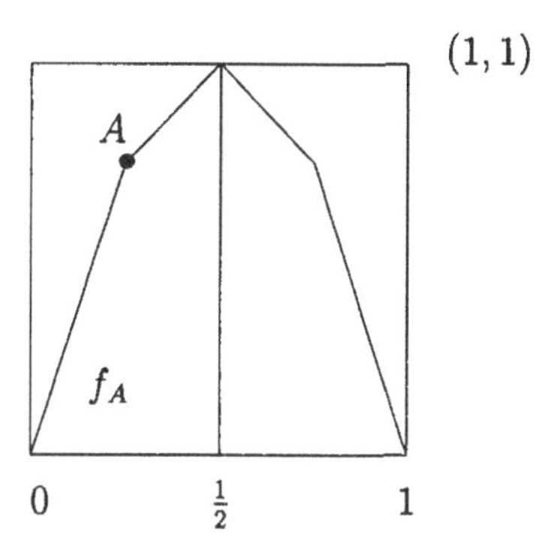

Figure 9.7: Barn map f_A

Naturally, we vary A with the restriction that the resulting map f_A be convex; we always have that $f_A(0) = f_A(1) = 0$, $f_A(1/2) = 1$, and f_A is piecewise linear. Informally, we refer to the point A as a "kink" in the graph of f_A. By varying both parameters a and A, one can "approximate" other unimodal maps. The open question is whether the monotonicity properties of these "kinked" maps (perhaps allowing for more than one kink) can be used to obtain the monotonicity of entropy results for one-parameter families of unimodal maps and in particular for the logistic family. A note of caution. It is known that a barn map can be uniformly approximated by polynomials [82, page 96]. However, a barn map cannot be approximated by a smooth map having negative Schwarzian derivative [131, Appendix A.1], and indeed logistic maps are such smooth maps. The Schwarzian derivative is defined in Definition 10.5.1.

To begin, one considers the case of one kink, namely barn maps. We have the following open question.

Question 9.6.1. Let

$$\mathcal{K} = \{A \mid a \mapsto h_{top}(af_A) \text{ is nondecreasing } \}.$$

Characterize the parameters $A \in \mathcal{K}$. What can be said about the set $\mathcal{K}$?

On can use an algorithm from [22] to obtain a numerical estimate for the map $a \mapsto h_{top}(f_a)$ (this algorithm is easily programmed). For example, using this algorithm, numerical estimates indicate that when $A = (0.26, 0.69)$ the map $a \mapsto h_{top}(f_A)$ is not monotone increasing [13].

If one obtained an answer to Question 9.6.1, the next step would be to re-ask the question with the number of kinks increased. Lastly, can the monotonicity results obtained be carried over to the logistic family (using "real" techniques)?

Chapter 10

Symmetric Tent Maps

We begin this chapter (Sections 10.1 and 10.2) with a few more combinatoric tools and then move to a detailed study of ω-limit sets arising from symmetric tent maps $\{T_a\}_{a\in[0,2]}$. Next, we investigate the *phase portrait* (Section 10.3) for this family of maps. From the phase portrait we are able to better understand the asymptotic dynamics of the turning point as the parameter a is varied. Recall that a unimodal map that is not renormalizable, has no attracting periodic points, and has no wandering intervals is topologically conjugate to a symmetric tent map (Theorem 3.4.27), and hence many of the results of this chapter hold for a boarder class of maps than the family of symmetric tent maps. For example, later in the chapter we introduce S-unimodal maps, many of which satisfy these three properties.

In Section 10.4 we use properties of the phase portrait along with a known measure result [35] to prove that the set $\tilde{\mathcal{D}} = \{a \in [\sqrt{2}, 2] \mid \overline{\{c_{S_k(a)}(a)\}_{k\geq 0}} = [c_2(a), c_1(a)]\}$ has full Lebesgue measure. The proof is a nice use of bounded distortion. Next (Section 10.5) we introduce S-unimodal maps and discuss three conditions: slow recurrence, finite criticality, and Collet-Eckmann; the first is a combinatoric property, the second topological, and the last metric. We see that the combinatoric and topological conditions give the metric one, and that Lebesgue almost every $a \in [1, 2]$ is such that the kneading sequence of T_a is slowly recurrent. A graphical consequence of critically finite is provided. In the next section we briefly discuss *attractors* for S-unimodal maps. Lastly, we discuss the combinatorics of renormalization, providing examples within the symmetric tent family.

When more than one parameter value, a, is being used, we may write $D_n(a)$ for the levels in the Hofbauer tower for T_a, $Q_a(k)$ for the kneading map of T_a, $S_k(a)$ for the cutting times of T_a and $c_n(a)$ for $T_a^n(c)$. Unless stated otherwise, $c = c(a) = c_0(a) = 0.5$.

Chapter 3 and Sections 6.1, 6.2, and 9.4 contain background material for this chapter.

10.1 Preliminary Combinatorics

We recall the notion of *itinerary* provided in Definition 5.1.3 and the *parity-lexicographical ordering* (plo) discussed in Section 5.2. Let $f : J \to J$ be unimodal with turning point c. For each $x \in I$, the itinerary of x under the map f is given by:

$$I(x, f) = \langle I_0(x), I_1(x), I_2(x), \ldots \rangle,$$

where

$$I_j(x) = \begin{cases} 0 & \text{if } f^j(x) < c \\ * & \text{if } f^j(x) = c \\ 1 & \text{if } f^j(x) > c. \end{cases}$$

The plo ($\preceq$) works as follows: Let $v \neq w$ be itineraries and find the first position where v, w differ; compare in that position using the usual ordering $0 < * < 1$ if the number of 1's preceding this position is even and use the ordering $0 > * > 1$ otherwise. We need to take into account parity since, every time we apply the right side of the unimodal map, we reverse orientation. Hence, if we have done so an odd number of times (i.e., an odd number of 1's), then one must reverse orientation.

Lemma 10.1.1. *Let $f : J \to J$ be a unimodal map with turning point c. Fix $x < y \in J$. Then $I(x, f) \preceq I(y, f)$.*

Proof. If $I(x, f) = I(y, f)$, then we are done. Thus, let $k = \min\{j \mid I_j(x) \neq I_j(y)\}$. If the number of 1's preceding the kth position is even (odd), then $f^k(x) \leq f^k(y)$ ($f^k(x) \geq f^k(y)$) and hence $I(x, f) \preceq I(y, f)$ ($I(x, f) \succeq I(y, f)$). □

Lemma 10.1.2. *Let $f : J \to J$ be a strictly unimodal map (recall Definition 3.4.11) with turning point c. Assume there are no homtervals and let $x \neq y \in J$. Then $I(x, f) \neq I(y, f)$.*

Proof. Suppose to the contrary that $I(x, f) = I(y, f)$. Without loss of generality, assume $x < y$. It then follows from Lemma 10.1.1 and $I(x, f) = I(y, f)$ that $I(z, f) = I(x, f) = I(y, f)$ for all $z \in (x, y)$, and therefore $*$ cannot be in $I(x, f)$, as f is strictly unimodal. Thus, $[x, y]$ is a homterval, a contradiction. □

Definition 10.1.3. Let $\mathcal{A}_{0,1,*}$ denote the collection of all one-sided infinite strings of 0's and 1's and all finite such strings followed by a $*$.

Exercise 10.1.4. ◇ *Let $f : J \to J$ be strictly unimodal with turning point c, and let $A \in \mathcal{A}_{0,1,*}$. If $I(c_2, f) \preceq A \preceq I(c_1, f)$, then there exists $x \in J$ with $I(x, f) = A$.* **HINT:** *Suppose to the contrary that no such x exists. Set $L_A = \{y \mid I(y, f) \prec A\}$ and $R_A = \{y \mid I(y, f) \succ A\}$. Show L_A and R_A are open sets, contradicting $L_A \cup R_A = J$. See [56, Theorem II.3.8].*

Proposition 10.1.5. *Let $a \in [\sqrt{2}, 2]$ and $A \in \mathcal{A}_{0,1,*}$. Then, there exists $x \in [c_2(a), c_1(a)]$ with $I(x, T_a) = A$ if and only if $I(c_2(a), T_a) \preceq A \preceq I(c_1(a), T_a)$. Moreover, such an x is unique.*

Proof. The result follows from Lemmas 10.1.1 and 10.1.2 and Exercise 10.1.4. □

In fact, Proposition 10.1.5 holds for all $a \in [1, 2]$, with $[c_2(a), c_1(a)]$ being replaced by the nonwandering set Ω.

Proposition 10.1.5 allows one to identify each $x \in [c_2(a), c_1(a)]$ by its itinerary $I(x, T_a)$. Moreover, we know precisely which itineraries arise, as they must lie between $I(c_2(a), T_a)$ and $I(c_1(a), T_a)$ in the parity-lexicographical ordering. Identifying points by their itineraries is a tool we frequently implement. For example, given a and n, the domain of each monotone lap of T_a^n has a unique finite code (of length n) associated to it, namely, the first n entries of $I(z, T_a)$ for any z in the domain. See Figure 10.1 for examples of this coding.

Exercise 10.1.6. *Determine the codes of length 5 that occur for each of the maps given in Figure 10.1.*

Exercise 10.1.7. *Fix $a \in [\sqrt{2}, 2]$ and $z \in [c_2(a), c_1(a)]$ such that $I(z, T_a)$ does not contain an $*$. Suppose $\lim_{n\to\infty} z_n = z$. Prove that $\lim_{n\to\infty} I(z_n, T_a) = I(z, T_a)$ in $\{0, 1\}^{\mathbb{N}}$. What happens when $*$ appears in $I(z, T_a)$?*

Exercise 10.1.8. *Let $T = T_2$. Construct, with itineraries, $z \in [0, 1]$ such that $\omega(z, T) = [0, 1]$, that is, such that the forward orbit of z is dense in $[0, 1]$.* **HINT:** *Set $\mathcal{L} = \cup_{n\geq 1} \{0, 1\}^n$ and enumerate $\mathcal{L}$ as $\{l_1, l_2, l_3, \cdots\}$. Form $e \in \{0, 1\}^{\mathbb{N}}$ by concatenating elements from $\mathcal{L}$. Show $I(0, T_2) \preceq e \preceq I(1, T_2)$ and hence there exists $z \in [0, 1]$ with $I(z, T_2) = e$. Using itineraries, prove $\omega(z, T_2) = [0, 1]$.*

In Section 9.4 we saw that $h_{top}(T_a) = \log(a) = \lim_{n\to\infty} l(T_a^n)$ (Exercise 9.4.8). Hence, the map $a \mapsto h_{top}(T_a)$ is a strictly increasing function. Thus, for $a \neq b$ we have that the kneading sequences of T_a and T_b are distinct, that is, $I(c_1(a), T_a) \neq I(c_1(b), T_b)$; otherwise (using the coding of laps and Proposition 10.1.5), $l(T_a^n) = l(T_b^n)$ for all n and therefore $h_{top}(T_a) = h_{top}(T_b)$.

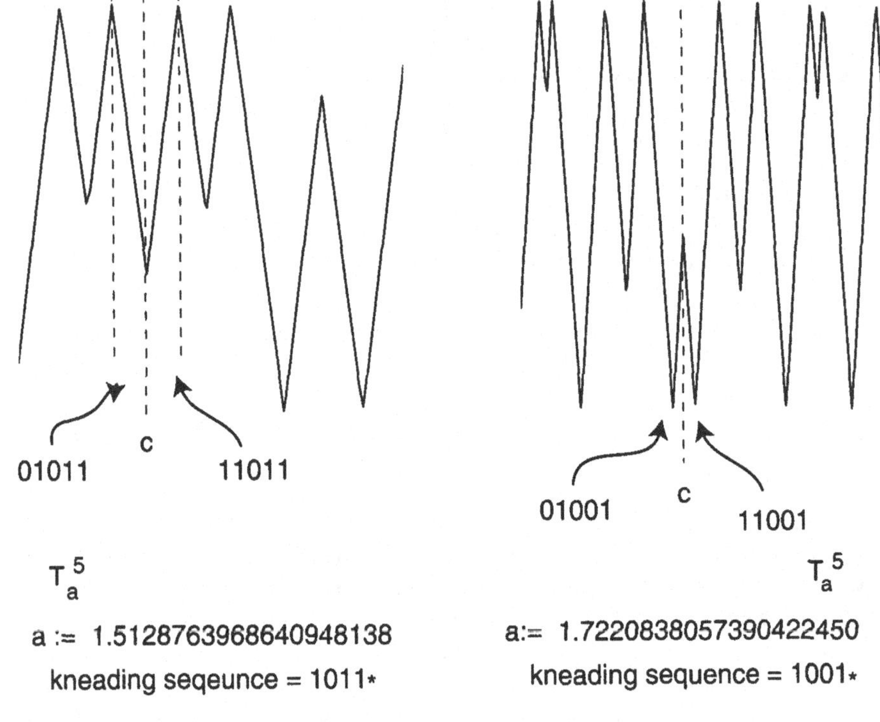

Figure 10.1: Coding of laps

Exercise 10.1.9. *Prove that the map $a \mapsto I(c_1(a), T_a)$ is strictly increasing, that is, $a < b$ implies that $I(c_1(a), T_a) \prec I(c_1(b), T_b)$.* **HINT:** *Suppose to the contrary that $a < b$ with $I(c_1(a), T_a) \succeq I(c_1(b), T_b)$. Then, using Proposition 10.1.5, conclude that $l(T_a^n) \geq l(T_b^n)$ for all n and hence $h_{top}(T_a) \geq h_{top}(T_b)$, a contradiction.*

Remark 10.1.10. For $\sqrt{2} < a^m \leq 2$, with $m \in \{2, 2^2, 2^3, \dots\}$, the map T_a is m times renormalizable (recall Proposition 3.4.26). Moreover, there exists a unique $b \in (\sqrt{2}, 2]$ such that T_a^m is topologically conjugate to T_b. Indeed, let $J \ni c$ be maximal such that $T_a^m(J) \subset J$ and $T_a^m|J$ is again unimodal. Set $g = T_a^m|J$. Then, $I(c, g) = I(c, T_b)$; see Section 10.7. The uniqueness of b follows from the map $b \mapsto h_{top}(T_b) = \log(b)$ being strictly increasing.

Remark 10.1.10 tells us that, to study the ω-limit sets for the one-parameter family of symmetric tent maps, it suffices to restrict our attention to $a \in [\sqrt{2}, 2]$. Note that $I(c_1(\sqrt{2}), T_{\sqrt{2}}) = 101^\infty$ and $I(c_1(2), T_2) = 10^\infty$.

Next, we look at another characterization of which sequences from $\mathcal{A}_{0,1,*}$ are indeed kneading sequences for some unimodal map. The admissibility condition for a kneading map, given in Definition 6.1.12, also provides a characterization.

Definition 10.1.11. Let $A = A_0, A_1, A_2, \ldots \in \mathcal{A}_{0,1,*}$. We say A is *shift maximal* provided

$$A_j, A_{j+1}, \ldots \preceq A$$

for all $j \geq 1$. Some authors use the term *admissible* in place of *shift maximal.*

Exercise 10.1.12. *Prove that $I(c_1(a), T_a)$ is shift maximal for all $a \in [1,2]$.* **HINT:** *Use Lemma 10.1.1 and the fact that $c_j(a) \leq c_1(a)$ for all j.*

Remark 10.1.13. In fact, if $A \in \mathcal{A}_{0,1,*}$ is shift maximal, then there exists some unimodal map g with $I(c_1, g) = A$; recall Remark 6.1.14. The proof of this fact is beyond the scope of the text; see [87].

Remark 10.1.14. Recalling Section 6.1, we have the following observations.

1. Given an admissible kneading map $Q : \mathbb{N} \to \mathbb{N} \cup \{0\}$, one can define $A \in \mathcal{A}_{0,1,*}$ by setting $A = e_1, e_2, \ldots$, where $e_1 = 1$ and for $j \geq 1$ we have

$$e_{S_{j-1}+1}, \ldots, e_{S_j} = e_1, \ldots, e_{S_{Q(j)}-1}, \bar{e}_{S_{Q(j)}}. \qquad (10.1)$$

   Then, indeed A is shift maxmimal.

2. Given $A \in \mathcal{A}_{0,1,*}$ shift maximal and containing no $*$, one can produce $Q : \mathbb{N} \to \mathbb{N} \cup \{0\}$ by breaking A into blocks of maximal length agreeing with initial blocks of A, as in (10.1). Then, Q is indeed admissible.

Note that we made no use of a unimodal map in this remark; it is purely a combinatoric remark. Can you provide a proof?

Exercise 10.1.15. *Prove that, with the exception of 0^∞, $*$, 1^∞, $1*$, and 10^∞, each shift maximal sequence begins as 10^m1 for some $m \in \mathbb{N}$.*

Exercise 10.1.16. *Let $A \in \mathcal{A}_{0,1,*}$ be shift maximal and suppose A begins as 10^m1. Prove that if 10^n appears in A, then $n \leq m$.* **HINT:** *Suppose to the contrary and simply compare strings using the plo.*

Lemma 10.1.17. *Fix $a \in [1,2]$ and suppose $I(c_1(a), T_a)$ is infinite, say $I(c_1(a), T_a) = e_1, e_2, e_3, \ldots$. Let $\{S_k\}_{k \geq 0}$ be the sequence of cutting times for T_a. Then $e_1, e_2, \ldots, e_{S_k - 1}, *$ is shift maximal for all $k \geq 1$.*

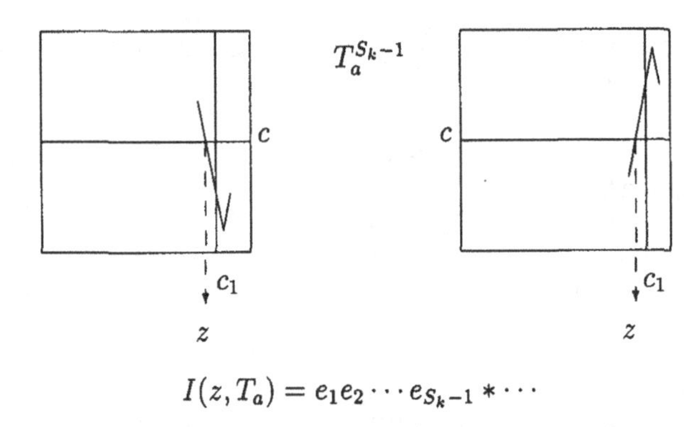

$$I(z, T_a) = e_1 e_2 \cdots e_{S_k-1} * \cdots$$

Figure 10.2: Piece of $T_a^{S_k-1}$

Proof. Exercises 6.1.1 and 6.1.4 imply that one of the two pictures in Figure 10.2 hold. Let z be as in Figure 10.2.

We see that each of $e_1, e_2, \dots, e_{S_k-1}, 0$, $e_1, e_2, \dots, e_{S_k-1}, 1$, and $e_1, e_2, \dots, e_{S_k-1}, *$ are initial segments of itineraries. We have two cases to consider: $e_{S_k} = 0$ and $e_{S_k} = 1$. We do the first case; the second is similar. We have $I(z, T_a) = e_1, e_2, \dots, e_{S_k-1}, *$.

In this case, $T_a^{S_k-1}$ maps the closed interval $[z, c_1(a)]$ monotonically onto $[c_{S_k}, c]$. Hence, $T_a^j(z) \notin [z, c_1(a)]$ for all $1 \leq j < S_k - 1$ (otherwise, $T_a^{S_k-1-j}(T_a^j(z)) = c$ with $S_k - 1 - j < S_k - 1$, contradicting the monotonicity of $T_a^{S_k-1}$ on $[z, c_1(a)]$). The result now follows. □

Lemma 10.1.18. *Fix $a \in (\sqrt{2}, 2]$ and suppose $I(c_1(a), T_a)$ is infinite, say $I(c_1(a), T_a) = e_1, e_2, e_3, \dots$. Let $\{S_k\}_{k \geq 0}$ be the sequence of cutting times for T_a. Then, the following hold:*

1. *For all $k \geq 1$, we have $e_1, e_2, \dots, e_{S_k-1}* \prec e_1, e_2, \dots$.*

2. *There exists $k_0 \in \mathbb{N}$ such that, if $k > k_0$, then $e_1, e_2, \dots, e_{S_k-1}* \succ 101^\infty$.*

3. *Let k_0 be as in (2) and $k \geq k_0$. Then there exists $b_k \in (\sqrt{2}, 2]$ with $I(c_1(b_k), T_{b_k}) = e_1, e_2, \dots, e_{S_k-1}, *$.*

Proof. Item (1) follows immediately from Exercise 6.1.4. For item (2), note that as $a > \sqrt{2}$ we have $I(c_1(a), T_a) \succ 101^\infty$. Let $m \in \mathbb{N}$ be the index of the first position where $I(c_1(a), T_a)$ and 101^∞ differ. Any k_0 with $S_{k_0} > m$ works.

Now for item (3). Let k_0 be as in item (2) and $k > k_0$. Notice that $I(c_2(a), T_a) \preceq e_1, e_2, \dots, e_{S_k-1}* \preceq I(c_1(a), T_a)$. Thus we may set

$$b_k = \inf \{b \in (\sqrt{2}, 2] \mid I(c_2(b), T_b) \preceq e_1, e_2, \dots, e_{S_k-1}* \preceq I(c_1(b), T_b)\}.$$

Since $k \geq k_0$, it follows that $b_k > \sqrt{2}$. We claim that $I(c_1(b_k), T_{b_k}) = e_1, e_2, \ldots, e_{S_k-1}*$. □

Exercise 10.1.19. *In the proof of Lemma 10.1.18, prove that* $I(c_1(b_k), T_{b_k}) = e_1, e_2, \ldots, e_{S_k-1}*$.

Lemma 10.1.18 provides a way to approximate an infinite (nonperiodic or nonpreperiodic) kneading sequence by finite kneading sequences. More precisely, let a and k_0 be as in Lemma 10.1.18. Then we have

$$b_{k_0+1} < b_{k_0+2} < b_{k_0+3} < \ldots < a$$

with $\lim_{j\to\infty} b_{k_0+j} = a$ and $I(c_1(b_k), T_{b_k})$ agreeing with $I(c_1(a), T_a)$ through the first $S_k - 1$ positions.

10.2 ω-Limit Sets

What are possible ω-limit sets for symmetric tent maps? Due to renormalization, it suffices to consider T_a with $a \in [\sqrt{2}, 2]$ (see Remark 10.1.10). Since any ω-limit set is a proper closed invariant subset of the core, we know from Remark 3.4.17 that an ω-limit set is either the core or is nowhere dense. We can say more. Figure 10.3 provides an overview of the options for $\omega(x, T_a)$.

We saw in Proposition 3.2.9 that if the orbit of x is finite, then $\omega(x, T_a)$ is a periodic cycle. If the orbit of x is dense in the core, then $\omega(x, T_a) = [c_2(a), c_1(a)]$. As mentioned in Remark 6.2.13, the set $\{a \in [\sqrt{2}, 2] \mid \omega(c, T_a) = [c_2(a), c_1(a)]\}$ has full Lebesgue measure.

Suppose the orbit of x is infinite but not dense in the core. Then $\omega(x, T_a)$ is nowhere dense and hence totally disconnected; otherwise, $\omega(x, T_a)$ contains an interval and, since T_a is leo, $\omega(x, T_a)$ would contain $[c_2(a), c_1(a)]$ (recall Remark 3.4.17). Either all points in the orbit of x are isolated with respect to the orbit itself, that is, for every $n \geq 0$ there exists an open set $U_n \ni T_a^n(x)$ such that $T_a^j(x) \notin U_n$ for all $j \neq n$, or at most finitely many (possibly none) points in the orbit of x are isolated with respect to the orbit itself. In the latter case, we have that $\omega(x, T_a)$ is a Cantor set. In the former, if $\omega(x, T_a)$ is not countable, then it contains a Cantor set (use Exercise 3.2.18). Hence, we have Figure 10.3.

Exercise 10.2.1. *Suppose the orbit of x is infinite but not dense in the core, and that at most finitely many points in the orbit of x are isolated with respect to the orbit. Prove $\omega(x, T_a)$ is a Cantor set.* **HINT:** *As noted*

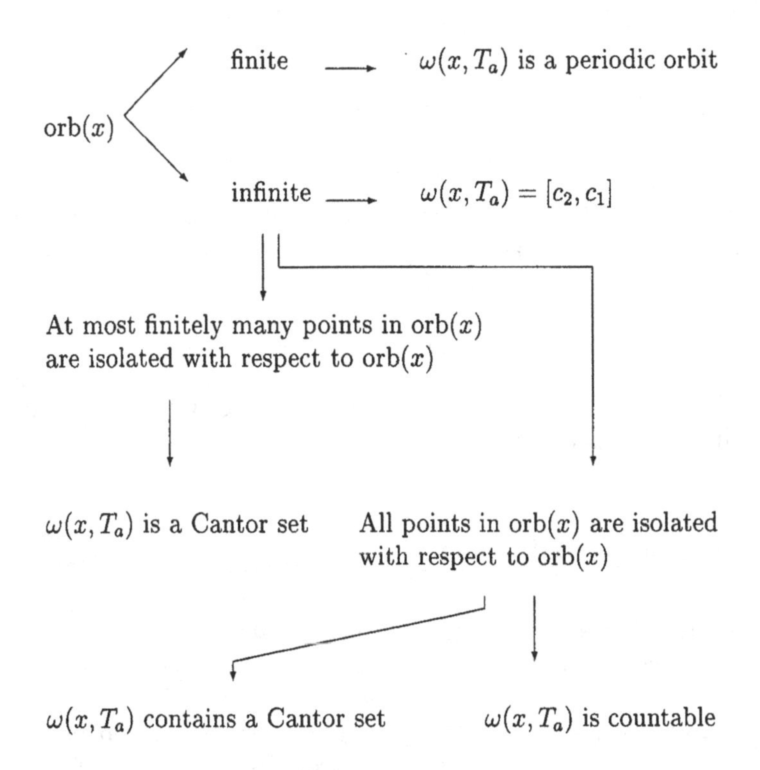

Figure 10.3: ω-limit sets for T_a

above, we have that $\omega(x, T_a)$ is totally disconnected. It is closed and therefore compact. Hence, it remains to show it is perfect. Use that at most finitely many points in the orbit are not recurrent, to conclude perfect.

Example 10.2.2. *Choose $a \in [\sqrt{2}, 2]$ such that $\omega(c, T_a) = [c_2(a), c_1(a)]$. As we have noted, the set of such a's has full Lebesgue measure in $[\sqrt{2}, 2]$. Set $A = I(c_1(a), T_a)$. Then A is shift maximal and hence begins with 10^m1 for some m. Set $B = 10^{2m}A$. Then, again, B is shift maximal and indeed there exists a unique $b > a$ in $[\sqrt{2}, 2]$ such that $I(c_1(b), T_b) = B$. Verify each of the following.*

1. *The turning point c is not recurrent.*

2. *At most finitely many points in the orbit of c, under T_b, are isolated with respect to the orbit of c.*

3. *$\omega(c, T_b)$ contains periodic orbits.*

4. *$\omega(c, T_b)$ is not a minimal set.*

5. *$\omega(c, T_b)$ is a Cantor set.*

6. *$Q_b(k)$ is bounded and hence T_b is longbranched (recall Proposition 6.2.6; in fact, longbranched follows immediately from item (1)).*

7. *$T_b|\omega(c, T_b)$ is not one-to-one (recall Remark 3.2.14).*

This completes Example 10.2.2.

Exercise 10.2.3. *Set $A = 10000011101111101111111011111111110\cdots$, that is, after the initial 1, groups of 1's of length $2n+1$ for $n \geq 1$ appear with exactly one 0 between groups. Prove A is shift maximal. Suppose T_a is such that $I(c_1(a), T_a) = A$. Prove:*

1. *All points in the orbit of $c_1(a)$ are isolated with respect to the orbit.*

2. *$\omega(c_1(a), T_a)$ is countable.*

3. *The nonzero fixed point of T_a is contained in $\omega(c_1(a), T_a)$.*

HINT: *If $y = \lim_{k\to\infty} c_{n_k}$, then $\lim_{k\to\infty} I(c_{n_k}(a), T_a) = I(y, T_a)$ in $\{0,1\}^{\mathbb{N}}$.*

Example 10.2.4. *We provide an example where all points in the orbit of c_1 are isolated with respect to the orbit and where $\omega(c, T_a)$ is uniquely expressed as a Cantor set and a countable set (as in Exercise 3.2.18).*

Let $\Sigma = \{e = \langle e_i \rangle_{i\geq 0} \in \{0,1\}^{\mathbb{N}} \mid e_i e_{i+1} \neq 00$ for all $i\}$. Note that Σ is a Cantor set. Let $\{v_i\}$ be an enumeration of all finite subwords from elements of Σ. For example: $v_1 = 0, v_2 = 1, v_3 = 01, v_4 = 10, v_5 = 11, v_6 = 010, v_7 = 011, v_8 = 101, v_9 = 110, v_{10} = 111, \ldots$.

For $n \geq 1$ define $w_n \in \{0,1\}^{4n+5}$ by

$$w_n = (1)^{2n+1}000(1)^{2n+1};$$

for example, $w_1 = 111000111$, $w_2 = 1111100011111$, and $w_3 = 111111100011$ 11111.

Next, define $A \in \{0,1\}^{\mathbb{N}}$ as follows:

$$A = 1000000 v_1 w_1 v_2 w_2 v_3 w_3 \ldots.$$

Check that indeed A is shift maximal. Moreover, there exists $a \in [\sqrt{2}, 2]$ with $I(c_1(a), T_a) = A$. Verify the following:

1. *If $e \in \Sigma$, then $I(c_2(a), T_a) \preceq e \preceq I(c_1(a), T_a)$.*

2. For each $n \geq 1$, $I(c_2(a), T_a) \preceq (1)^n000(1)^\infty \preceq I(c_1(a), T_a)$.

Use Proposition 10.1.5 to identify points in $[c_2(a), c_1(a)]$ as follows.

1. *For each $e \in \Sigma$, let $z_e \in [c_2(a), c_1(a)]$ be the unique point such that $I(z_e, T_a) = e$.*

2. *For each $n \geq 1$, let $z_n \in [c_2(a), c_1(a)]$ be the unique point such that $I(z_n, T_a) = (1)^n000(1)^\infty$.*

Let p denote the nonzero fixed point of T_a, and set $K = \{z_e \mid e \in \Sigma\}$ and $H = \{T_a^j(z_n) \mid j \geq 0 \text{ and } n \geq 1\} \setminus \{T_a^{-1}(p)\}$. Verify each of the following.

1. *K is a Cantor set.*

2. *$\omega(c(a), T_a) = K \cup H$ and $K \cap H = \emptyset$.*

3. *H is countable and not closed.*

4. *$T_a(K) \subset K$.*

This completes Example 10.2.4.

The construction given in Example 10.2.4 is not unique. One could replace Σ by a shift space of choice. We now have examples of all branches in Figure 10.3

Exercise 10.2.5. *Let $a \in [\sqrt{2}, 2]$ be such that all points in the orbit of the turning point c are isolated with respect to the orbit, such as in Exercise 10.2.3 and Example 10.2.4. Prove $Q(k)$ is bounded and hence T_a is longbranched.*

Exercise 10.2.6. ◇ *Provide an example where c is recurrent (and not periodic) for the map T_a and where T_a is longbranched. Notice that $Q(k)$ is bounded and $\omega(c, T_a)$ is a Cantor set.*

Exercise 10.2.7. ◇ *Suppose $a \in [\sqrt{2}, 2]$ with $\omega(c, T_a) = [c_2(a), c_1(a)]$. Can $Q(k)$ be bounded?*

10.3 Phase Portrait

In this section we investigate the *phase portrait* for the one-parameter family of symmetric tent maps. Such portraits are useful in analyzing how dynamical behaviors change in one-parameter families of maps. As noted in the previous section, the parameter space is precisely $[\sqrt{2}, 2]$. Recall that $I(c_1(\sqrt{2}), T_{\sqrt{2}}) = 101^{\infty}$. We begin with some definitions. Set

$$\mathcal{P} = \{a \in [\sqrt{2}, 2] \mid c \text{ is either periodic or eventually periodic under } T_a\}.$$

Definition 10.3.1. For $a \in [\sqrt{2}, 2]$ and $n \in \mathbb{N}$, set $\xi_n(a) = T_a^n(c)$. Let $\omega_n(a)$ be the maximal open interval in the parameter space containing a such that ξ_n is monotone on $\omega_n(a)$. Notice that $\omega_n(a)$ is not defined for $a \in \mathcal{P}$ when c is periodic for T_a and n is large, since no such open interval exists. See Figure 10.4

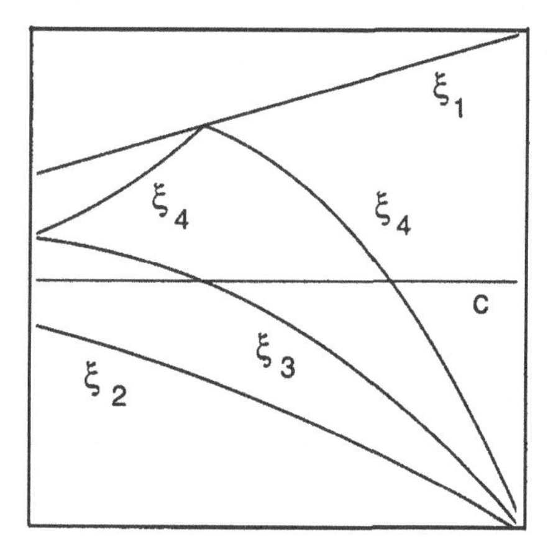

Figure 10.4: Phase portrait: ξ_1 to ξ_4

The interested reader can obtain proofs of the next two lemmas from [35, 59, 152]; statements of both lemmas are easily understood and are very useful. Although the proofs consist entirely of technical calculations and we omit them, we strongly suggest the reader obtain the references and work through proofs of these two lemmas.

Lemma 10.3.2. *There exist positive constants α and β such that for all $k \geq 2$ and all $a \in [\sqrt{2}, 2]$*

$$\alpha a^k \leq |\xi_k'(a)| \leq \beta a^k,$$

where ξ_k' is defined.

Lemma 10.3.3. *Fix $\epsilon_0 > 0$ and $a_0 \in (\sqrt{2}, 2]$. Then there exists $K_0 \in \mathbb{N}$ such that, for all $k \geq K_0$,*

$$\frac{|\xi_k'(b)|}{|\xi_k'(a)|} \leq 1 + \epsilon_0$$

whenever a, b belong to the same lap of $\xi_k|[a_0, 2]$.

Let $a \in [\sqrt{2}, 2]$ and suppose c is not periodic under T_a. Then $\xi_k'(a)$ exists for all k and Lemma 10.3.2 tells us that $\lim_{k\to\infty} \xi_k'(a) = \infty$. For $n \in \mathbb{N}$, one defines the *distortion* of T_a on $\omega_n(a)$ as

$$\sup_{b,d\in\omega_n(a)} \left\{ \frac{|\xi_n'(b)|}{|\xi_n'(d)|} \right\}.$$

Thus, Lemma 10.3.3 tells us that, for large n, the distortion of T_a on $\omega_n(a)$ is approximately 1 and hence is bounded. Therefore, for large n, we have that $\xi_n|\omega_n(a)$ is "close to linear," that is, when looking at the graph of ξ_n on $\omega_n(a)$ for large n, the graph looks linear.

Lemma 10.3.4. *The set $\mathcal{P}$ is dense in $[\sqrt{2}, 2]$.*

Proof. Suppose to the contrary that there exists an open interval $U = [\alpha, \beta] \subset [\sqrt{2}, 2]$ such that $\mathcal{P} \cap U = \emptyset$. Using Lemma 10.3.2, choose $n \in \mathbb{N}$ such that

$$|\xi_n'(a)| > \frac{2}{\beta - \alpha} \tag{10.2}$$

for all $a \in U$ (since $\mathcal{P} \cap U = \emptyset$, we have that $\xi_n'(a)$ exists for all $a \in U$). By the Mean Value Theorem, there exists $b \in U$ with

$$|\xi_n'(b)| = \left| \frac{\xi_n(\beta) - \xi_n(\alpha)}{\beta - \alpha} \right|. \tag{10.3}$$

However, $|\xi_n(\beta) - \xi_n(\alpha)| \leq 2$ and formula (10.3) give that $|\xi_n'(b)| \leq \frac{2}{\beta-\alpha}$, contradicting formula (10.2). □

Definition 10.3.5. For each $a \in [\sqrt{2}, 2]$, let z_a denote the nonzero fixed point of T_a. Thus,

$$z_a = \frac{2a}{2a+1},$$

and $a \mapsto z_a$ is a continuous map on $[\sqrt{2}, 2]$.

Definition 10.3.6. We call a parameter $a \in [\sqrt{2}, 2]$ *prefixed*, provided there exists an $m \in \mathbb{N}$ such that $T_a^m(c) = z_a$. Thus, if a is prefixed, then the kneading sequence of T_a is eventually 1^∞. Note that $\sqrt{2}$ is prefixed.

Exercise 10.3.7. ♣ *[32, 59] Prove that the prefixed parameters are dense in* $[\sqrt{2},2]$.

Lemma 10.3.8. *Let* $b \in (\sqrt{2},2]$ *be a prefixed parameter and for* $n \in \mathbb{N}$ *let* $\omega_n(b) = (\alpha_n, \beta_n)$. *Then there exists* $M \in \mathbb{N}$ *such that*

$$\xi_n(\omega_n(b)) = \begin{cases} (c_2(\alpha_n), c_1(\beta_n)) & \text{if } \xi_n \text{ is increasing} \\ (c_2(\beta_n), c_1(\alpha_n)) & \text{if } \xi_n \text{ is decreasing} \end{cases}$$

for all $n \geq M$.

Proof. We have that $\omega_n(b)$ is a nondegenerate interval for all n, since b is prefixed. Fix $m \in \mathbb{N}$ and consider ξ_m for $a \in \omega_m(b)$. See Figure 10.5. Without loss of generality, assume ξ_m is monotone increasing on $\omega_m(b)$. Then ξ_{m+l} is monotone increasing on $\omega_{m+l}(b)$ for l even and monotone decreasing for l odd. Say $\omega_m(b) = (\alpha, \beta)$.

We claim there exists j odd such that $\xi_{m+j}(\omega_{m+j}(b)) \ni c$. Otherwise, ξ_{m+l} is monotone on (b, β) for all l odd, contradicting $\lim_n \ \xi_n'(b) = \infty$. Similary, there exists k odd such that $\xi_{m+j+k}(\omega_{m+j+k}(b)) \ni c$. Set $M = m+j+k+1$. See Figure 10.5. □

From Lemma 10.3.8 we see that the phase portrait for the symmetric tent family has long stretches contained in it, namely the intervals $\xi_n(\omega_n(b))$ for b prefixed and n large. We make use of these stretches to show that once an itinerary occurs, it occurs as the tail of a kneading sequence for a dense set of parameters (see Exercise 10.3.10).

Remark 10.3.9. Fix $b \in (\sqrt{2},2)$ and $x \in [c_2(b), c_1(b)]$. Let $a \in (\sqrt{2},2]$ with $a > b$. Then,

$$I(c_2(a),T_a) \preceq I(c_2(b),T_b) \preceq I(x,T_b) \preceq I(c_1(b),T_b) \preceq I(c_1(a),T_a). \quad (10.4)$$

Formula (10.4) follows easily from the definition of $\preceq$ and Proposition 10.1.5. Hence, for every $x \in [c_2(b), c_1(b)]$ and each $a > b$ (with $a \in (\sqrt{2},2]$), there exists a unique $z_{a,x} \in [c_2(a), c_1(a)]$ with $I(z_{a,x}, T_a) = I(x, T_b)$. In less precise terms, once an itinerary exists, it continues to exist. Indeed, the map $a \mapsto z_{a,x}$ is continuous in a.

Notice that if $\xi_n(a) = z_{a,x}$ for some $n \in \mathbb{N}$ and $a \in (b,2]$, then the kneading sequence of T_a begins as some finite string (of length $n-1$) followed by $I(x,T_b)$. In the case that $I(x,T_b)$ is not finite, then $I(x,T_b)$ is the tail of the kneading sequence for T_a.

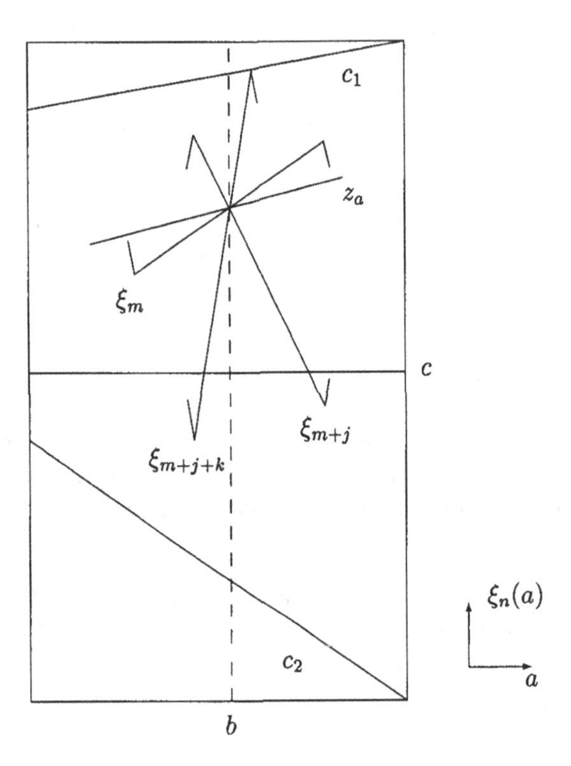

Figure 10.5: Prefixed parameters

Exercise 10.3.10. *Fix $b \in (\sqrt{2}, 2)$ and $x \in [c_2(b), c_1(b)]$. Let $U \subset (b, 2)$ be an open interval. Prove there exists $a \in U$ and $n \in \mathbb{N}$ such that $\xi_n(a) = z_{a,x}$. Less formally, once an itinerary occurs, it occurs as the tail of a kneading sequence for a dense set of parameters.* **HINT:** *Use Exercise 10.3.7, Lemma 10.3.8, and Remark 10.3.9.*

Exercise 10.3.11. *Fix $b \in (\sqrt{2}, 2)$ and set $x = c_1(b)$. Choose $a > b$ and $n \in \mathbb{N}$ such that $\xi_n(a) = z_{a,x}$. Note that if c is periodic (preperiodic) for T_b, then it is also periodic (preperiodic) for T_a. Suppose c is not periodic or preperiodic for T_b. If $\omega(c, T_b) = [c_2(b), c_1(b)]$, then we saw in Example 10.2.2 that $\omega(c, T_a)$ is a Cantor set. What happens in the other cases from Figure 10.3?*

Next we see how to determine cutting times from the phase portrait. Let $a \in (\sqrt{2}, 2] \setminus \mathcal{P}$ be given and suppose that $n = S_k(a) = S_k$ is a cutting time for T_a; let $Q_a(k) = Q(k)$. Then, one of the pictures in Figure 10.6 holds (see Exercise 10.3.12). From Lemma 10.3.3, we see that for large n the graph of ξ_n is almost linear and hence for ease we draw linear functions in Figure

10.6. We have that $\omega_{S_k}(a) = (u, v)$. The point b in Figure 10.6 is such that $\xi_{S_k}(b) = c = \frac{1}{2}$. We have that S_k is a cutting time for all $a' \in (b, v)$, $D_{S_k}(a') = [T_{a'}^{S_k}(c), T_{a'}^{S_{Q(k)}}(c)]$ for all $a' \in (u, v) \setminus \{b\}$, and S_k is not a cutting time for all $a' \in (u, b)$. For further discussion see [152, Chapter 3].

Exercise 10.3.12. $\diamondsuit$ *Fix $a \in (\sqrt{2}, 2] \setminus \mathcal{P}$ and let the kneading sequence of T_a be given by $e_1, e_2, e_3, \ldots$. Show n is a cutting time ($n = S_k$) for T_a with $e_1, e_2, \ldots, e_{S_k - 1}, * \succ 101^\infty$ if and only if $\xi_n(\omega_n(a)) \ni c$ and $a > b$, where b is the unique point in $\omega_n(a)$ such that $\xi_n(b) = c$.*

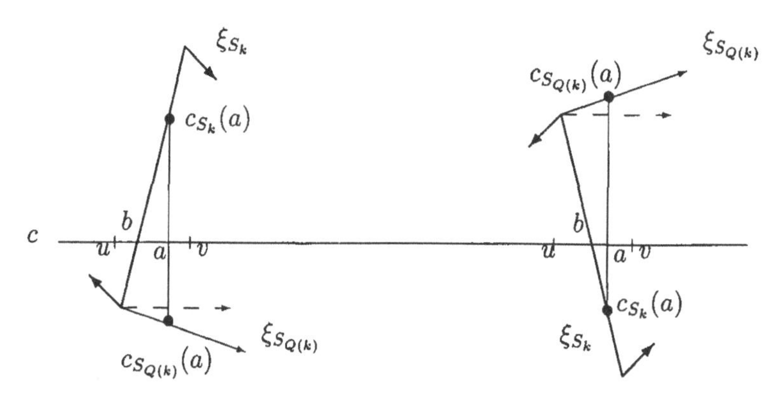

Figure 10.6: Plotting $\xi_n(a)$ at cutting time $n = S_k$

Using Exercise 10.3.12, we can prove that the set of parameters a with $\lim_{k \to \infty} Q_a(k) = \infty$ is dense in $[\sqrt{2}, 2]$ and is uncountable. Set

$$\mathcal{I} = \{a \in [\sqrt{2}, 2] \mid \lim_{k \to \infty} Q_a(k) = \infty\}.$$

We know $\mathcal{I}$ has zero Lebesgue measure as it sits in the complement of $\{a \in [\sqrt{2}, 2] \mid \omega(c, T_a) = [c_2(a), c_1(a)]\}$; recall Remark 6.2.13.

Lemma 10.3.13. *[30] The set $\mathcal{I}$ is dense in $[\sqrt{2}, 2]$ and is uncountable.*

Proof. Let $U \subset [\sqrt{2}, 2]$ be an open interval. Choose $a_1 \in U \setminus \mathcal{P}$ and a cutting time $n_1 = S_{k_1}(a_1)$ such that $\omega_{n_1}(a_1) \subset U$. We can make such a choice due to $\mathcal{P}$ being countable and Lemma 10.3.2. Let $\epsilon_1 > 0$, and set $J_1 = \{a \in \omega_{n_1}(a_1) \mid n_1 = S_{k_1}(a) \text{ and } |c - c_{n_1}(a)| < \epsilon_1\}$. Then, for each $a \in J_1$, we have that n_1 is a cutting time for T_a and $|c - c_{n_1}(a)| < \epsilon_1$. Notice that J_1 is an open subinterval of U; recall Exercise 10.3.12.

Fix $a_2 \in J_1 \setminus \mathcal{P}$. Then $n_1 = S_{k_1}(a_2)$. Set $n_2 = S_{k_1+1}(a_2)$. Choose $0 < \epsilon_2 < \frac{\epsilon_1}{2}$ such that $J_2 := \{a \in \omega_{n_2}(a_2) \mid n_2 = S_{k_1+1}(a) \text{ and } |c - c_{n_2}(a)| < \epsilon_2\} \subset J_1$ and such that the sets J_1 and J_2 share no boundary points. Then (again use Exercise 10.3.12) for each $a \in J_2$ we have that $n_1 = S_{k_1}(a)$, $n_2 = S_{k_1+1}(a)$, $|c - c_{n_1}(a)| < \epsilon_1$, and $|c - c_{n_2}(a)| < \epsilon_2$. Also, $\bar{J}_2$ is a proper closed subinterval of $\bar{J}_1$.

Fix $a_3 \in J_2 \setminus \mathcal{P}$. Then $n_1 = S_{k_1}(a_3)$ and $n_2 = S_{k_1+1}(a_3)$. Set $n_3 = S_{k_1+2}(a_3)$. Choose $0 < \epsilon_3 < \frac{\epsilon_2}{2}$ such that $J_3 := \{a \in \omega_{n_3}(a_3) \mid n_3 = S_{k_1+2}(a) \text{ and } |c - c_{n_3}(a)| < \epsilon_3\} \subset J_2$ and the sets J_2 and J_3 share no boundary points. Then, for each $a \in J_3$ we have that $n_1 = S_{k_1}(a)$, $n_2 = S_{k_1+1}(a)$, $n_3 = S_{k_1+2}(a)$, $|c - c_{n_1}(a)| < \epsilon_1$, $|c - c_{n_2}(a)| < \epsilon_2$, and $|c - c_{n_3}(a)| < \epsilon_3$. Also, $\bar{J}_3$ is a proper closed subinterval of $\bar{J}_2$.

Continue this process and set

$$a_* = \cap_{n \geq 1} J_n.$$

Remember that if $\liminf_{k\to\infty} Q_{a_*}(k) < \infty$, then (by Exercise 6.1.7) there exists some $\delta > 0$ such that for infinitely many k we have $|c - c_{S_{k-1}}| > \delta$. Hence, we have that $\lim_{k\to\infty} Q_{a_*}(k) = \infty$ with $a_* \in U$. By varying the choices of $\{a_i\}$ and hence of the sequences of cutting times $\{n_i\}$, one can show that $\mathcal{I}$ is uncountable. □

Proposition 10.3.21 tells us that "long stretches" in the phase portrait (as we saw for prefixed parameters) are common. The next few exercises and lemmas provide a further understanding of the phase portrait and are used to prove Proposition 10.3.21. These long stretches in the phase portrait are useful for measure results (see [30, 35]) and for producing behaviors, such as in Remark 10.3.9 and Exercise 10.3.10.

Definition 10.3.14. For $a \in [\sqrt{2}, 2]$ and $n \in \mathbb{N}$ set

$$\begin{aligned} \alpha_n(a) &= \inf\{y \in [c_2(a), c_1(a)) \mid T_a^{n-1}|[y, c_1(a)] \text{ is monotone }\} \\ \beta_n(a) &= \sup\{y \in [c_1(a), 1] \mid T_a^{n-1}|[c_1(a), y] \text{ is monotone }\}. \end{aligned}$$

Next, set $r_n(a) = T_a^{n-1}(\alpha_n)$, $s_n(a) = T_a^{n-1}(\beta_n)$, $H_n(a) = [\alpha_n(a), \beta_n(a)]$, and $V_n(a) = \langle r_n(a); s_n(a)\rangle$. Notice that the nth level of the Hofbauer tower is precisely $\langle c_n; r_n\rangle$, that is, $D_n(a) = \langle c_n; r_n\rangle$. Hence,

$$r_n(a) = \begin{cases} c_{n-S_k(a)} & \text{if } S_k(a) < n < S_{k+1}(a) \text{ for some } k \\ c_{S_{Q(k)}(a)} & \text{if } n = S_k(a) \text{ for some } k. \end{cases}$$

See Figure 10.7. If a is fixed, for ease, we write α_n, β_n, $\cdots$, V_n.

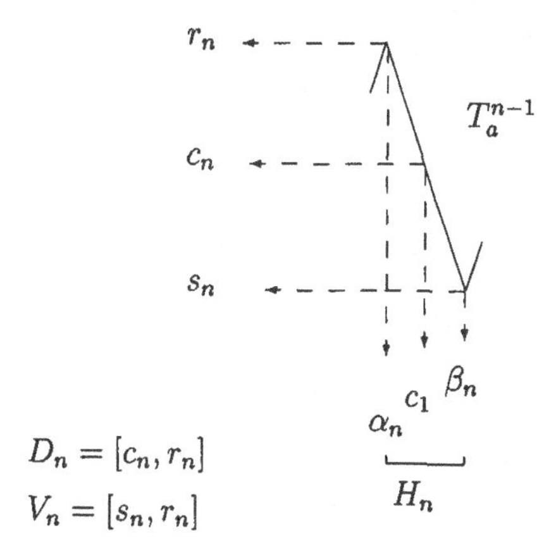

Figure 10.7: Lap of T_a^{n-1} containing c_1

Note that in Definition 10.3.14 we could have simply written that $r_n(a) = c_{n-S_k(a)}$ for $S_k(a) < n \leq S_{k+1}(a)$, since $S_{k+1}(a) - S_k(a) = S_{Q_a(k)}(a)$ (recall formula (6.2)).

Exercise 10.3.15. *Fix $a \in [\sqrt{2}, 2]$. Prove*

$$V_{n+1} = \begin{cases} T_a(V_n) & \text{if } c \notin V_n \\ T_a(W_n) & \text{if } c \in V_n, \end{cases}$$

where W_n is the closure of the component of $V_n \setminus \{c\}$ containing c_n and $V_1 = [0, 1]$.

Exercise 10.3.16. *Fix $a \in [\sqrt{2}, 2]$. Prove*

$$r_{n+1} = \begin{cases} T_a(r_n) & \text{if } c \notin \langle r_n; c_n \rangle \\ c_1 & \text{if } c \in \langle r_n; c_n \rangle \end{cases}$$

and

$$s_{n+1} = \begin{cases} T_a(s_n) & \text{if } c \notin \langle s_n; c_n \rangle \\ c_1 & \text{if } c \in \langle s_n; c_n \rangle. \end{cases}$$

We have seen that the functions $\xi_n(a)$ are continuous in a (although they are not differentiable for all a and n). In contrast to this fact, we have that the functions $r_n(a)$ and $s_n(a)$ are piecewise continuous. The discontinuities occur precisely at endpoints of the intervals $\omega_n(a)$.

Exercise 10.3.17. *[152] Let $a \in [\sqrt{2}, 2]$ be such that c is not periodic for T_a. Hence, $\omega_n(a)$ is a nondegenerate open interval for all n. Fix n and say $\omega_n(a) = (\alpha, \beta)$. Prove*

$$\lim_{a \downarrow \alpha} r_n(a) = \xi_n(\alpha)$$

and

$$\lim_{a \uparrow \beta} s_n(a) = \xi_n(\beta).$$

HINT: *Induct on n and use the recursive formulas in Exercise 10.3.16.*

Exercise 10.3.18. *Let $b \in [\sqrt{2}, 2]$ be such that c is not periodic for T_b. Again, $\omega_n(b)$ is a nondegenerate open interval for all n. Notice that for $e \neq d \in \omega_n(b)$, the strings $I(c_1(e), T_e)$ and $I(c_1(d), T_d)$ agree for the first $n-1$ positions. Conclude that $r_n(a) = \xi_m(a)$ for all $a \in \omega_n(b)$, where*

$$m = n - S_k(b) \text{ with } S_k(b) < n \leq S_{k+1}(b) \text{ for some } k.$$

The next lemma tells us that the "slope" of ξ_n becomes increasingly steeper than the "slope" of r_n or s_n as n increases; see Figure 10.8.

Lemma 10.3.19. *[152] Let $a \in [\sqrt{2}, 2]$ be such that c is not periodic for T_a. Then*

$$\lim_{n\to\infty} \frac{r_n'(a)}{\xi_n'(a)} = 0 \qquad \text{and} \qquad \lim_{n\to\infty} \frac{s_n'(a)}{\xi_n'(a)} = 0.$$

Proof. From Exercise 10.3.18, Lemma 10.3.2, and Lemma 10.3.3 we have that $|r_n'(a)| = |\xi_m'(a)| \leq \beta a^m$ and $|\xi_n'(a)| \geq \alpha a^n$, where m is as in Exercise 10.3.18. Hence,

$$\frac{|r_n'(a)|}{|\xi_n'(a)|} \leq \frac{\beta a^m}{\alpha a^n} = \frac{\alpha}{\beta}\frac{1}{a^{n-m}} = \frac{\alpha}{\beta}\frac{1}{a^{S_k(a)}},$$

where $S_k(a) < n \leq S_{k+1}(a)$ for some k. However, $S_k(a) \to \infty$ as $n \to \infty$ and hence the result follows. See Figure 10.8. □

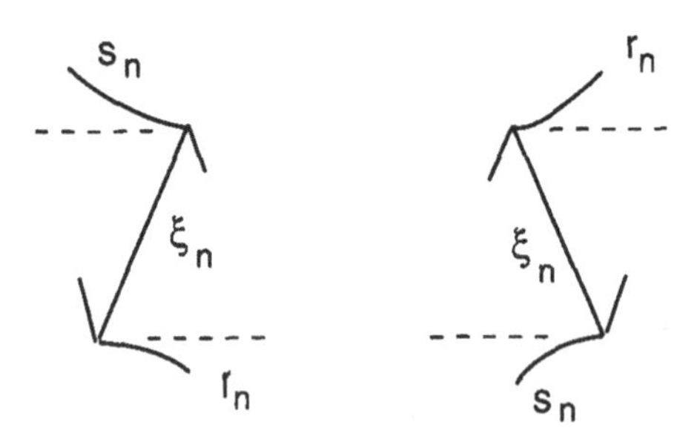

Figure 10.8: Graphs of ξ_n, r_n, and s_n

Lemma 10.3.20. *Let $a \in [\sqrt{2}, 2]$ be such that $\omega(c, T_a) = [c_2(a), c_1(a)]$. Then, there exist infinitely many n's with $V_n = [c_2(a), c_1(a)]$.*

Proof. Let p be the nonzero fixed point of T_a and $K \in \mathbb{N}$. As p is repelling, we may choose preimages of c closest to p:

$$c_{-2} < c_{-4} < c_{-6} < \ldots < p < \ldots < c_{-5} < c_{-3} < c_{-1}.$$

Note that $\lim_{n\to\infty} c_{-n} = p$. Let $v > K$ be even. Since $\omega(c, T_a) = [c_2(a), c_1(a)]$, there exists $c_m \in (c_{-v}, c_{-v-1})$. Take m minimal. Then, since the endpoints of $T_a^{m-1}(H_m) = V_m$ are in orb(c) and m is minimal, we have that $V_m \supseteq [c_{-v}, c_{-v-1}]$. However, $T_a^{v+2}([c_{-v}, c_{-v-1}]) = [c_2(a), c_1(a)]$. Hence, setting $k = m - 1 + v + 2$, we have $V_{k+1} = [c_2(a), c_1(a)]$ with $k+1 > K$. □

Proposition 10.3.21. *Let $a \in [\sqrt{2}, 2]$ be such that $\omega(c, T_a) = [c_2(a), c_1(a)]$. Then there exist arbitrarily large n's such that ξ_n maps $\omega_n(a) = (\alpha, \beta)$ monotonically onto $(c_2(\alpha), c_1(\beta))$ or $(c_2(\beta), c_1(\alpha))$.*

Proof. The proof is immediate from Exercise 10.3.17, Lemma 10.3.19, and Lemma 10.3.20. □

Proposition 10.3.21 appears in [43, Lemma 4] with a cosmetically different proof. Lemma 10.3.20 also appears in [43, Lemma 3], although it is stated in terms of the action $(\hat{I}, \hat{T}_a)$ on the Hofbauer tower (recall Definition 6.1.15).

Definition 10.3.22. Fix $a \in [\sqrt{2}, 2]$. We call $n \in \mathbb{N}$ a *co-cutting time* provided $c \in \langle c_n; s_n \rangle$.

Exercise 10.3.23. *Let $a \in [\sqrt{2}, 2]$ be such that $\omega(c, T_a) = [c_2(a), c_1(a)]$. Suppose n is such that ξ_n maps $\omega_n(a) = (\alpha, \beta)$ monotonically onto $(c_2(\alpha), c_1(\beta))$ or $(c_2(\beta), c_1(\alpha))$. Prove that n is either a cutting or co-cutting time.*

For more information on co-cutting times see [39, 40, 152].

10.4 Measure Results

We remarked in Section 6.2 that the set of parameters $\mathcal{D}$ has full Lebesgue measure in $[\sqrt{2}, 2]$; recall (from formula (6.7)) that

$$\mathcal{D} = \{a \in [\sqrt{2}, 2] \mid \overline{\{c_n\}_{n\geq 0}} = [c_2(a), c_1(a)]\}.$$

The proof of this result (see [35]) is beyond the scope of the text. In this section we use the fact that $\mathcal{D}$ has full measure to prove

$$\tilde{\mathcal{D}} = \{a \in [\sqrt{2}, 2] \mid \overline{\{c_{S_k(a)}(a)\}_{k\geq 0}} = [c_2(a), c_1(a)]\}$$

has full Lebesgue measure in $[\sqrt{2}, 2]$. In both proofs there are two key components: long stretches in the phase portrait (as discussed in the previous section) and bounded distortion (Lemma 10.3.3). Obtaining the long stretches needed to prove $\mathcal{D}$ has full measure is beyond the scope of the text, although the more advanced reader is encouraged to see [35] for the details. However, using that $\mathcal{D}$ has full measure along with Proposition 10.3.21, one can more easily obtain the long stretches needed to prove $\tilde{\mathcal{D}}$ has full measure. Once one has obtained the long stretches, both proofs are essentially the same and illustrate a standard application of bounded distortion.

Note that $\tilde{\mathcal{D}} \subset \mathcal{D}$. We believe this containment is proper and encourage the reader to provide an explicit example of $a \in \mathcal{D} \setminus \tilde{\mathcal{D}}$; recall Question 6.4.8. Porosity results for the set $\mathcal{D}$ can be found in [30]. We do not discuss these stronger results, as they are beyond the scope of the text. We let $|A|$ denote the Lebesgue measure of a set A.

Proposition 10.4.1. *The set $\tilde{\mathcal{D}}$ has full Lebesgue measure in $[\sqrt{2}, 2]$.*

Proof. Let $U \subset [0,1]$ be an open interval such that $U \subset [c_2(a), c_1(a)]$ for all $a \in [\sqrt{2}, 2]$ (recall Figure 10.4) and let $\epsilon = |U|$. Set

$$\mathcal{B} = \{a \in \mathcal{D} \mid c_{S_k(a)}(a) \notin U \text{ for all } k \geq 0\}.$$

Suppose to the contrary that $|\mathcal{B}| > 0$. Then we may choose $b \in \mathcal{B}$ to be a Lebesgue point of density of $\mathcal{B}$. It follows from Lemmas 10.3.2 and 10.3.3 that

$$\lim_{n\to\infty} |\omega_n(b)| = 0,$$

and hence, as b is a point of density of $\mathcal{B}$, we have that

$$\lim_{n\to\infty} \frac{|\omega_n(b) \cap \mathcal{B}|}{|\omega_n(b)|} = 1.$$

Choose $M \in \mathbb{N}$ such that $n \geq M$ implies

$$\frac{|\omega_n(b) \cap \mathcal{B}|}{|\omega_n(b)|} \geq 1 - \frac{\epsilon}{4} \tag{10.5}$$

and (use Lemma 10.3.3)

$$\sup_{a,\tilde{a}\in\omega_n(b)} \left\{ \frac{|\xi_n'(a)|}{|\xi_n'(\tilde{a})|} \right\} < 1 + \frac{\epsilon}{2}.$$

Fix $n \geq M$. Then,

$$\int_{\omega_n(b)\cap\mathcal{B}} |\xi_n'| \geq |\omega_n(b) \cap \mathcal{B}| \inf_{a\in\omega_n(b)\cap\mathcal{B}} |\xi_n'(a)| \tag{10.6}$$

and

$$\int_{\omega_n(b)} |\xi_n'| \leq |\omega_n(b)| \sup_{a \in \omega_n(b)} |\xi_n'(a)|. \tag{10.7}$$

Thus,

$$\begin{aligned}
\left(\int_{\omega_n(b)} |\xi_n'|\right) \frac{|\omega_n(b) \cap \mathcal{B}|}{|\omega_n(b)|} &\leq |\omega_n(b) \cap \mathcal{B}| \sup_{a \in \omega_n(b)} |\xi_n'(a)| \quad \text{(use 10.7)} \\
&\leq \frac{\sup_{a \in \omega_n(b)} |\xi_n'(a)|}{\inf_{a \in \omega_n(b) \cap \mathcal{B}} |\xi_n'(a)|} \int_{\omega_n(b) \cap \mathcal{B}} |\xi_n'| \quad \text{(use 10.6)}.
\end{aligned}$$

Hence,

$$\begin{aligned}
\frac{|\omega_n(b) \cap \mathcal{B}|}{|\omega_n(b)|} &\leq \frac{\sup_{a \in \omega_n(b)} |\xi_n'(a)|}{\inf_{a \in \omega_n(b) \cap \mathcal{B}} |\xi_n'(a)|} \frac{\int_{\omega_n(b) \cap \mathcal{B}} |\xi_n'|}{\int_{\omega_n(b)} |\xi_n'|} \\
&\leq \frac{\sup_{a \in \omega_n(b)} |\xi_n'(a)|}{\inf_{a \in \omega_n(b)} |\xi_n'(a)|} \frac{\int_{\omega_n(b) \cap \mathcal{B}} |\xi_n'|}{\int_{\omega_n(b)} |\xi_n'|} \\
&\leq \frac{|\xi_n(\omega_n(b) \cap \mathcal{B})|}{|\xi_n(\omega_n(b))|} (1 + \frac{\epsilon}{2}).
\end{aligned} \tag{10.8}$$

Say $\omega_n(b) = (\alpha, \beta)$. Since $b \in \mathcal{D}$, using Proposition 10.3.21, we may choose n large such that $\xi_n(\omega_n(b)) = (c_2(\alpha), c_1(\beta))$ or $\xi_n(\omega_n(b)) = (c_2(\beta), c_1(\alpha))$. Hence, $U \subset \xi_n(\omega_n(b))$. Without loss of generality, assume $\xi_n(\omega_n(b)) = (c_2(\alpha), c_1(\beta))$. Then,

$$|\xi_n(\omega_n(b) \cap \mathcal{B})| \leq |c_2(\alpha) - c_1(\beta)| - \epsilon \tag{10.9}$$

and

$$|\xi_n(b)(\omega_n(b))| = |c_2(\alpha) - c_1(\beta)|. \tag{10.10}$$

Combining formulas (10.5), (10.8), (10.9), and (10.10) we have

$$\begin{aligned}
1 - \frac{\epsilon}{4} &\leq \frac{|\omega_n(b) \cap \mathcal{B}|}{|\omega_n(b)|}(1 + \frac{\epsilon}{2}) \\
&\leq \frac{|c_2(\alpha) - c_1(\beta)| - \epsilon}{|c_2(\alpha) - c_1(\beta)|}(1 + \frac{\epsilon}{2}).
\end{aligned} \tag{10.11}$$

Using the general function $x \mapsto \frac{x-\epsilon}{x}$, one can check that formula (10.11) is impossible. Hence $|\mathcal{B}| = 0$.

We now have that

$$|\{a \in \mathcal{D} \mid c_{S_k(a)}(a) \in U \text{ for some } k\}| = |\mathcal{D}|.$$

Lastly, as U was arbitrary and $\mathcal{D}$ has full measure, the proposition follows. □

Exercise 10.4.2. *Provide a proof for Proposition 10.4.1 using Proposition 6.2.12 (in place of Proposition 10.3.21) to obtain sufficiently long stretches in the phase portrait.*

10.5 Slow Recurrence and the CE Condition

We discuss three conditions: *slow recurrence*, *finite criticality*, and *Collet-Eckmann* (CE). Slow recurrence is a combinatoric condition for kneading sequences [152, 110], finite criticality is a topological condition for unimodal maps [130], and Collet-Eckmann is a metric condition for unimodal maps [57]. For S-unimodal maps (see Definition 10.5.1) we have that slow recurrence implies Collet-Eckmann [152] and that finite criticality characterizes Collet-Eckmann maps [130]. Hence, we have combinatoric and topological conditions guaranteeing a metric condition.

In measurable dynamics, one often works with invariant measures that are absolutely continuous with respect to Lebesgue measure. It is known that S-unimodal maps satisfying the Collet-Eckmann condition admit such measures. Much energy has gone into determining whether maps satisfy the Collet-Eckmann condition; for a history of this work see [115, Chapter V]. It is somewhat surprising to have nonmetric properties, such as slow recurrence and finite criticality, guaranteeing the metric condition of Collet-Eckmann.

Results involving the existence and uniqueness of invariant measures are beyond the scope of the text. However, given the combinatoric and topological flavor of the text, we include statements of results from [152] and [130]. See [91] for a general discussion on why one investigates invariant measures, and see [115] for a unified treatment of many results on invariant measures. We begin with some definitions. Recall that a map $f : \mathbb{R} \to \mathbb{R}$ is called C^3 provided f', f'', and f''' exist and are continuous.

Definition 10.5.1. Let $f : I \to I$ be a C^3 unimodal map with turning point c. We call f *S-unimodal* provided each of the following three properties hold.

1. The map f has *negative Schwarzian derivative*, that is, for $x \neq c$ we have
$$Sf(x) = \frac{f'''(x)}{f'(x)} - \frac{3}{2}\left(\frac{f''(x)}{f'(x)}\right)^2 < 0,$$
denoted $Sf < 0$.

2. The turning point is *nonflat* , that is, there exist $l, L > 1$ such that
$$\frac{|x-c|^{l-1}}{L} \leq |f'(x)| \leq L|x-c|^{l-1}$$
for $x \in I$.

3. We have $|f'(0)| > 1$.

The Schwarzian derivative was introduced into dynamics by Singer [158]. One can compute that $S(f \circ g)(x) = Sf(g(x)) \times (g'(x))^2 + Sg(x)$ and hence, if both f and g have negative Schwarzian derivative, then so also does $f \circ g$, that is, a negative Schwarzian derivative is preserved under composition. In particular, all iterates of a map with negative Schwarzian derivative have negative Schwarzian derivative. Singer proved that if $Sf < 0$, then every attracting periodic orbit (recall Definition 3.1.6) in the core attracts a critical point of f, see for example, [64, Chapter 1.11]. In the results discussed in this and the next section, the role played by a negative Schwarzian derivative is to control the nonlinearity of the branches of f^n. Recent work of Kozlovski [101] shows that, for many purposes, the condition $Sf < 0$ can be weakened to: f is C^2. See also [76].

Definition 10.5.2. [57] Let f be S-unimodal with turning point c. We say f satisfies the *Collet-Eckmann (CE) condition* provided there exists $\kappa > 0$ and $\lambda > 1$ such that

$$|(f^n)'(f(c))| > \kappa\lambda^n$$

for all $n \geq 1$.

As an aside, we remark that if f and g are both C^3 unimodal maps that are topologically conjugate with conjugacy h, then h need not preserve a negative Schwarzian derivative, the order of a critical point, or the CE condition. However, if f and g have negative Schwarzian derivative and a nonflat critical point, then indeed CE is preserved by h [115].

Definition 10.5.3. Let f be unimodal with kneading sequence (recall Definition 6.1.3) $e = e_1, e_2, e_3, \ldots$. Assume $e \in \{0,1\}^{\mathbb{N}}$. For each $n \in \mathbb{N}$ set

$$\mathcal{R}(n) = \min\{j \geq 1 \mid e_{n+j} \neq e_j\}.$$

Exercise 10.5.4. *Let $f : I \to I$ be unimodal with kneading sequence $e = e_1, e_2, e_3, \ldots \in \{0,1\}^{\mathbb{N}}$. Recall the sequences $\{z_k\}_{k\geq 0}$ and $\{\hat{z}_k\}_{k\geq 0}$ given in (and below) formula (6.1). For ease of notation, let z_{-1} denote the left endpoint of I and $\hat{z}_{-1}$ denote the right endpoint of I. Then, as $e \in \{0,1\}^{\mathbb{N}}$, we have that $c_n \neq z_k, \hat{z}_k$ for any k. Prove*

$$c_n \in (z_{k-1}, z_k) \cup (\hat{z}_k, \hat{z}_{k-1}) \iff \mathcal{R}(n) = S_k.$$

Conclude that $\mathcal{R}(n) \in \{S_k\}_{k\geq 0}$ for all n.

Definition 10.5.5. [152] Let $e \in \{0,1\}^{\mathbb{N}}$ be the kneading sequence for some unimodal map f. We say e is *slowly recurrent* provided

$$\lim_{l\to\infty} \limsup_{i\to\infty} \frac{1}{i} \sum_{j=1}^{i} \begin{cases} \mathcal{R}(j) & \text{if } \mathcal{R}(j) \geq l \\ 0 & \text{otherwise} \end{cases} = 0.$$

Let $\mathcal{SR}$ denote the collection of all slowly recurrent kneading sequences.

Set

$$\mathcal{E} = \{\log(a) \mid a \in [1,2] \text{ and } I(c_1(a), T_a) \in \mathcal{SR}\},$$

that is, $\mathcal{E}$ consists of the values of topological entropy of symmetric tent maps T_a for which the kneading sequence of T_a is slowly recurrent.

Theorem 10.5.6. *[152] The kneading sequence of T_a is slowly recurrent for Lebesgue almost every $a \in [1,2]$, that is, the set*

$$\{a \mid I(c_1(a), T_a) \in \mathcal{SR}\}$$

has full Lebesgue measure in $[1,2]$.

Theorem 10.5.7. *[152] Let f be a S-unimodal map whose kneading sequence is slowly recurrent. Then, f satisfies the CE condition.*

Calculating topological entropy from lapnumbers (recall Theorem 9.4.1), we have that the topological entropy of a unimodal map is contained in $[0, \log(2)]$. A value of topological entropy is called *Collet-Eckmann* provided every S-unimodal map with this entropy value satisfies the CE condition. It follows from Theorems 10.5.6 and 10.5.7 that Lebesgue almost every value of topological entropy is indeed Collet-Eckmann. Thus, we see that the combinatoric condition of slow recurrence guarantees the metric condition of CE for S-unimodal maps. Next we discuss the topological condition of finite criticality given in [130] and state the result of [130] for the unimodal setting (Theorem 10.5.10). For S-unimodal maps, we have the equivalence of finite criticality (topological condition) with the CE condition (metric condition).

For $f : I \to I$ unimodal, $n \in \mathbb{N}$, and $U \subset I$ an open interval, we have that $f^{-n}(U)$ is either empty or consists of finitely many pairwise disjoint intervals (not all of which need to be open). In the event that $x \in f^{-n}(U)$, let $Comp(x, f^{-n}(U))$ denote the component (i.e., interval) in $f^{-n}(U)$ containing x. We let $\#A$ denote the cardinality of a set A.

Definition 10.5.8. [130] Let $f : I \to I$ be unimodal with turning point c. We call f *critically finite* provided there exist $M > 0$, $P > 0$, and $\delta > 0$ such that for every $x \in I$ there exists a strictly increasing sequence of positive integers $\{n_i\}_{i \geq 1}$ such that for each i we have $n_i \leq Pi$ and

$$\#\{j \mid 0 \leq j \leq n_i \text{ and } c \in Comp(f^j(x), f^{-(n_i-j)}(B(f^{n_i}(x), \delta)))\} \leq M.$$

The notion "critically finite" is also called "topologically Collet-Eckmann." It was originally formulated for complex maps (see [140, 141]), where it is actually easier to formulate.

One can give a "graphical consequence" of critically finite. For $x \in I$, consider the graph of f^{n_i}, and in particular consider the horizontal δ-band about the point $(x, f^{n_i}(x))$. See Figure 10.9. In this figure, we have that $c \in Comp(f^j(x), f^{-(n_i-j)}(B(f^{n_i}(x), \delta)))$ for $j = n_i - l,\ n_i - m,\ n_i - q$. Thus, critically finite is a restriction on the number of turns (at different heights) the graph of f^{n_i} may have about the point $(x, f^{n_i}(x))$ and within the δ-band (before the graph leaves the δ-band).

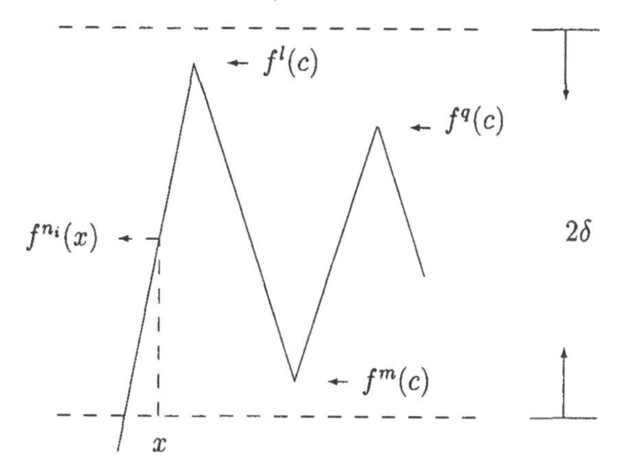

Figure 10.9: Graphical consequence of critically finite

The bound on the number of turns (at different heights) is not a characterization of critically finite. Indeed, it can be that

$$c \in Comp(f^j(x), f^{-(n_i-j)}(B(f^{n_i}(x), \delta)))$$

without a turn resulting in the δ-band about the point $(x, f^{n_i}(x))$ before the graph leaves the δ-band. For example, in Figures 10.10 and 10.11 we have that $c \in Comp(f^j(x), f^{-(n_i-j)}(B(f^{n_i}(x), \delta)))$ for $j = n_i - l$ and $n_i - m$. Yet the turn resulting from $j = n_i - m$ does not occur before the graph leaves the δ-band.

Exercise 10.5.9. *Let f be a longbranched (recall Definition 6.2.3) unimodal map. Prove there exists $\gamma > 0$ and $M > 0$ such that for all x and all n we have that the number of turns within a γ-band about the point $(x, f^n(x))$ before the graph leaves the band is bounded by M. Why can you* not *conclude that every longbranched unimodal map is critically finite?* **HINT:** *See Figures 10.10 and 10.11.*

The result of [130] restricted to the unimodal setting is Theorem 10.5.10.

Theorem 10.5.10. *[130] Let f be an S-unimodal map. Then f satisfies the CE condition if and only if f is critically finite.*

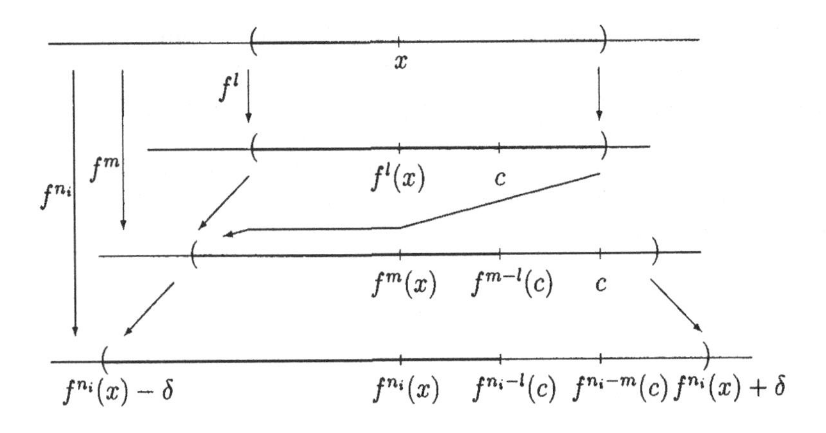

Figure 10.10: $c \in Comp(f^j(x), f^{-(n_i-j)}(B(f^{n_i}(x),\delta)))$ for $j = n_i - l$ and $n_i - m$

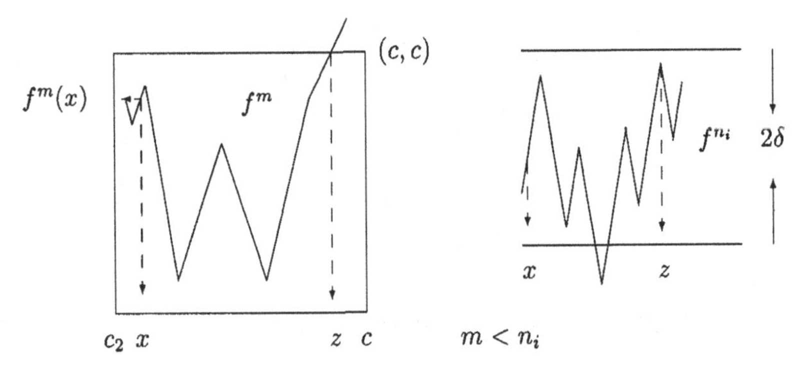

Figure 10.11: Turn at z after the graph has left and reentered the δ-band

10.6 Attractors

In studying asymptotic behaviors for S-unimodal maps one considers *attractors*. We have both *metric* and *topological* attractors (defined below) and a classification of such attractors for S-unimodal maps. It turns out that each topological attractor is indeed a metric attractor; however, there are metric attractors that are not topological attractors. We only briefly touch here on these topics and send the reader to [23, 46, 94, 111, 113, 115] for further reading. Recall that $|E|$ denotes the Lebesgue measure of the set E.

Definition 10.6.1. Let $f : I \to I$ be an S-unimodal map. For $A \subset I$ set $\mathcal{B}(A) = \{x \in I \mid \omega(x, f) \subset A\}$. We call $\mathcal{B}(A)$ the *basin of* A. We say

1. A is a *metric attractor* provided $|\mathcal{B}(A)| > 0$ and there is no proper subset of A with this property;

2. A is a *topological attractor* provided $\mathcal{B}(A)$ is a residual set (a set is *residual* provided its complement is the countable union of nowhere dense sets) and there is no proper subset of A with this property.

It can be shown that A is closed.

Proposition 10.6.2 [77] provides a classification of topological attractors for S-unimodal maps.

Proposition 10.6.2. *[77] Let $f : I \to I$ be an S-unimodal map and $A \subset I$ be a topological attractor. Then exactly one of the following hold:*

1. *The set A is a finite set (and hence a periodic orbit).*

2. *The set A consists of finitely many disjoint closed intervals.*

3. *The set A is a solenoidal attractor (i.e., $A = \omega(c, f)$, with f being infinitely renormalizable).*

The family $g_a(x) = ax(1 - x)$ consists of S-unimodal maps. If a is such that $\frac{1}{2}$ is periodic under g_a, then $\omega(\frac{1}{2}, g_a)$ is a topological attractor of type 1 from Proposition 10.6.2. The 2^∞ map g_{a_*} from this family, discussed in Chapter 5, is such that $\omega(\frac{1}{2}, g_{a_*})$ is a topological attractor of type 3 from Proposition 10.6.2; more generally, any infinitely renormalizable map from this family is such that the ω-limit set of $\frac{1}{2}$ is a topological attractor of type 3. The next section provides examples of type 2 from Proposition 10.6.2.

A classification of metric attractors is given in [24]. Indeed, a metric attractor is one of the three types in Proposition 10.6.2 or is an *absorbing Cantor set.* We say a set A is an absorbing Cantor set provided it is a Cantor set, it is a metric attractor, and it is not solenoidal; thus it is a metric attractor and not a topological attractor. It is known that there are no absorbing Cantor sets as attractors for the logistic family $g_a(x) = ax(1-x)$. However, there are S-unimodal maps with absorbing Cantor sets as metric attractors. Moreover, if f is S-unimodal with an absorbing Cantor set A, then the Lebesgue measure of A is 0, A is a minimal set, $A = \omega(c, f)$, and the turning point is persistently recurrent.

10.7 Combinatorics and Renormalization

We begin with some combinatorics and then connect to renormalization. The material in this section is not used elsewhere in the text. For further reading see [33, 56, 162]

Definition 10.7.1. Let $n, m \in \mathbb{N}$, $P = P_1, P_2, \ldots, P_n \in \{0,1\}^n$, and $Q = Q_1, Q_2, \ldots, Q_m \in \{0,1\}^m$. We say P has *odd parity* (*even parity*) provided there are an odd number (even number) of 1's appearing in P. The *star product* of P and Q is defined as

$$P \star Q = \begin{cases} P\tilde{Q}_1 P\tilde{Q}_2 \ldots P\tilde{Q}_m P & \text{if the parity of } P \text{ is odd} \\ PQ_1PQ_2 \ldots PQ_mP & \text{if the parity of } P \text{ is even,} \end{cases}$$

where

$$\tilde{Q}_i = \begin{cases} 1 & \text{if } Q_i = 0 \\ 0 & \text{if } Q_i = 1. \end{cases}$$

One can generalize the star product to sequences P and Q, where P has finite length and $Q \in \{0,1\}^{\mathbb{N}}$. For example, let $P = 1$ and $Q = 10000\ldots = 10^\infty$. Then $P \star Q = 10111111\ldots = 101^\infty$.

Example 10.7.2. *Let $P = 1$ and $Q = 10$. Then $P \star Q = 10111$. Let T_b be the symmetric tent map with the kneading sequence $10111* = (P \star Q)*$, and let T_a be the symmetric tent map with the kneading sequence $10* = Q*$. For ease of notation, set $c_i(a) = T_a^i(c)$ and $c_i(b) = T_b^i(c)$ for $i \geq 1$.*

We show that T_b is renormalizable with $n = n_b = 2$ and restrictive interval $J = J_b = [c_2(b), c_4(b)]$ (recall Definition 3.4.24). Moreover, $T_b^{n_b}|J_b$ is affinely topologically conjugate to T_a (recall Remark 3.3.7).

As in Exercise 3.1.4, we see that a is the root of $x^3 - 2x^2 + 1 = 0$ that sits in $(\sqrt{2}, 2]$ and b is the root of $x^6 - 2x^5 + 2x^3 - 2x^2 + 2x - 1 = 0$ that sits in $(1, \sqrt{2}]$. Approximations are $a \approx 1.6180339887498948482$ and $b \approx 1.2720196495140689643$. Precisely, a is the golden mean $\frac{1+\sqrt{5}}{2}$. We have that the turning point $c = \frac{1}{2}$ is periodic of period 3 under T_a and of period 6 under T_b.

Observe that $a \in (\sqrt{2}, 2]$, $b \in (\sqrt[4]{2}, \sqrt{2})$, and $b^2 = a$.

The first three iterates of the map T_a appear in Figure 10.12, and the first two iterates of T_b appear in Figure 10.13. From Figure 10.13 we see that $T_b^2|J_b$ is unimodal and hence that T_b is renormalizable with the restrictive interval $J = J_b = [c_2(b), c_4(b)]$.

Set $f = T_a|[c_2(a), c_1(a)]$, $g_1 = T_b^2|[c_2(b), c_4(b)]$, and $g_2 = T_b^2|[c_3(b), c_1(b)]$. Then for $i = 1, 2$ the map g_i is topologically conjugate to f. Thus, for

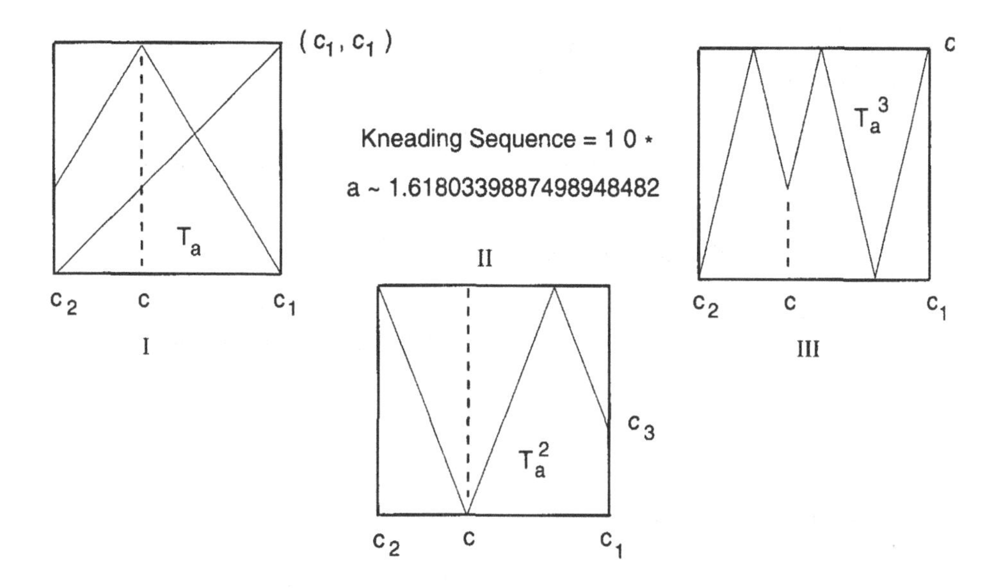

Figure 10.12: First three iterates of T_a

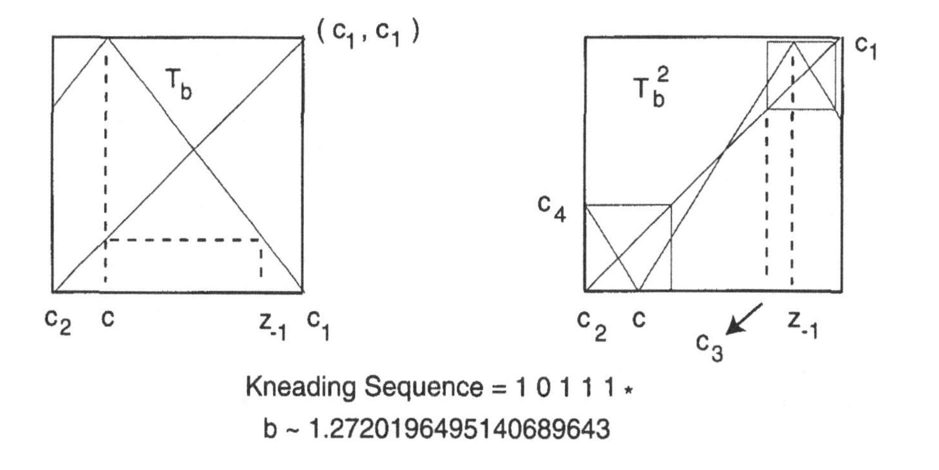

Figure 10.13: First two iterates of T_b

example, if one plotted the graphs of g_2, g_2^2, and g_2^3, one would obtain, respectively, the graphs I, II, and III from Figure 10.12. Similarly for g_1, g_1^2, and g_1^3 (with a change of orientation).

Can you obtain the affine conjugacy between g_1 and T_a as described in Remark 3.3.7?

Example 10.7.2 is a specific case from the next proposition. For ease of notation we set $P \star P = (P)^2$, $P \star (P \star P) = (P)^3, \ldots$.

Proposition 10.7.3. *Let $m \geq 2$ be a positive integer and $b \in (2^{1/2^m}, 2^{1/2^{m-1}}]$. Let Q_b be the kneading sequence of the map $T_{b^{2m-1}}$. Then each of the following hold.*

1. *The kneading sequence of T_b is $(1)^{m-1} \star Q_b$.*

2. *The map T_b is renormalizable with $T_b^{2^{m-1}}|J$ topologically conjugate to $T_{b^{2m-1}}$, where J is the restrictive interval containing the turning point $c = 1/2$.*

Recursively, we have that T_b is $m-1$ times renormalizable.

It is known that, for $\sqrt{2} < b \leq 2$, the kneading sequence of T_b has no star structure (i.e., cannot be written as some $P \star Q$). From Proposition 10.7.3 we see that for $1 < b \leq \sqrt{2}$ there exists $m \in \mathbb{N}$ such that the kneading sequence of T_b is of the form $(1)^m \star Q$ with $b^{2^m} > \sqrt{2}$ and Q the kneading sequence of $T_{b^{2m}}$. Moreover, T_b is finitely renormalizable for $1 < b \leq \sqrt{2}$. There are no infinitely renormalizable symmetric tent maps.

The family $g_a(x) = ax(1-x)$, $a \in [0,4]$, contains finitely and infinitely renormalizable maps (along with nonrenormalizable maps). Infinitely renormalizable unimodal maps have an "infinite star structure." For example, let A denote the kneading sequence of the 2^∞ map from Chapter 5. Then, informally, $\lim_{n\to\infty}(1)^n = A$. Hence, we view A as 1 starred with itself infinitely many times.

One can write the connection between the star structure and renormalization as follows. Let f be a unimodal map with the kneading sequence $P \star Q$, where P has length n (not including the $*$). Then f is renormalizable and $f^{n+1}|J$ is topologically conjugate (where J is the restrictive interval containing the turning point of f) to a unimodal map with the kneading sequence Q. For a detailed study of these topics see [56].

At the end of Section 3.4 we asked the question: Which symmetric tent maps are topologically conjugate to a quadratic map? As remarked in Section 3.4, if T_a is not renormalizable (i.e., has slope in $(\sqrt{2}, 2]$) and the turning point is not periodic, then there is exactly one quadratic map g_a topologically conjugate to T_a. Indeed, let T_a be nonrenormalizable with a turning point not periodic and let A denote the kneading sequence of T_a. Then there exists exactly one g_a with the kneading sequence A. The maps T_a and g_a are topologically conjugate with the conjugacy given by the map h, which takes "itineraries to itineraries," that is, $I(x, T_a) = I(h(x), g_a)$. Again, see [56] for more detail.

We close this section with a formula to compute topological entropy when working with start products. The formula is part of the folklore, and hence we cannot provide an explicit reference. Although the formula is stated in terms of symmetric tent maps, it holds for more general unimodal maps.

Exercise 10.7.4. $\diamondsuit$ *Let T_a, T_b, T_s be symmetric tent maps with kneading sequences P, Q and $P \star Q$, respectively. Then the topological entropy of T_s is given by*

$$h(T_s) = \max\left(h(T_a), \frac{h(T_b)}{|P|}\right).$$

Chapter 11

Unimodal Maps and Rigid Rotations

In this chapter we present results of [48] proving that, given any $\rho \in [0,1]\backslash\mathbb{Q}$, there exists a unimodal map f such that (S^1, R_ρ) is a factor of $(\omega(c,f), f)$ (recall Remark 3.3.2 for the definition of factor). One might ask whether one can obtain the stronger result of conjugacy? As S^1 is not homeomorphic to a Cantor set, and in this setting $\omega(c,f)$ is indeed a Cantor set, a conjugacy is not possible. See [48, 47] for further details and results.

Chapter 3 and Sections 6.1, 7.2, and 8.3 contain background material for this chapter.

11.1 Adding Machines in Unimodal Maps

Given a unimodal map f with turning point c and kneading map $Q(k)$, we construct an adding machine $(\Omega, \mathbf{P})$ from the sequence of cutting times $\{S_k\}$. In the event that $\lim_{k\to\infty} Q(k) = \infty$, we have that $(\omega(c,f), f)$ is a factor of $(\Omega, \mathbf{P})$ (Theorem 11.1.15). The condition $\lim_{k\to\infty} Q(k) = \infty$ is not required to define the adding machine $(\Omega, \mathbf{P})$, but rather comes into play for the continuity of the map $\mathbf{P}$. In this section we provide only the information on the adding machine $(\Omega, \mathbf{P})$ needed to obtain Theorem 11.1.15; see Section 13.3 for a more detailed discussion of $(\Omega, \mathbf{P})$.

Let $S_0 < S_1 < S_2 < S_3, \ldots$ be the sequence of cutting times for some unimodal map f (recall it is always the case that $S_0 = 1$ and $S_1 = 2$). We do not assume that $\lim_{k\to\infty} Q(k) = \infty$. Let n be a nonnegative integer. Then, there is a unique k such that

$$S_k \le n < S_{k+1}.$$

Let $\langle n\rangle_k$ and $r_k(n)$ be integers such that

$$n = \langle n\rangle_k \; S_k + r_k(n),$$

with $0 \le r_k(n) < S_k$. Iterating we get an expansion for n in terms of cutting times:

$$n = \langle n\rangle_0 \; S_0 + \langle n\rangle_1 \; S_1 + \cdots \langle n\rangle_k \; S_k.$$

Set $\langle n\rangle_j = 0$ for $j > k$. Then we get

$$\langle n\rangle = \{\langle n\rangle_j\}_{j\ge 0} \in \{0,1\}^{\mathbb{N}}.$$

Example 11.1.1. *Let $S_0 = 1,\ S_1 = 2,\ S_2 = 3,\ S_3 = 5,\ S_4 = 8, \ldots,$ that is, the Fibonacci sequence. Then,*

$$\begin{aligned} 4 &= 1\; S_0 + 0\; S_1 + 1\; S_2 \\ 13 &= 0\; S_0 + 0\; S_1 + 0\; S_2 + 0\; S_3 + 0\; S_4 + 1\; S_5 \\ 12 &= 1\; S_0 + 0\; S_1 + 1\; S_2 + 0\; S_2 + 1\; S_4. \end{aligned}$$

Hence,

$$\begin{aligned} \langle 4\rangle &= 1\,0\,1\,\overline{0} \\ \langle 13\rangle &= 0\,0\,0\,0\,0\,1\,\overline{0} \\ \langle 12\rangle &= 1\,0\,1\,0\,1\;\overline{0}. \end{aligned}$$

Definition 11.1.2. Let $x = x_0, x_1, x_2, x_3, \ldots$ and $y = y_0, y_1, y_2, y_3, \ldots$ be elements of $\{0,1\}^{\mathbb{N}}$. Define $\rho(x,y)$ by

$$\rho(x,y) = \sum_{i\ge 0} \frac{|x_i - y_i|}{2^i}.$$

Then, ρ is a metric on $\{0,1\}^{\mathbb{N}}$.

Definition 11.1.3. Set

$$\langle \mathbb{N}\rangle = \{\langle n\rangle \mid n \in \mathbb{N}\}.$$

Let Ω denote the closure of $\langle \mathbb{N}\rangle$ in $\{0,1\}^{\mathbb{N}}$; thus $\langle \mathbb{N}\rangle \subset \Omega \subset \{0,1\}^{\mathbb{N}}$.

Remark 11.1.4. We know, from Section 3.6, that $(\{0,1\}^{\mathbb{N}}, \rho)$ is a compact metric space. As Ω is a closed subset of $\{0,1\}^{\mathbb{N}}$, we have that (Ω, ρ) is also a compact metric space (recall Exercise 1.1.28).

Remark 11.1.5. From the construction of Ω we have

$$\Omega = \{\omega = \omega_0, \omega_1, \ldots \mid \sum_{i=0}^{j} \omega_i S_i < S_{j+1} \;\text{ for all }\; j \ge 0\}.$$

Theorem 11.1.6. *[48] We have the following characterization of* Ω:

$$\begin{aligned}\Omega &= \{\omega = \omega_0, \omega_1, \dots \mid \textit{for all } j \geq 0, \textit{ if } Q(j+1) \leq i \leq \\ &\quad j-1 \textit{ and } \omega_j = 1, \textit{ then } \omega_i = 0\}.\end{aligned} \tag{11.1}$$

Proof. Let $\omega = \omega_0, \omega_1, \dots \in \Omega$ and suppose $\omega_j = 1$ for some j. Then $\sum_{i=0}^{j} \omega_i S_i < S_{j+1}$ gives

$$\sum_{i=0}^{j-1} \omega_i S_i < S_{j+1} - S_j = S_{Q(j+1)}. \tag{11.2}$$

It follows from (11.2) that if $Q(j+1) \leq i \leq j-1$, then $\omega_i = 0$.

Let $\omega = \omega_0, \omega_1, \dots \in \{0,1\}^{\mathbb{N}}$ be contained in the right-hand side of (11.1). We show $\omega \in \Omega$ by showing

$$\sum_{i=0}^{j} \omega_i S_i < S_{j+1} \tag{11.3}$$

holds for all $j \geq 0$. We induct on j. For $j = 0$, we have $\omega_0 S_0 = \omega_0 < S_{0+1} = 2$. Fix $k \geq 1$ and assume (11.3) holds for all $j < k$. We show (11.3) holds for $j = k$. Set $k' = \max\{Q(k+1) - 1, 0\}$; note that $k' < k$. If $\omega_k = 0$, then $\sum_{i=0}^{k} \omega_i S_i = \sum_{i=0}^{k-1} \omega_i S_i < S_k < S_{k+1}$ by our induction hypothesis. Suppose $\omega_k = 1$ and $Q(k+1) > 0$. Then

$$\begin{aligned}\sum_{i=0}^{k} \omega_i S_i &= \sum_{i=0}^{k'} \omega_i S_i + S_k \\ &= \sum_{i=0}^{k'} \omega_i S_i + \left(S_{k+1} - S_{Q(k+1)}\right) \qquad \text{(see formula (6.2))} \\ &= \sum_{i=0}^{k'} \omega_i S_i + (S_{k+1} - S_{k'+1}) \\ &< S_{k'+1} + S_{k+1} - S_{k'+1} = S_{k+1} \qquad \text{(induction hypothesis).}\end{aligned}$$

Lastly, suppose $\omega_k = 1$ and $Q(k+1) = 0$. Then, as ω is in the right-hand side of (11.1), we have $\sum_{i=0}^{k} \omega_i S_i = S_k$. The result follows from $S_k < S_{k+1}$. □

Exercise 11.1.7. *Let the sequence of cutting times be the Fibonacci sequence. Show that* Ω *consists of all sequences from* $\{0,1\}^{\mathbb{N}}$ *that do NOT have two consecutive 1's.*

Let's summarize what we have so far in this section. We begin with a sequence of cutting times (gotten from some unimodal map)

$$S_0 = 1 < S_1 = 2 < S_3 < S_4 < S_5 < \cdots.$$

Next, for each $n \in \mathbb{N}$ we obtain the unique $\{S_k\}$-expansion of n, denoted $\langle n \rangle$, with $\langle n \rangle \in \{0,1\}^{\mathbb{N}}$. We then let Ω be the closure of $\langle \mathbb{N} \rangle$ in $\{0,1\}^{\mathbb{N}}$, where $\{0,1\}^{\mathbb{N}}$ is a metric space with the metric given in Definition 11.1.2. Now, we are going to define a self-map $\mathbf{P} : \Omega \to \Omega$ that we call *add and carry* on Ω. Of course, $\mathbf{P}(\langle n \rangle) = \langle n+1 \rangle$, but we have to define $\mathbf{P}$ on all of Ω. The dynamical system $(\Omega, \mathbf{P})$ is called an *$\{S_k\}$-odometer* or, more simply, an *adding machine.* (Remember that $\Omega \subset \{0,1\}^{\mathbb{N}}$.)

Definition 11.1.8. [48] Let a sequence of cutting times $\{S_k\}$ be given. For $\omega = \omega_0, \omega_1, \omega_2, \ldots \in \Omega$ and $j \in \mathbb{N}$, set

$$\omega(j) = \sum_{i \geq 0}^{j} \omega_i S_i.$$

Let $\Omega_0 = \{\omega \in \Omega \mid$ there exists $M_\omega \in \mathbb{N}$ such that for all $j \geq M_\omega$, $\omega(j) < S_{j+1} - 1\}$. For $\omega \in \Omega_0$ and $j \geq M_\omega$, set

$$\mathbf{P}(\omega) = \langle \omega(j)+1 \rangle_0, \langle \omega(j)+1 \rangle_1, \ldots, \langle \omega(j)+1 \rangle_j, \omega_{j+1}, \omega_{j+2}, \ldots.$$

It is not difficult to show that this definition is independent of the choice of $j \geq M_\omega$. For $\omega \in \Omega \setminus \Omega_0$, set $\mathbf{P}(\omega) = \bar{0}$.

Remark 11.1.9. As mentioned, the map $\mathbf{P} : \Omega \to \Omega$ is continuous if and only if $\lim_{k\to\infty} Q(k) = \infty$. We prove this fact in Section 13.3. Also see [47, 48].

Definition 11.1.10. Let $\{S_k\}$ be a sequence of cutting times for some unimodal map. For $S_{k-1} < n \leq S_k$, set $\beta(n) = n - S_{k-1}$. It follows from formulas (6.2) and (6.3) that $\beta(n) \leq S_{Q(k)} < S_k$.

Exercise 11.1.11. *Let $\{S_k\}$ be a sequence of cutting times for some unimodal map f, and let $\{D_n\}$ be the levels in the associated Hofbauer tower. Suppose that $S_{k-1} < n \leq S_k$ for some k. Note that $D_n = \langle c_n; c_{\beta(n)} \rangle$. Prove that $D_{\beta(n)} \supset D_n$.*

Exercise 11.1.12. *Let $\{S_k\}$ be a sequence of cutting times for some unimodal map. Fix m, p such that $m \geq S_p$. Let $i \in \mathbb{N}$ be minimal such that $\beta^i(m) < S_p$. Prove there exists $r \geq p$ such that $m = S_r + \beta^i(m)$ and hence $c_m \in D_{S_r + \beta^i(m)}$.* **HINT:** *Consider $\beta^{i-1}(m)$ and use Exercise 11.1.11.*

Remark 11.1.13. Let f be unimodal with kneading map $Q(k)$. Assume that $\lim_{k\to\infty} Q(k) = \infty$ and that there are no homtervals. It follows from Exercise 6.2.10 that $\lim_{n\to\infty} |D_n| = 0$, where the D_n's are from the associated Hofbauer tower.

Remark 11.1.14. Let $\{S_k\}$ be a sequence of cutting times for some unimodal map. Let $\omega = \omega_0, \omega_1, \ldots \in \Omega$ be such that ω is not eventually 0. Say, $\omega_j = 1 \iff j \in \{q_i\}_{i\geq 0}$. We show in Remark 13.4.14 that the intervals $\{D_{\omega(q_i)}\}_{i\geq 0}$ are nested.

Theorem 11.1.15. *[48] Let f be unimodal with kneading map $Q(k)$ and cutting times $\{S_k\}$. Suppose that $\lim_{k\to\infty} Q(k) = \infty$ and that there are no homtervals. Define $\pi : \langle \mathbb{N} \rangle \to \{c_n\}_{n\geq 0}$ by*

$$\pi(\langle n \rangle) = c_n.$$

Then π is uniformly continuous and has a continuous extension $\pi : \Omega \to \omega(c, f)$ such that the following diagram commutes. Moreover, $\pi^{-1}(c) = \langle 0 \rangle$.

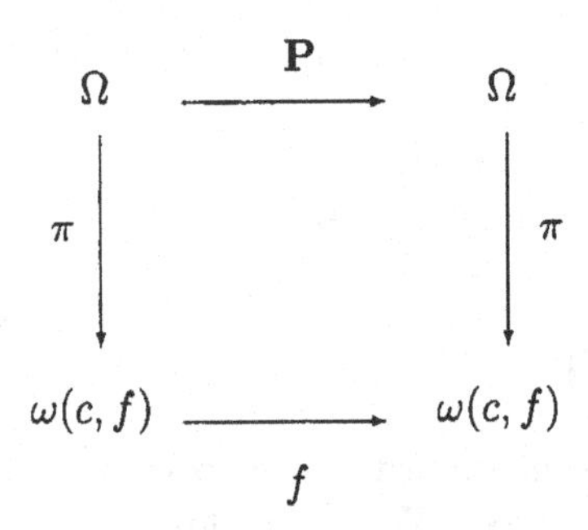

Proof. Fix $\epsilon > 0$. As $\lim_{k\to\infty} Q(k) = \infty$, we have that $\lim_{n\to\infty} |D_n| = 0$ (where the D_n's are the levels in the associated Hofbauer tower). Choose p such that $|D_n| < \frac{\epsilon}{2}$ for all $n \geq S_p$. Let m be arbitrary and choose l such that $\langle l \rangle$ agrees with $\langle m \rangle$ through the $(p-1)$ position. We show $|c_l - c_m| < \epsilon$ and hence conclude that π is uniformly continuous.

Let $i \geq 0$ be minimal such that $\beta^i(m) < S_p$. Set $k = \beta^i(m)$. We break it into cases.

Case 1: Assume $m, l \geq S_p$. Since $\langle l \rangle$ agrees with $\langle m \rangle$ through the $(p-1)$ position, there exist (use Exercise 11.1.12) $r, r' \geq p$ such that

$$c_m \in D_{S_r+k} \quad \text{and} \quad c_l \in D_{S_{r'}+k}.$$

As D_{S_r+k} and $D_{S_{r'}+k}$ share a common boundary point (namely c_k) and $r, r' \geq p$ (thus $|D_{S_r+k}|, |D_{S_{r'}+k}| < \frac{\epsilon}{2}$), we have $|c_m - c_l| < \epsilon$.

Case 2: Assume $m < S_p$ and $l \geq S_p$. In this case, $m = k$. As $l \geq S_p$, there exists $r \geq p$ such that $c_l \in D_{S_r+k} = D_{S_r+m}$. By construction of the Hofbauer tower, we know $c_m \in D_{S_r+m}$. As in Case 1, we have $|c_m - c_l| < \epsilon$.

Case 3: Assume $l < S_p$ and $m \geq S_p$. The argument is similar to Case 1.

Case 4: Assume $l, m < S_p$. Since $\langle l \rangle$ agrees with $\langle m \rangle$ through the $(p-1)$ position, we have $l = m$.

We now show that $\pi^{-1}(c) = \langle 0 \rangle$. Suppose to the contrary that $\langle 0 \rangle \neq \omega = \omega_0, \omega_1, \ldots \in \Omega$ is such that $\pi(\omega) = c$. As $\pi(n) = c_n$ for all n, we have an infinite sequence $\{q_i\}_{i \geq 0}$ with $\omega_j = 1 \iff j \in \{q_i\}$. From Remark 11.1.14 we have that $\pi(\omega) = c \in D_{\omega(q_1)} = D_{S_{q_0}+S_{q_1}}$ and hence that $S_{q_0} + S_{q_1} = S_t$ for some t. However, $\sum_{i=0}^{j} \omega_j S_j < S_{j+1}$ for all j gives that $S_{q_0} + S_{q_1} < S_{q_1+1}$, contradicting $S_{q_0} + S_{q_1} = S_t$. □

Remark 11.1.16. We make two observations. First, it is in Theorem 11.1.15 that we use $\lim_{k \to \infty} Q(k) = \infty$; we had not imposed this condition earlier. Second, the map π is not necessarily one-to-one; we investigate when π is one-to-one in Chapter 13. When π is not one-to-one, we have only that $(\omega(c, f), f)$ is a factor of $(\Omega, \mathbf{P})$, that is, the systems are not topologically conjugate.

Remark 11.1.17. In summary, we now have the following facts. The map $\pi : \Omega \mapsto \omega(c)$ is such that $\pi(\langle n \rangle) = c_n$. Suppose $e = (e_0, e_1, \ldots) \in \Omega$ with $e_j = 1$ infinitely often; hence $e \neq \langle n \rangle$ for any n. Say $e_j = 1 \iff j \in \{q_i\}_{i \geq 0}$ and set $\omega(i) = \sum_{j \leq i} S_j$. Then $\pi(e) = \cap_{i \geq 0} D_{\omega(q_i)}$; see Figure 11.1. In Figure 11.1 the levels for $D_{S_{q_1}}$, $D_{S_{q_1+1}}$, $D_{S_{q_2}}$, and $D_{S_{q_2+1}}$ are shown; however, the actual intervals are not shown.

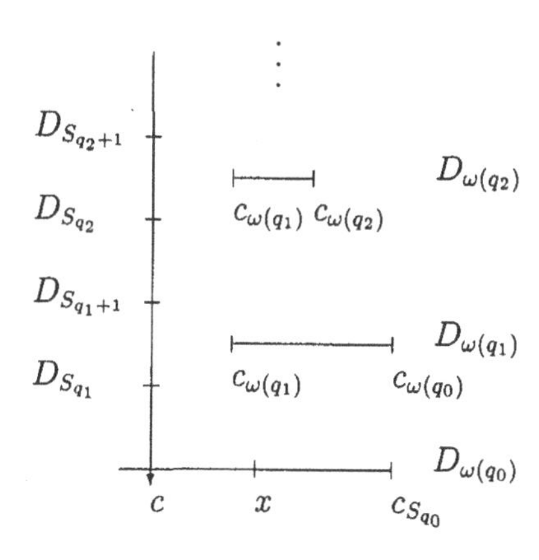

Figure 11.1: $x = \pi(e) = \cap_{i \geq 0} D_{\omega(q_i)}$

Given a sequence of cutting times $\{S_k\}$ we defined the $\{S_k\}$-odometer (or adding machine) $(\Omega, \mathbf{P})$. More generally, given an increasing sequence of positive integers $\{G_k\}$, one can define an $\{G_k\}$-odometer; that is, one need not start with cutting times [48, 75]. In Sections 11.2 and 11.3, we connect all this machinery up with irrational rotations on a circle and prove that for any irrational rotation there exists a unimodal map f such that $(\omega(c,f),f)$ factors onto the irrational rotation [48]. An extensive reference on adding machine is [8].

11.2 Rigid Rotations in Unimodal Maps – I

We are interested in the case when rigid rotations of a circle occur as factors of unimodal systems. In this and the next subsection, we present results of [48] showing that any irrational rotation can be obtained as a factor of a unimodal system. Given an irrational $\rho \in [0,1]$, Theorem 11.2.6 provides conditions on a kneading map to obtain the irrational rotation as a factor of the associated unimodal map. An explicit algorithm for constructing such a kneading map is given in the next section.

Remark 11.2.1. Throughout this section, we assume $\lim_{k\to\infty} Q(k) = \infty$ for all unimodal maps used. We want the add and carry operator $\mathbf{P}$ to be continuous (see Section 13.3).

For $x \in \mathbb{R}$, let $\| x \|$ denote the distance of x to the closest point in $\mathbb{Z}$, round(x) denote the closest integer to x (where round$(n+\frac{1}{2}) = n+1$), and $\{\{x\}\} = x-$round(x). Thus, $\{\{x\}\} \in [-\frac{1}{2},\frac{1}{2})$ and $\| x \| \in [0,1)$. Note that $\{\{x\}\} \mod 1 = \{\{x\}\}$ for $\{\{x\}\} \in [0,\frac{1}{2})$ and $\{\{x\}\} \mod 1 = 1+\{\{x\}\}$ for $\{\{x\}\} \in [-\frac{1}{2},0)$. For $\rho \in [0,1]$, let $R_\rho(x) = x+\rho \mod 1$, that is, the rigid rotation of the unit circle S^1 by angle ρ. Recall that $\#A$ denotes the cardinality of A.

Definition 11.2.2. [48] Let $\rho \in [0,1]$, $\{S_k\}_{k\geq 0}$ be a sequence of cutting times for some unimodal map and Ω the associated odometer. Define $\Pi_\rho : \Omega \to S^1$ by

$$\Pi_\rho(\omega) = \left[\sum_{k\geq 0} \omega_k\{\{S_k\rho\}\}\right] \mod 1.$$

An immediate question is whether the map Π_ρ is well defined, that is, is $\Pi_\rho(\omega) \in S^1$ for all $\omega \in \Omega$? Proposition 11.2.4 provides a condition under which Π_ρ is well defined, and Section 11.3 provides a kneading map (and hence a sequence of cutting times $\{S_k\}$) for which Π_ρ is indeed well defined. Before getting to Proposition 11.2.4, we give an alternative way to compute Π_ρ.

Exercise 11.2.3. *Let $\rho \in [0,1]$, $\{S_k\}_{k\geq 0}$ be a sequence of cutting times for some unimodal map, and Ω be the associated odometer. Suppose Π_ρ is well defined. Show:*

$$\Pi_\rho(\omega) = \lim_{l\to\infty} \left[\left(\sum_{k=0}^{l} \omega_k S_k \rho \right) \mod 1 \right].$$

Proposition 11.2.4. *[48] Let $\rho \in [0,1]$, $\{S_k\}_{k\geq 0}$ be a sequence of cutting times for some unimodal map and Ω the associated odometer. If $\sum_{k\geq 0} \| \rho S_k \| < \infty$, then Π_ρ is well defined, continuous, and $\Pi_\rho \circ P = R_\rho \circ \Pi_\rho$, that is, the following diagram commutes.*

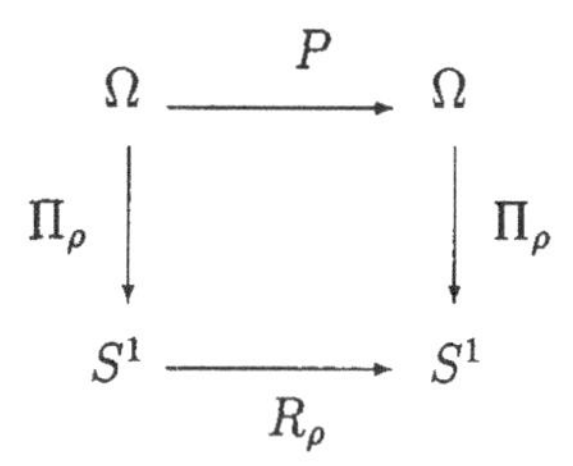

Exercise 11.2.5. ♣ *Prove Proposition 11.2.4.*

Theorem 11.2.6 ([48, Theorem 2]). *Let $\rho \in [0,1]$ and $\{S_k\}$ be such that $\sum_{k\geq 0} \| \rho S_k \| < \infty$. Recall the map π from Theorem 11.1.15. Then,*

$$\#\Pi_\rho(\pi^{-1}(x)) = 1$$

for each $x \in \omega(c,f)$ (f is a unimodal map generating the cutting times $\{S_k\}$).

Denoting this unique element by $\pi_\rho(x)$, we obtain a continuous factor map $\pi_\rho : (\omega(c,f), f) \to (S^1, R_\rho)$; that is, the following diagram commutes.

$$\begin{array}{ccc} & (\Omega, P) & \\ \pi \swarrow & & \searrow \Pi_\rho \\ (\omega(c), f) & \xrightarrow{\ \pi_\rho\ } & (S^1, R_\rho) \end{array}$$

Proof. Let $M = \{x \in \omega(c,f) \mid \#\Pi_\rho(\pi^{-1}(x)) > 1\}$.

Claim 1: If $x \in M$, then $\overline{orb(x,f)} \subset M$. (The proof of this claim is left to Exercise 11.2.7.)

Since $\lim_{k\to\infty} Q(k) = \infty$, the set $\omega(c,f)$ is minimal and therefore (using Claim 1 and the forward invariance of an ω-limit set) $M = \emptyset$ or $M = \omega(c,f)$. By Theorem 11.1.15, $c \notin M$ and therefore $M = \emptyset$. □

Exercise 11.2.7. ♣ *Prove Claim 1 from Theorem 11.2.6.* **HINT:** *Show first that the forward orbit (under f) of $x \in M$ is a subset of M. Then argue that $\overline{orb(x,f)} \subset M$.*

11.3 Rigid Rotations in Unimodal Maps – II

Fix $\rho \in [0,1]\setminus\mathbb{Q}$ and let $[0, a_1, a_2, a_3, \dots]$ be its continued fraction expansion. Again, convergents are denoted $\frac{p_i}{q_i}$, that is, $\frac{p_i}{q_i} = [0, a_1, \dots, a_i]$ (see Exercise 7.2.3). Throughout this section ρ is fixed. We follow [48] to construct a kneading map Q and hence a unimodal map f such that (S^1, R_ρ) is a factor of $(\omega(c,f), f)$; see Corollary 11.3.21, Theorem 11.3.22, and Remark 11.3.23. We point out that, in this construction, it is not the case that $(\omega(c,f), f)$ is conjugate to (S^1, R_ρ). For more detail see [47, 48].

Recall from number theory [81]:

1. $\frac{p_0}{q_0} = \frac{0}{1}$ and $\frac{p_1}{q_1} = \frac{1}{a_1}$.
2. $p_i = a_i p_{i-1} + p_{i-2}$ and $q_i = a_i q_{i-1} + q_{i-2}$.
3. $p_i q_{i+1} - p_{i+1} q_i = (-1)^{i+1}$.
4. $\frac{p_{2i}}{q_{2i}} < \rho < \frac{p_{2i+1}}{q_{2i+1}}$ and hence, $|\rho - \frac{p_j}{q_j}| \le \frac{1}{q_j q_{j+1}}$.

Exercise 11.3.1. *Fix $\rho \in [0,1] \setminus \mathbb{Q}$ with convergents $\frac{p_i}{q_i}$'s. Prove the following:*

1. $\{\{q_k\rho\}\} < 0 \iff k$ *is odd.*
2. $\{\{q_k\rho\}\} = (-1)^k \parallel q_k\rho \parallel$ *for* $k \ge 1$.

Definition 11.3.2. Define a kneading map Q as follows:

$$Q(n) = 0 \qquad \text{for} \qquad n < q_1,$$

$n_0 := 0$, $n_1 := q_1 - 1$, inductively set $n_{k+1} = n_k + a_{k+1}$, and

$$Q(n) = \begin{cases} n_k & \text{if } n_k < n < n_{k+1} \\ n_{k-1} & \text{if } n = n_{k+1}. \end{cases}$$

Exercise 11.3.3. *[48] Prove each of the following.*

1. $S_{n_k} = q_k$ *for all* k.
2. $S_{n_k+j} = (j+1)q_k$ *for* $1 \leq j < a_{k+1}$.
3. $\lim_{k\to\infty} Q(k) = \infty$.
4. Q *is admissible.* **HINT**: *Show* $Q(Q^2(n)+1) \leq n_{k-2} < n_{k-1} \leq Q(n+1)$ *for* $n_k \leq n < n_{k+1}$.

Throughout the rest of this section, Ω is the adding machine generated from the sequence of cutting times $\{S_k\}$ described in Exercise 11.3.3 and obtained from Definition 11.3.2. Exercise 11.3.5 shows that the map Π_ρ is well defined.

Exercise 11.3.4. *[48] Let* $\omega = \omega_0, \omega_1, \ldots \in \Omega$. *Prove for each* k

1. *There is at most one* $n_k \leq n < n_{k+1}$ *such that* $\omega_n = 1$. **HINT:** *Notice that* $Q(n+1) \leq n_k$ *for every* $n_k \leq n < n_{k+1}$.
2. *If* $\omega_n = 1$ *for some* $n_k \leq n < n_{k+1}$, *then* $\omega_{n_{k+1}+a_{k+2}-1} = \omega_{n_{k+2}-1} = 0$.

Exercise 11.3.5. *[48] Let* $\omega = \omega_0, \omega_1, \ldots \in \Omega$. *Prove*

$$\begin{aligned}\sum_k \omega_k \parallel \rho S_k \parallel &\leq \sum_k \max\{\parallel \rho j q_k \parallel \mid 0 \leq j \leq a_{k+1}\} \\ &\leq \sum_k \frac{1}{q_k} < \infty.\end{aligned}$$

HINT: *Use* $\parallel \rho j q_k \parallel \leq j|\rho q_k - p_k| \leq j q_k|\rho - \frac{j}{q_k}| \leq \frac{j}{q_{k+1}}$, $S_{n_k+j} = (j+1)S_{n_k}$ *for* $0 \leq j < a_{k+1}$ *and Exercise 11.3.4.*

Definition 11.3.6. [48] We say $\omega \in \Omega$ is *eventually maximal* provided there exists $K \in \mathbb{N}$ such that either

- $\omega_{n_{2k}-1} = 1$ for all $k \geq K$ or
- $\omega_{n_{2k+1}-1} = 1$ for all $k \geq K$.

Exercise 11.3.7. *[48] Show* $\mathbf{P}^m(\omega) = \langle 0 \rangle$ *for some* $m \geq 1$ *if and only if* ω *is eventually maximal.* **HINT:** *For the "only if" part, set* $m = S_{n_{2K}} - \omega(n_{2K}-1) - 1$.

Definition 11.3.8. [48] Define $\gamma, \tilde{\gamma} \in \Omega$ by

$$\begin{aligned}\gamma_j &= 1 \iff j \in \{n_{2i} - 1\}_{i\geq 1} \\ \tilde{\gamma}_j &= 1 \iff j \in \{n_{2i+1} - 1\}_{i\geq 0}.\end{aligned}$$

Exercise 11.3.9. *Note that γ and $\tilde{\gamma}$ are eventually maximal. Prove* $\mathbf{P}(\gamma) = \mathbf{P}(\tilde{\gamma}) = (\langle 0 \rangle)$ *and* $\mathbf{P}^{-1}(\langle 0 \rangle) = \{\gamma, \tilde{\gamma}\}$.

Definition 11.3.10. [48] Define Σ by

$$\Sigma = \{(b_i)_{i \geq 0} \mid 0 \leq b_0 < a_1, \ 0 \leq b_i \leq a_{i+1}, \ \text{ and } \ b_i \neq 0 \Rightarrow b_{i+1} < a_{i+2}\}.$$

Proposition 11.3.11. *[48] There is a bijection between Ω and Σ, denoted* $h : \Omega \to \Sigma$.

Exercise 11.3.12. *Use Exercise 11.3.4 to prove Proposition 11.3.11.* **HINT:** *Let $\omega = \omega_0, \omega_1, \ldots \in \Omega$. Define a bijection as follows:*

- $\omega_j = 0$ *for* $n_i \leq j < n_{i+1}$ *implies* $b_i = 0$.
- $\omega_j = 1$ *for* $n_i \leq j < n_{i+1}$*, say* $j = n_i + l$*, then* $b_i = l + 1$.

Exercise 11.3.13. *[48] For $\omega \in \Omega$ and $h(\omega) = (b_i)_{i \geq 0}$, prove that ω is eventually maximal if and only if there exists $K \in \mathbb{N}$ such that either*

- $b_{2k} = a_{2k+1}$ *(and thus,* $b_{2k-1} = 0$*) for all* $k \geq K$
- $b_{2k+1} = a_{2k+1}$ *(and thus,* $b_{2k} = 0$*) for all* $k \geq K$.

We refer to such $h(\omega)$'s as eventually maximal *in* Σ.

Definition 11.3.14. [48] Define a mapping $\Pi'_\rho : \ \Sigma \to S^1$ by

$$\Pi'_\rho(b) = \left[\sum_{i \geq 0} (-1)^i \ b_i \parallel \rho q_i \parallel \right] \quad \text{mod } 1.$$

Exercise 11.3.15. *Prove* $\Pi'_\rho \circ h = \Pi_\rho$. **HINT:** *Use Exercise 11.3.1.*

Exercise 11.3.16. *Prove* $\Pi'_\rho(h(\gamma)) = \Pi'_\rho(h(\tilde{\gamma})) = 1 - \rho$.

The proof of Theorem 11.3.17 is best understood if one completely understands the details behind Figure 11.2. Section 8.3 provides these details.

Theorem 11.3.17. *[48] The mapping Π'_ρ is a continuous mapping of Σ to* $[0, 1]$. *It is one-to-one except on eventually maximal sequences, where it is two-to-one.*

Proof. We do the case when $0 < \rho < \frac{1}{2}$. Observe, from Figure 11.2, that for $n \geq 0$ we have

$$\begin{aligned} \parallel \rho q_n \parallel &= \parallel \rho q_{n+2} \parallel + a_{n+2} \parallel \rho q_{n+1} \parallel \\ &= \sum_{i \geq 0} a_{n+2+2i} \parallel \rho q_{n+1+2i} \parallel . \end{aligned} \tag{11.4}$$

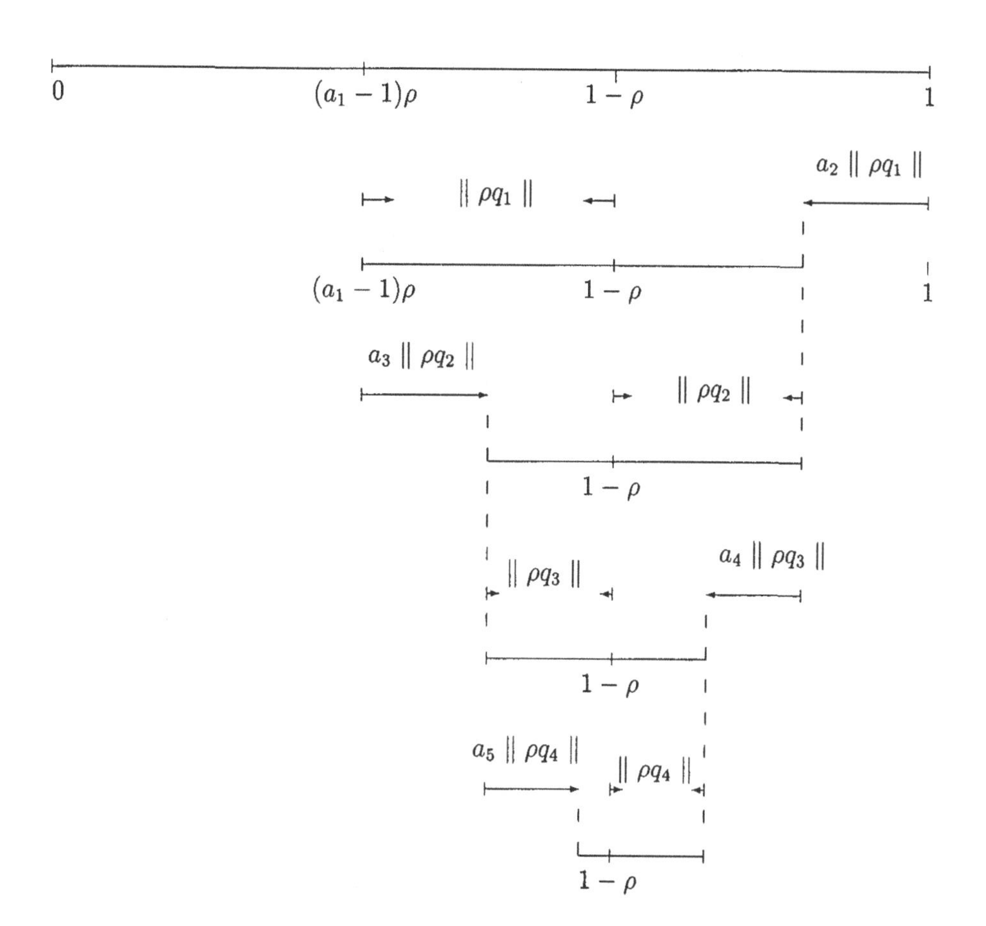

Figure 11.2: Case $\rho < \frac{1}{2}$ (not to scale)

Suppose $b \neq b' \in \Sigma$ are such that $\Pi'_\rho(b) = \Pi'_\rho(b')$. Let n be the first entry where b and b' differ and assume n is even and $b'_n = b_n + 1$. Other cases are similar. Set $x = \sum_{i \leq n} (-1)^i\, b_i \parallel \rho q_i \parallel$ and $x' = \sum_{i \leq n} (-1)^i\, b'_i \parallel \rho q_i \parallel = x + \parallel \rho q_n \parallel$. Then, use (11.4),

$$\begin{aligned} \Pi'_\rho(b) &\leq \max\{\Pi_\rho{}'(c) \mid c_i = b_i \text{ for } i \leq n\} \qquad (11.5) \\ &= x + \sum_{i \geq 1} a_{n+2i+1} \parallel \rho q_{n+2i} \parallel \\ &= x + \parallel \rho q_{n+1} \parallel \end{aligned}$$

and

$$\begin{aligned} \Pi'_\rho(b') &\geq \min\{\Pi_\rho{}'(c) \mid c_i = b'_i \text{ for } i \leq n\} \qquad (11.6) \\ &= x' - (a_{n+2}-1) \parallel \rho q_{n+1} \parallel + \sum_{i\geq 1} (-1)^i \, a_{n+2i+2} \parallel \rho q_{n+1+2i} \parallel \\ &= x + \parallel \rho q_n \parallel - (a_{n+2}-1) \parallel \rho q_{n+1} \parallel - \parallel \rho q_{n+2} \parallel \\ &= x + \parallel \rho q_{n+1} \parallel . \end{aligned}$$

From (11.5), (11.6), and $\Pi'_\rho(b) = \Pi'_\rho(b')$ we have the following:

$$\begin{aligned} \Pi'_\rho(b) &= \max\{\Pi'_\rho(c) \mid c_i = b_i \text{ for } i \leq n\} \\ \Pi'_\rho(b') &= \min\{\Pi'_\rho(c) \mid c_i = b'_i \text{ for } i \leq n\}. \end{aligned}$$

From the definition of Π'_ρ and Exercise 11.3.13, the right-hand sides of (11.5) and (11.6) are realized by eventually maximal sequences. Thus, $\Pi'_\rho(b) = \Pi'_\rho(b')$ only if b and b' are maximal from entry n, that is, b and b' are eventually maximal. More precisely, b and b' must be the eventually maximal sequences with Π'_ρ value equal to the right-hand sides of (11.5) and (11.6). We also see that Π'_ρ is one-to-one except on eventually maximal sequences.

The case $\frac{1}{2} < \rho < 1$ is similar; see Figure 11.3. □

Exercise 11.3.18. *Prove formulas (11.5) and (11.6).*

Theorem 11.3.19. *Let $P' : \Sigma \to \Sigma$ be defined as $P' = h \circ P \circ h^{-1}$ and let R_ρ denote the rigid rotation $R_\rho(x) = x + \rho \mod 1$. Then $\Pi'_\rho \circ P' = R_\rho \circ \Pi'_\rho$, that is, the following diagram commutes.*

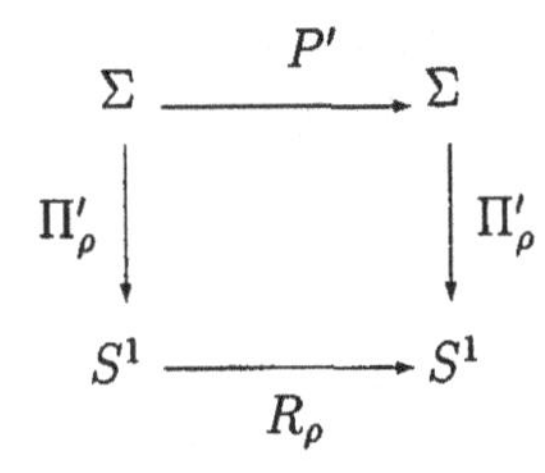

Exercise 11.3.20. *Prove Theorem 11.3.19. See [167].*

Corollary 11.3.21. *The following diagram commutes.*

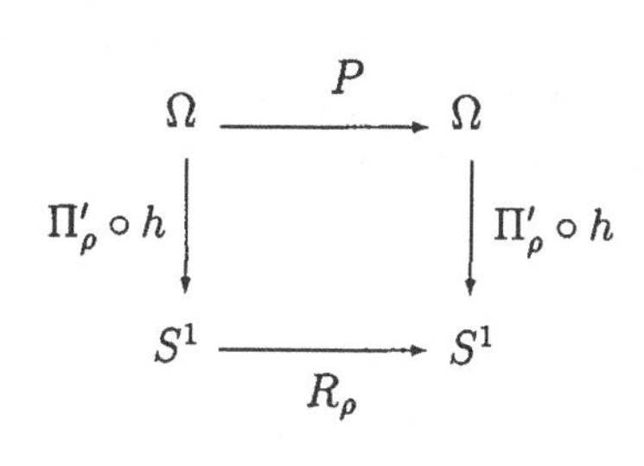

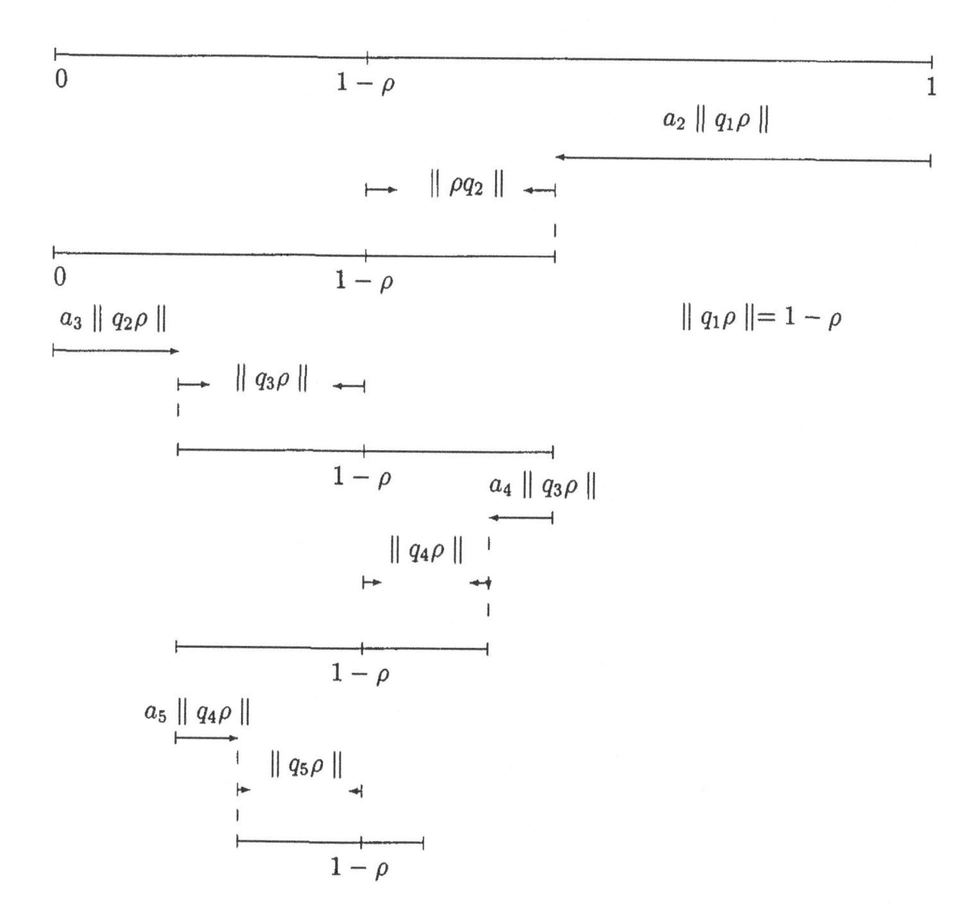

Figure 11.3: Case $\rho > \frac{1}{2}$ (not to scale)

Theorem 11.3.22. *[48] Let $\rho \in [0,1] \setminus \mathbb{Q}$ and $\{S_k\}$ the cutting times from Definition 11.3.2. Then $\Pi_\rho : \Omega \to S^1$ is continuous and one-to-one except on preimages of 0, where it is two-to-one.*

Proof. Use Corollary 11.3.21 and $\Pi'_\rho \circ h = \Pi_\rho$. □

Remark 11.3.23. Let $\rho \in [0,1] \setminus \mathbb{Q}$ and $\{S_k\}$ be the cutting times from Definition 11.3.2. Then $(\omega(c,f), f)$ factors over (S^1, R_ρ) via $\Pi_\rho \circ \pi^{-1}$, that is, the following diagram commutes.

$$\begin{array}{ccc} \omega(c) & \xrightarrow{\ f\ } & \omega(c) \\ \Pi_\rho \circ \pi^{-1} \downarrow & & \downarrow \Pi_\rho \circ \pi^{-1} \\ S^1 & \xrightarrow[R_\rho]{} & S^1 \end{array}$$

Chapter 12

β-Transformations, Unimodal Maps, and Circle Maps

In earlier chapters the interval maps studied were continuous. Now we look at a family of discontinuous maps: the β-transformations $x \mapsto \beta x \bmod 1$ for a real number $\beta > 1$ (the slope). These maps were introduced as a tool in number theory and later became a frequent example in ergodic theory; see Chapter 4. In this chapter, we study the connection between β-transformations and unimodal maps, in particular tent maps. We discuss the "flip-half-of the-graph" trick that turns tent maps into a version of the β-transformation. In Section 12.4, we start investigating *isomorphisms* (a measurable analog of conjugacy) between tent maps and β-transformations. For slope $\beta = 2$, it is quite rewarding to understand the isomorphism explicitly. In Section 12.5, we give a relation between the isomorphism and Ledrappier's example. Finally we discuss the (non)existence of isomorphisms for other slopes.

Chapter 3 and Sections 4.2, 8.1, 8.2 contain background material for this chapter.

12.1 β-Transformations and β-Expansions

Let us start with the angle doubling map $f_2 : [0,1) \to [0,1)$ defined by $f_2(x) = 2x \mod 1$, or, in other words,

$$f_2(x) = \begin{cases} 2x & \text{if } x < \frac{1}{2}, \\ 2x-1 & \text{if } x \geq \frac{1}{2}. \end{cases}$$

Similar to Definition 5.1.3, we can define itineraries: For $x \in [0,1)$, the *itinerary* $I(x, f_2) = I_0(x, f_2), I_1(x, f_2) \dots$, where $I_i(x, f_2) = 0$ if $f_2^i(x) < \frac{1}{2}$ and $I_i(x, f_2) = 1$ if $f_2^i(x) \geq \frac{1}{2}$.

Exercise 12.1.1. *Show that the itinerary of x coincides with the binary expansion, that is, $x = \sum_i I_i(x, f_2)2^{-i-1}$. (For the limit case we find $\lim_{x\nearrow 1} I(x, f_2) = 11111...$, so here also $1 = \sum_i I_i(1, f_2)2^{-i-1}$ is justified.)*

Exercise 12.1.2. *Find a map $g : [0, 1) \to [0, 1)$ such that the itinerary of a point x is exactly its decimal expansion. Do the same for the expansion in base 17 (i.e., we want $x = \sum_i I_i(x, g)(17)^{-i-1}$ with $I_i(x, g) \in \{0, 1, \dots, 16\}$).*

The β-transformation is a variation of the theme in the above exercises. Choose $\beta > 1$. One can choose β to be an integer; however, we obtain nothing new. Hence, it is more interesting to let β be noninteger. The β-transformation is defined as

$$f_\beta : [0, 1) \to [0, 1), \quad f_\beta(x) = \beta x \mod 1. \tag{12.1}$$

Renyi [144], Adler [1], and Smorodinsky [159] were among the first to study β-transformations. A comprehensive treatment of the map was made by William Parry [133]. More recent accounts were given in [17, 61, 96, 108, 153]. Figure 12.1 shows f_β for some $\beta \in (2, 3)$.

$$f_\beta(x) = \begin{cases} \beta x & \text{if } 0 < x < \frac{1}{\beta}, \\ \beta x - 1 & \text{if } \frac{1}{\beta} \leq x < \frac{2}{\beta}, \\ \beta x - 2 & \text{if } \frac{2}{\beta} \leq x < 1. \end{cases}$$

Figure 12.1: Graph of f_β for some $\beta \in (2, 3)$

Let $b = \lfloor \beta \rfloor$ and, if $\beta \in \mathbb{N}$, then $b = \beta - 1$. (Recall that $\lfloor x \rfloor$ denotes the integer part of x.) The map f_β has $b + 1$ branches; the first branch maps the interval $[0, \frac{1}{\beta})$ onto the whole interval $[0, 1)$, with slope β, the second branch maps the interval $[\frac{1}{\beta}, \frac{2}{\beta})$ onto $[0, 1)$, and so on. Only the last branch maps $[\frac{b}{\beta}, 1)$ into a part of $[0, 1)$. We define symbolic dynamics as before. So give the first interval $[0, \frac{1}{\beta})$ the symbol 0, the second interval $[\frac{1}{\beta}, \frac{2}{\beta})$ gets 1, and so on up to the last interval $[\frac{b}{\beta}, 1)$, which gets the symbol b. Define the itinerary $I(x, f_\beta)$ of a point accordingly.

Before proving Theorem 12.1.3, we need to say exactly what is meant by the expansion of x in base β. What we want is to write x as a sum

$$x = \sum_{i=0}^{\infty} a_i \beta^{-i-1},$$

where each $a_i \in \{0, 1, \ldots, b = \lfloor\beta\rfloor\}$. We do this as greedily as possible, that is, we take a_0 as large as possible so that $a_0\beta^{-1} \le x$. Since $x < 1$ and, by the choice of b, $0 \le a_0 \le b$. Then continue with the remainder $x - a_0\beta^{-1}$, that is, find a_1 as large as possible such that $a_1\beta^{-2} \le x - a_0\beta^{-1}$, and so on.

Theorem 12.1.3. *The itinerary of a point $x \in [0, 1)$ for the β-transformation f_β is exactly the expansion of x in base β.*

Proof. Take $x \in [0, 1)$ and let a_0 be the first digit in the β-expansion of x. Then $a_0\beta^{-1} \le x < (a_0 + 1)\beta^{-1}$. This means that x belongs to the $a_0 + 1$-st interval, and the first symbol of the itinerary is $I_0(x, f_\beta) = a_0$.

Write $x_0 = x$, $x_1 = f_\beta(x)$, $x_2 = f_\beta^2(x)$, and so on. Then $x_1 = \beta x_0 - a_0$, or, in other words, $x_1\beta^{-1} = x_0 - a_0\beta^{-1}$ is the remainder after subtracting the first term in the β-expansion.

By the definition of a β-expansion, we have $a_1\beta^{-2} \le x_1\beta^{-1} < (a_1+1)\beta^{-2}$. But multiplying these inequalities by β again, we also see that $I_1(x, f_\beta) = a_1$. Continue in this manner. □

In the sequel it is convenient to extend f_β to the point 1. We do this by

$$f_\beta(1) := \lim_{x \nearrow 1} f_\beta(x).$$

Exercise 12.1.4. *Find $\beta \in [1, 2]$ such that the itinerary of 1 for f_β is* $110101010\ldots = 1(10)^\infty$.

12.2 Flip-Half-of-the-Graph Trick

In this section we discuss a *flip-half-of-the-graph* trick in low-dimensional dynamics. This trick is used in Sections 12.3 and 12.4. For our discussion here, we use tent maps. Recall the symmetric family of tent maps T_β from Definition 3.1.3. First we rescale the tent map as follows:

$$\tau_\beta : [0, 1] \to [0, 1], \ \tau_\beta(x) = 1 - \beta|x - \tfrac{1}{2}|.$$

The turning point of τ_β is $c = \frac{1}{2}$. See Figure 12.2.

Exercise 12.2.1. *Verify that τ_β is topologically conjugate to the symmetric tent map T_β restricted to $[\hat{T}_\beta(c), T_\beta(c)]$. (Recall that $\hat{x}$ is the unique point different from x such that $T_\beta(x) = T_\beta(\hat{x})$; we use $\hat{T}_\beta(c) = \widehat{T_\beta(c)}$.) Find the conjugacy h.*

Next, for $\beta \in (1,2]$ we define a map $\varphi_\beta : [0,1] \to [0,1]$ by

$$\varphi_\beta(x) = \begin{cases} \beta x + 1 - \frac{\beta}{2} & \text{if } x < \frac{1}{2}, \\ \beta x - \frac{\beta}{2} & \text{if } x \geq \frac{1}{2}. \end{cases}$$

The map φ_β is obtained from τ_β by "flipping" the right half of τ_β's graph; see Figure 12.2. This flipping is simply given by $\varphi_\beta(x) = 1 - \tau_\beta(x)$ for $x \geq \frac{1}{2}$. Notice that φ_β is piecewise increasing with slope β, similar to f_β (the β-transformation from formula (12.1) with $\beta \in (1,2]$. A map such as φ_β is sometimes called a *Lorenz map* [79, Section 2.3], since they form a very simplified version of a famous meteorological model of Lorenz.

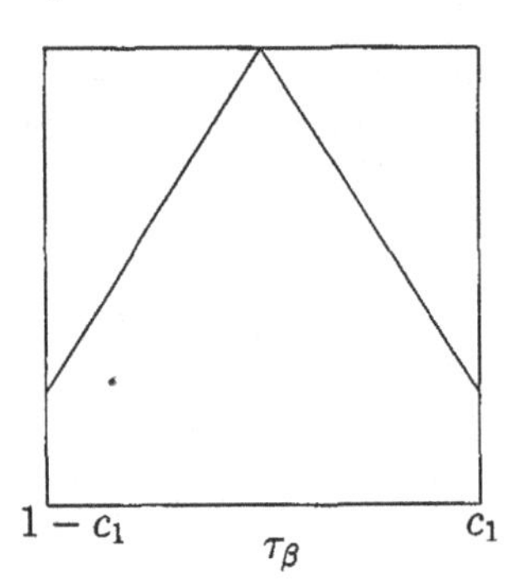

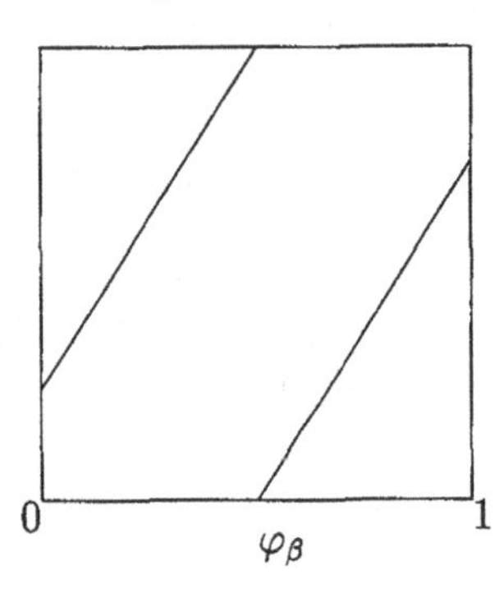

Figure 12.2: Graphs of τ_β and φ_β

Exercise 12.2.2. *Show that $\tau_\beta \circ \varphi_\beta = \tau_\beta \circ \tau_\beta$, so τ_β and φ_β are two-to-one semiconjugate and τ_β itself is the semiconjugacy.*

Exercise 12.2.3. *Show that for $n \geq 0$ and x such that $\tau_\beta^i(x) \neq c$ for all $0 \leq i < n$ we have:*

$$\tau_\beta^n(x) = \begin{cases} \varphi_\beta^n(x) & \text{if } \tau_\beta^n \text{ is locally increasing at } x, \\ 1 - \varphi_\beta^n(x) & \text{if } \tau_\beta^n \text{ is locally decreasing at } x. \end{cases} \tag{12.2}$$

Recall that $I(x, \tau_\beta) = I_0(x, \tau_\beta), I_1(x, \tau_\beta), \dots \in \{0,1\}^{\mathbb{N}}$ is the itinerary of the point x with respect to τ_β. Similarly define the itinerary $I(x, \varphi_\beta) = I_0(x, \varphi_\beta),\ I_1(x, \varphi_\beta), \dots \in \{0,1\}^{\mathbb{N}}$, where $I_i(x, \varphi_\beta) = 0$ if $\varphi_\beta^i(x) < \frac{1}{2}$ and $I_i(x, \varphi_\beta) = 1$ if $\varphi_\beta^i(x) \geq \frac{1}{2}$.

Lemma 12.2.4. *Fix $n \geq 1$ and suppose that $\tau_\beta^i(x) \neq c$ for all $i \leq n$. Then $I_n(x, \tau_\beta) = I_n(x, \varphi_\beta)$ if and only if τ_β^n is locally increasing at x. Furthermore,*

$$I_n(x, \varphi_\beta) = \begin{cases} 0 & \text{if } \#\{0 \leq i \leq n \mid I_i(x, \tau_\beta) = 1\} \text{ is even}, \\ 1 & \text{if } \#\{0 \leq i \leq n \mid I_i(x, \tau_\beta) = 1\} \text{ is odd}. \end{cases} \tag{12.3}$$

Proof. The first statement follows directly from Exercise 12.2.3. Indeed, if τ_β^n is locally increasing at x, then $\tau_\beta^n(x) = \varphi_\beta^n(x)$, so that $I_n(x,\tau_\beta) = I_n(x,\varphi_\beta)$. If τ_β^n is locally decreasing at x, then $\tau_\beta^n(x) = 1 - \varphi_\beta^n(x)$; thus $I_n(x,\tau_\beta) = 1 - I_n(x,\varphi_\beta)$.

For the second statement, τ_β^n is locally increasing at x precisely when $\#\{0 \le i < n \mid \tau_\beta^i(x) > c\} = \#\{0 \le i < n \mid I_i(x,\tau_\beta) = 1\}$ is even. (If $n = 0$, then we assume the cardinality is even by default.) In this case we have $\tau_\beta^n(x) = \varphi_\beta^n(x)$, so either $\varphi^n(x) > c$ and

$$1 = I_n(x,\varphi_\beta) \text{ with } \#\{0 \le i \le n \mid I_i(x,\tau_\beta) = 1\} \text{ being odd,}$$

or $\varphi_\beta^n(x) < c$ and

$$0 = I_n(x,\varphi_\beta) \text{ with } \#\{0 \le i \le n \mid I_i(x,\tau_\beta) = 1\} \text{ being even.}$$

A similar argument works if τ_β^n is locally decreasing at x. □

Definition 12.2.5. Let $\Sigma = \{0,1\}^{\mathbb{N}}$; we define a map $\psi : \Sigma \to \Sigma$. For $s = s_1, s_2, \ldots \in \Sigma$ denote $\psi(s)$ as $\psi(s) = u_1, u_2, \ldots$. Then,

$$u_n = \begin{cases} 0 & \text{if } \#\{0 \le i \le n \mid s_i = 1\} \text{ is even,} \\ 1 & \text{if } \#\{0 \le i \le n \mid s_i = 1\} \text{ is odd.} \end{cases}$$

Notice that it would have sufficed to say: Let $\psi : \Sigma \to \Sigma$ be defined as $\psi = I(\cdot\,,\tau_\beta) \circ I(\cdot\,,\varphi_\beta)^{-1}$; see formula (12.3) and Exercise 12.2.7.

Exercise 12.2.6. *Show that ψ is a homeomorphism of Σ into itself. Find a formula for ψ^{-1}.*

Exercise 12.2.7. *Show that $\psi \circ I(\cdot,\varphi_\beta) = I(\cdot,\tau_\beta)$.* **HINT:** *For each n, consider the intervals of monotonicity of τ_β^n and φ_β^n. It follows from Exercise 12.2.3 that these intervals are the same. Next, check the itinerary (up to length n) for each of these intervals.*

Exercise 12.2.8. *Find all the fixed points of ψ. For periodic points of period > 1, see Exercise 12.5.8.*

Exercise 12.2.9. *Show that every element $s \in \Sigma$ is recurrent under ψ and that $\omega_\psi(s)$ is minimal for every s.* **HINT:** *Examine how ψ acts on finite blocks $e_1e_2\ldots e_N$ and use uniform recurrence.*

12.3 A Relation Between Unimodal Maps and Circle Maps

In this section we explore the flip-half-of-the-graph trick some more, but this time for arbitrary unimodal maps. The initial idea comes from [72]; more details can be found in [39].

Our goal is to exhibit a collection of unimodal maps $\{g_a\}_{a\in\Lambda}$ such that for each $a \in \Lambda$ (here, Λ is an index set yet to be defined) we have:

1. $\omega(c, g_a)$ is a minimal Cantor set.

2. There is a continuous map $h : \omega(c, g_a) \to S^1$ and an irrational angle ρ such that $h \circ g_a = R_\rho \circ h$.

3. The map g_a is longbranched; more precisely, $Q(k) \leq 1$ for all k. Moreover, there is a precise relation between ρ and the cutting times of g_a, namely

$$\rho = \lim_{k\to\infty} \frac{k}{S_k}.$$

4. The map g_a is not one-to-one on $\omega(c, g_a)$. Instead, $g_a^3(c)$ has two preimages in $\omega(c, g_a)$.

Remark 12.3.1. Let $\rho \in [0,1] \setminus \mathbb{Q}$. In Section 11.3 an algorithm was given to construct a kneading map Q such that, for a unimodal map f with kneading map Q (for example a symmetric tent map), we have (S^1, R_ρ) being a factor of $(\omega(c, f), f)$. The results of this section provide another way to obtain irrational rigid rotations as factors of unimodal maps. Again, $\omega(c, g_a)$ is a Cantor set, and hence we cannot have a conjugacy.

Let g be an arbitrary unimodal map, restricted to $[\hat{g}(c), g(c)]$. (Again, recall that $\hat{x}$ is the unique point different from x such that $g(x) = g(\hat{x})$; we use $\hat{g}(c) = \widehat{g(c)}$.) We assume, for simplicity, that g is symmetric. Rescale $[\hat{g}(c), g(c)]$ to $[0,1]$ and note that the critical point is $\frac{1}{2}$. A map φ_g is obtained from (the rescaled) g by flipping the right half of g's graph:

$$\varphi_g(x) = \begin{cases} g(x) & \text{if } x \leq \frac{1}{2}, \\ 1 - g(x) & \text{if } x > \frac{1}{2}. \end{cases}$$

We are doing the flip-half-of-the-graph trick from Section 12.2.

Let us identify the endpoints 0 and 1 of the interval $[0,1]$. So now φ_g is a circle map, and we can define $\varphi_g(\frac{1}{2}) = 0 = 1$. The only discontinuity that φ_g now has is at 0. In fact, φ_g is a degree one circle map as described in Section 8.2.

Exercise 12.3.2. *Check that if $g_a(x) = 1 - a(x - \frac{1}{2})^2$, then $\varphi_a := \varphi_{g_a}$ becomes $\varphi_a(x) = 1 - a(x - \frac{1}{2})^2$ for $x \leq \frac{1}{2}$ and $\varphi_a(x) = a(x - \frac{1}{2})^2$ for $x \geq \frac{1}{2}$. For which parameter values a is φ_a one-to-one (two-to-one)?*

Let Φ_g be the degree 1 lift of φ_g. We apply the "pour-water" algorithm from Definition 8.2.8 to the lift Φ_g to obtain $(\Phi_g)_u := \bar{\Phi}_g$ and its associated circle map $\bar{\varphi}_g$. Recall (Definition 8.2.4) that φ_g has a rotation interval $Rot(\varphi_g) = [\underline{\rho}_{\Phi_g}, \overline{\rho}_{\Phi_g}]$ and that $\rho(\bar{\varphi}_g) = \overline{\rho}_{\Phi_g}$. Figure 12.3 shows φ_g and $\bar{\varphi}_g$ for $g(x) = 1 - \beta|x - \frac{1}{2}|$ with $\beta = \frac{3}{2}$.

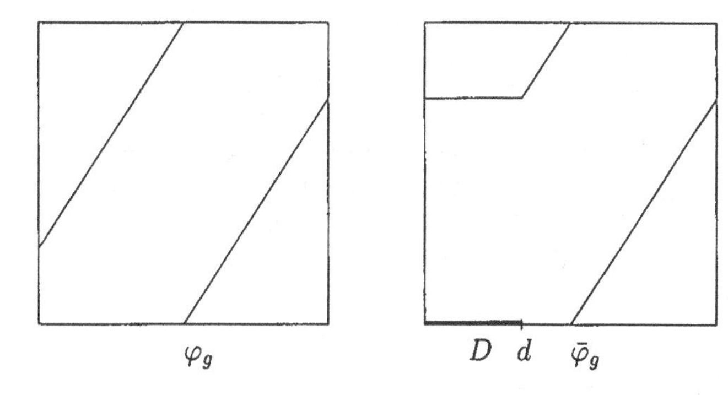

Figure 12.3: Graphs of φ_g and $\bar{\varphi}_g$ with interval D

Exercise 12.3.3. *Let φ_a be the family of "quadratic" circle maps from Exercise 12.3.2. Show that $a \mapsto \rho(a) := \rho(\bar{\varphi}_a)$ is a nondecreasing function, $\rho(2) = \frac{1}{2}$, and $\rho(4) = 1$. Repeat the exercise for $g_a(x) = 1 - a|x - \frac{1}{2}|$, and check that $\rho(1) = \frac{1}{2}$ and $\rho(2) = 1$.*

Thus, it is easy to find examples of families $\{\bar{\varphi}_a\}$ for which their rotation numbers $\rho(a) := \rho(\bar{\varphi}_a)$ increase (not necessarily strictly) from $\frac{1}{2}$ to 1 with the parameter a. Moreover, the rotation number depends continuously (recall Remark 8.2.13) on the map, and therefore continuously on a. It follows that there must be many parameter values a for which $\rho(a)$ is irrational. For such parameter values, the maps $\bar{\varphi}_a$ have no periodic points. It is precisely these parameter values that compose the index set Λ.

Theorem 12.3.4. *Let g_a be a family of symmetric unimodal maps and let φ_a and $\bar{\varphi}_a$ be the corresponding families of circle maps. Assume that $\rho(a) := \rho(\bar{\varphi}_a)$ is irrational. Then 1 has a twisted lift orbit for φ_a. Furthermore, $\omega(c, g_a)$ is a minimal Cantor set and g_a is longbranched.*

Proof. Fix a such that $\rho(a)$ is irrational. Let $D = [0, d]$ be the interval on which $\bar{\varphi}_a$ is constant. Then $\bar{\varphi}_a(c) = 1 \in \partial D$ (due to the identification of 0 and 1) and $\bar{\varphi}_a(D) = \bar{\varphi}_a^2(c)$. Since $\rho(a)$ is irrational, $\bar{\varphi}_a^n(c) \notin D$ for all $n \geq 2$ (otherwise $\bar{\varphi}_a^2(c)$ is periodic and $\rho(a)$ would have been rational).

As for any orbit of a nondecreasing degree one circle map, 1 has a twisted lift orbit under $\bar{\varphi}_a$ (recall Definition 8.2.11). The point 1 has a twisted lift orbit for φ_a as well, since $\bar{\varphi}_a^n(1) = \varphi_a^n(1)$ for all $n \geq 0$.

Next we show that the set $\omega(c, \bar{\varphi}_a)$ is a Cantor set. Indeed, we show that $\omega(c, \bar{\varphi}_a)$ is compact, totally disconnected, and perfect (recall Definition 1.1.39). It follows from Exercise 3.2.5 that $\omega(c, \bar{\varphi}_a)$ is compact. Since $\bar{\varphi}_a^n(c) \notin D$ for $n \geq 2$, we have $\omega(c, \bar{\varphi}_a) \cap (0, d) = \emptyset$ and hence $\omega(c, \bar{\varphi}_a)$ is not the full circle. Suppose that $[x, y]$ is a nondegenerate interval in $\omega(c, \bar{\varphi}_a)$. Without loss of generality, we may assume $[x, y]$ is maximal, that is, $[x - \epsilon, x]$ and $[y, y + \epsilon]$ are not contained in $\omega(c, \bar{\varphi}_a)$ for any $\epsilon > 0$. As $[x, y] \subset \omega(c, \bar{\varphi}_a)$, we may choose $m < n$ such that $\bar{\varphi}_a^m(c)$ and $\bar{\varphi}_a^n(c)$ belong to $[x, y]$. Thus, $\bar{\varphi}_a^{n-m}([x, y]) \cap [x, y] \neq \emptyset$, and therefore (using the maximality of $[x, y]$) $\bar{\varphi}_a^{n-m}([x, y]) \subset [x, y]$. Hence, $[x, y]$ contains a periodic point, and we have that $\rho(a)$ is rational, a contradiction. Thus $\omega(c, \bar{\varphi}_a)$ contains no intervals and therefore is totally disconnected.

Next take $y \in \omega(c, \bar{\varphi}_a)$ arbitrary. By Theorem 8.1.10, the points $\{\bar{\varphi}_a^j(c)\}$ with $j \geq 0$ have the same order as the points $\{j\rho(a) \pmod 1\}_{j \geq 0}$ in the circle. Moreover, $\{\bar{\varphi}_a^j(c)\}_{j \geq 0}$ accumulates on y. Therefore, for any $\epsilon > 0$ there exist $m < n$ such that $y - \epsilon < \bar{\varphi}_a^n(c) < \bar{\varphi}_a^m(c) < y$ (or $y < \bar{\varphi}_a^m(c) < \bar{\varphi}_a^n(c) < y + \epsilon$). It follows that $\bar{\varphi}_a^{n-m}([\bar{\varphi}_a^m(c), y]) = [\bar{\varphi}_a^n(c), \bar{\varphi}_a^{n-m}(y)] \subset [\bar{\varphi}_a^n(c), y]$; otherwise, that is, if $[\bar{\varphi}_a^n(c), \bar{\varphi}_a^{n-m}(y)] \supset [\bar{\varphi}_a^n(c), y]$, then $[\bar{\varphi}_a^m(c), y]$ contains a periodic point and $\rho(a) \in \mathbb{Q}$. Since $\epsilon > 0$ is arbitrary, we obtain that y is recurrent but not periodic. Therefore $\omega(c, \bar{\varphi}_a)$ contains no isolated points, that is, $\omega(c, \bar{\varphi}_a)$ is perfect. We now have that $\omega(c, \bar{\varphi}_a)$ is a Cantor set.

Claim 1: The set $\omega(c, \bar{\varphi}_a)$ is minimal.

We leave the proof of Claim 1 to the reader; see Exercise 12.3.5. Since $\bar{\varphi}_a^n(1) = \varphi_a^n(1)$ for all $n \geq 0$, $\omega(1, \varphi_a)$ is also a minimal Cantor set.

Now we consider $\omega(c, g_a)$. Recall (Exercise 12.2.2) that $g_a \circ g_a = g_a \circ \varphi_a$, so g_a is semiconjugate to φ_a (here we assume that g_a has been rescaled as in the definition of τ_β from Section 12.2). It immediately follows that $\omega(c, g_a) = \omega(1, g_a) = g_a(\omega(1, \varphi_a))$ is a Cantor set.

Claim 2: The set $\omega(c, g_a)$ is minimal.

The proof of Claim 2 is left to the reader (Exercise 12.3.5).

It remains to prove that g_a is longbranched. Let $d_1 \in [c, 1]$ be such that $\varphi_a(d_1) = d$. Since $\varphi_a^n(1) \notin [0, d]$ for all $n \geq 0$, we have that $\mathrm{orb}_{\varphi_a}(1) \cap [c, d_1] = \emptyset$. Now we use (12.2) from Exercise 12.2.3 to conclude:

- If $g_a^{n-1}(1) \in [1 - d_1, c]$, then g_a^{n-1} cannot be decreasing at 1.

- If $g_a^{n-1}(1) \in [c, d_1]$, then g_a^{n-1} cannot be increasing at 1.

In other words, any return $g_a^n(c)$ to $[1 - d_1, d_1]$ is even (recall Definition 6.3.3). It follows that, for any cutting time n, $g_a^n(c) \notin [1 - d_1, d_1]$, since cutting times are odd returns (recall Exercise 6.3.4).

Claim 3: The map g_a is longbranched.

The proof of Claim 3 is left to the reader (Exercise 12.3.5). □

Exercise 12.3.5. *Prove Claims 1–3 of Theorem 12.3.4.*

It follows from [35] that, for Lebesgue a.e. $a \in (1, 2]$, the tent map T_a is not longbranched. Hence we can conclude from Theorem 12.3.4 that for most tent maps T_a we have that $\rho(a) = \rho(\bar{\varphi}_a) \in \mathbb{Q}$.

Using any family of unimodal maps satisfying the hypotheses of Theorem 12.3.4, we have item (1) and part of item (3) from the goal of this section. We leave item (2) as an exercise and now establish item (4) and the remaining parts of item (3). Thus we assume $\{g_a\}$ is a family of unimodal maps satisfying the hypotheses of Theorem 12.3.4 and that we have a fixed g_a with $\rho(a) = \rho(\bar{\varphi}_a) \notin \mathbb{Q}$.

Exercise 12.3.6. *Establish item (2) of our goal.* **HINT:** *Use* $\rho = \rho(a) = \rho(\bar{\varphi}_a)$.

Exercise 12.3.7. *Check that the point d_1 from the proof of Theorem 12.3.4 has itinerary $I(d_1, g_a) = 11\nu_3\nu_4\nu_5\dots$, where $\nu = \nu_1\nu_2\nu_3\dots$ is the kneading sequence of g_a. Suppose that ν starts with* 1011 *and then show that the closest precritical points in $g_a^{-2}(c)$ belong to $(1 - d_1, d_1)$. Conclude that the kneading map $Q(k) \leq 1$ for all k.*

The next set of exercises is meant to prove that $Q(k) \leq 1$ for all g_a such that $\rho(\bar{\varphi}_a) \notin \mathbb{Q}$, not just the ones covered in Exercise 12.3.7. Moreover, we obtain a way to derive the rotation number $\rho = \rho(\bar{\varphi}_a)$ from the kneading sequence ν of g_a.

Exercise 12.3.8. *Let $\tilde{\xi} = I(1, \varphi_a)$ be the itinerary of* 1 *with respect to φ_a. In our definition, ν starts with index* 1*, while $\tilde{\xi}$ starts with index* 0*. After re-indexing $\xi_i = \tilde{\xi}_{i-1}$ for $i \geq 1$, show that $\xi = \psi(\nu)$, where ψ is given by Definition 12.2.5.*

Exercise 12.3.9. *Use the graphs of φ_a and $\bar{\varphi}_a$ to show that $\xi_1\xi_2 = 11$ and that ξ contains no two* 0*'s in a row (ξ is from Exercise 12.3.8).*

Exercise 12.3.10. *Define a sequence $(\tilde{S}_i)$ by $\tilde{S}_0 = 1$ and*

$$\tilde{S}_i = \min\{k > \tilde{S}_{i-1} \mid \xi_k = \xi_{k-\tilde{S}_{i-1}}\}.$$

Show that the sequence $(\tilde{S}_i)$ is exactly the sequence of cutting times for g_a. Using Exercise 12.3.9, show that $Q(k) \leq 1$ for all k.

Exercise 12.3.11. ♣ *[39] Show that $\rho(a) = \lim_n \frac{1}{n} \#\{1 \leq i \leq n \mid \xi_i = 1\}$. Conclude that $\rho(a) \geq \frac{1}{2}$ and that $\rho(a) = \lim_k k/S_k$.*

We now have item (3) of our goal.

Exercise 12.3.12. ♣ *[39] For this exericse only, assume that $\bar{\varphi}_a$ is such that $\rho(a) \in \mathbb{Q}$. In this case, $\bar{\varphi}_a^n(c) \in D$ for some $n \geq 2$. Show that $n-1 = S_k$ is a cutting time for g_a and that $\rho(a) = k/S_k$.*

Finally, we prove item (4) of our goal. The map $\bar{\varphi}_a$ is not a homeomorphism, of course, because of the flat piece in the graph. The following proposition shows that $\bar{\varphi}_a : \omega(c, \bar{\varphi}_a) \to \omega(c, \bar{\varphi}_a)$ is not one-to-one. The same holds for $\varphi_a|\omega(c, \varphi_a)$ and $g_a|\omega(c, g_a)$.

Proposition 12.3.13.

- *The restriction $\bar{\varphi}_a : \omega(c, \bar{\varphi}_a) \to \omega(c, \bar{\varphi}_a)$ is not one-to-one; there are two preimages of $\bar{\varphi}_a(1)$ in $\omega(c, \bar{\varphi}_a)$.*
- *The restriction $\varphi_a : \omega(c, \varphi_a) \to \omega(c, \varphi_a)$ is not one-to-one; there are two preimages of $\varphi_a(1)$ in $\omega(c, \varphi_a)$.*
- *The restriction $g_a : \omega(c, g_a) \to \omega(c, g_a)$ is not one-to-one; there are two preimages of $g_a^2(1)$ in $\omega(c, g_a)$.*

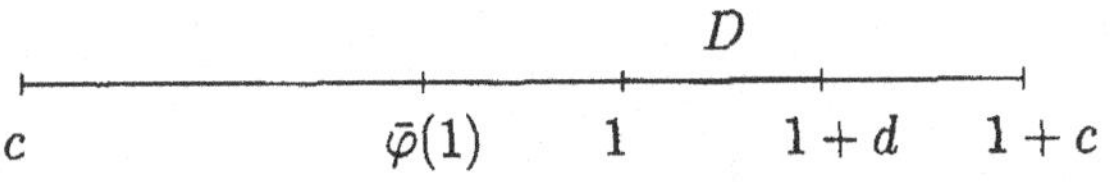

D

$c \qquad \bar{\varphi}(1) \qquad 1 \qquad 1+d \qquad 1+c$

Proof. The above picture shows the circle cut open at c, as well as the interval D and its image under $\bar{\varphi}_a$ (notice that $\bar{\varphi}_a(D) = \bar{\varphi}_a(1)$). Since there are no wandering intervals, we have $\omega(c, \bar{\varphi}_a) = S^1 \setminus \cup_{n \geq 0} \bar{\varphi}_a^{-n}((1, 1+d))$. In particular, $\mathrm{orb}_{\bar{\varphi}_a}(c)$ accumulates on both endpoints of D. Therefore, both 1 and $d = 1 + d$ belong to $\omega(c, \bar{\varphi}_a)$, and $\bar{\varphi}_a(1) = \bar{\varphi}_a(d)$ are the same point.

Because $\bar{\varphi}_a$ and φ_a coincide on $\mathrm{orb}_{\bar{\varphi}_a}(c)$, again 1 and $d \in \omega(c, \varphi_a)$, and $\varphi_a(1) = \varphi_a(d)$.

Now, for g_a, we have that $1 = g_a(c) \in \omega(c, g_a)$ and $g_a(1) \in \omega(c, g_a)$. Since $\mathrm{orb}_{\varphi_a}(c)$ accumulates on d, Exercise 12.2.3 implies that $\mathrm{orb}_{g_a}(c)$ accumulates on d and/or $\hat{d} = 1 - d$. Hence $\mathrm{orb}_{g_a}(c)$ accumulates on $g_a(d) = g_a(\hat{d})$ as well. But $g_a(d) = \hat{g}_a(1) = 1 - g_a(1)$. Therefore both $g_a(1)$ and $\hat{g}_a(1)$ belong to $\omega(c, g_a)$, and they have the same image. □

12.4 Comparing β-Transformations and Tent Maps

Choose $\beta \in (1, 2]$ and let T_β be the symmetric tent map restricted to its core. The β-transformation f_β and T_β have similar behaviors in many respects. For one thing, they have the same topological entropy $h_{top}(T_\beta) = h_{top}(f_\beta) = \log \beta$. As we saw in Chapter 9, topological entropy [5, 168] can be defined for piecewise monotone maps as the exponential growth rate of the *lapnumber*, that is, the number of branches:

$$h_{top}(g) = \lim_{n \to \infty} \frac{1}{n} \log \#\{\text{branches of } g^n\}.$$

Topological entropy is preserved under topological conjugacy. Yet, the tent map T_β and the β-transformation f_β are not topologically conjugate.

Exercise 12.4.1. *Give a simple argument why T_β and f_β are not topologically conjugate.*

In the case that the slope is 2, we can say more. For ease of notation, in the remainder of this section set $f := f_2$ and $T := T_2$. Remember that T is restricted to its core; however, for $\beta = 2$, the core of T_β is precisely $[0, 1]$.

To express the relation between f and T precisely, we need the following definition.

Definition 12.4.2. Let $f : X \to X$ and $g : Y \to Y$ be maps that are measurable with respect to measures m_X on X resp. m_Y on Y. We say that $h : X \to Y$ is an *isomorphism* between f and g if there exists $X_0 \subset X$ with $m_X(X_0) = 1$ and $Y_0 \subset Y$ with $m_Y(Y_0) = 1$ such that $h : X_0 \to Y_0$ is one-to-one and onto, $m_X(h^{-1}(B)) = m_Y(B)$ for all measurable sets $B \subset X$, and $h \circ f = g \circ h$. In this case we call f and g *isomorphic.*

In other words, f and g are isomorphic if there is a conjugacy h that need not be continuous but which is defined and one-to-one almost everywhere and which preserves the measures. For our purposes we will always think of Lebesgue measure m on $[0, 1]$ as the measure. Although measure theory (see Chapters 2 and 4 and [55]) is used in Definition 12.4.2, the reader unfamiliar with it can access the remainder of this section. Simply replace the phrase "m-a.e." by "except for at most countably many points." The fact that h preserves Lebesgue measure is not used in this section.

Exercise 12.4.3. *Show that if f and g are isomorphic, then f^k and g^k are also isomorphic for any fixed $k \geq 1$.*

Theorem 12.4.4. *The full tent map $T = T_2$ and the β-transformation $f = f_2$ are isomorphic.*

Proof. The easiest way to construct a map $h : [0,1] \to [0,1]$ such that $h \circ T = f \circ h$ is by means of the itineraries. Write $\Sigma = \{0,1\}^{\mathbb{N}}$. Note that $I(\cdot, T) : [0,1] \to \Sigma$ and $I(\cdot, f) : [0,1] \to \Sigma$ are both onto and one-to-one except for all dyadic rationals $m2^{-n} \in (0,1)$, $m, n \in \mathbb{N}$. This is a countable set of points, and, for both T and f, they are precisely backward orbits of $\frac{1}{2}$. Define $h(\cdot) = I(\cdot, f)^{-1} \circ I(\cdot, T)$. Then it is easy to see that h is well defined, onto, one-to-one and $h \circ T = f \circ h$, except again for the dyadic rationals. A little more work will show that h preserves Lebesgue measure. □

Exercise 12.4.5. *Show that $h(\cdot) = I(\cdot, T)^{-1} \circ \psi \circ I(\cdot, T)$, where ψ is as in Definition 12.2.5.* **HINT:** *Observe that $\tau_2 = T_2$ and $\varphi_2 = f_2$, and consult Exercise 12.2.7.*

In order to see how the map h (from the proof of Theorem 12.4.4) moves points around, we give a more geometric construction of h. The remainder of this section provides this geometric construction of h.

Let us start by defining partitions P_n from the open intervals of monotonicity of T^n. Thus,

$$\begin{aligned} P_1 &= \{(0, \tfrac{1}{2}), (\tfrac{1}{2}, 1)\}, \\ P_2 &= \{(0, \tfrac{1}{4}), (\tfrac{1}{4}, \tfrac{1}{2}), (\tfrac{1}{2}, \tfrac{3}{4}), (\tfrac{3}{4}, 1)\}, \\ \vdots & \qquad \vdots \qquad\quad \vdots \qquad\quad \vdots \\ P_n &= \{(0, \tfrac{1}{2^n}), (\tfrac{1}{2^n}, \tfrac{2}{2^n}), \dots, (\tfrac{2^n - 1}{2^n}, 1)\}. \end{aligned}$$

Note that (see Exercise 12.2.3) the intervals of monotonicity for T^n and f^n are the same. Each interval $J \subset P_n$ has (under T) a constant itinerary up to length n; denote this itinerary by $I(J,T) = I_0(J,T), \dots, I_{n-1}(J,T)$. Notice that the length of any $J \in P_n$ goes to 0 as $n \to \infty$. Although we do not use the next exercise in the geometric construction of h, it is a general remark on partitions P_n for leo unimodal maps (recall Exercise 6.2.10).

Exercise 12.4.6. *Let g be any full unimodal map (i.e., $g(c) = 1$ and $g(1) = g(0) = 0$). Construct the analogous partitions P_n. Show that if g is leo, then $\lim_{n\to\infty} \sup\{|J| \mid J \in P_n\} = 0$.*

We construct the map h as a pointwise limit of maps h_i, where each $h_i : [0,1] \to [0,1]$ is affine with slope ± 1 on the intervals $J \in P_i$. Each h_i is also one-to-one, except that we do not specify what happens to the endpoints of the J's, that is, for each $i \geq 1$, the points $n2^{-i}$ with $n = 0, 1, \ldots, 2^i$. Since $\{n2^{-i} \mid i \in \mathbb{N}, 0 \leq n \leq 2^i\}$ is only a countable set of points, the limit map h is well defined, except for at most these points.

Each h_i is an isomorphism, preserving Lebesgue measure. This is also true for the limit map h, which will not be very surprising once we have specified the h_i's. But a rigorous proof of this fact is beyond the scope of the text.

One important property that the h_i's have is that

$$h_{n+1}(J) \subset h_n(J) \text{ for all } J \in P_n \text{ and all } n. \tag{12.4}$$

This implies that, for m-a.e. $x \in [0, 1]$, the limit $h(x) := \lim_{n\to\infty} h_n(x)$ exists. Indeed, fix x such that x is not a boundary point of any $J \in P_n$ for all n. Fix $\epsilon > 0$. By the construction of the P_n's, there exists $n \in \mathbb{N}$ such that $\sup\{|J| \mid J \in P_n\} < \epsilon$. Take $J \in P_n$ so that $x \in J$. By (12.4) we have for any $s, t \geq n$ that $h_s(x)$ and $h_t(x) \in h_n(J)$, so $|h_s(x) - h_t(x)| < |h_n(J)| = |J| < \epsilon$. Therefore $\{h_i(x)\}_i$ is a Cauchy sequence. As Cauchy sequences in $[0, 1]$ converge, the limit $h(x)$ exists.

Let us now specify the h_i's. First take

$$h_1(x) = \begin{cases} x & \text{if } x < \frac{1}{2}, \\ \frac{3}{2} - x & \text{if } x > \frac{1}{2}. \end{cases}$$

In other words, h_1 is the identity, except that the second half of its graph is flipped. See Figure 12.4.

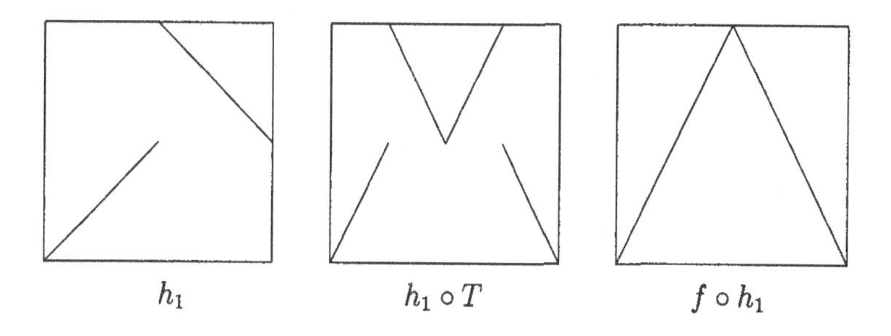

Figure 12.4: Graph of h_1 and its compositions

For $n \geq 1$, $J \in P_n$, and $x \in J$ set

$$F_J(x) = \sup J - (x - \inf J).$$

Then

$$h_1(x) = \begin{cases} x & \text{if } x \in (0, \frac{1}{2}), \\ F_{(\frac{1}{2},1)}(x) & \text{if } x \in (\frac{1}{2}, 1). \end{cases}$$

Continue the construction of the h_i's by induction: If $J \in P_i$ and $x \in J$, then

$$h_i(x) = \begin{cases} h_{i-1}(x) & \text{if } I_{i-1}(J,T) = 0, \\ h_{i-1} \circ F_J(x) & \text{if } I_{i-1}(J,T) = 1. \end{cases}$$

Exercise 12.4.7. *Show that h_n is orientation preserving (i.e., increasing) on $J \in P_n$ if and only if $\#\{0 \leq i < n \mid I_i(J,T) = 1\}$ is even.*

Figures 12.5 and 12.6 show the graphs of h_2, h_3 and also the graphs of their compositions with T and f.

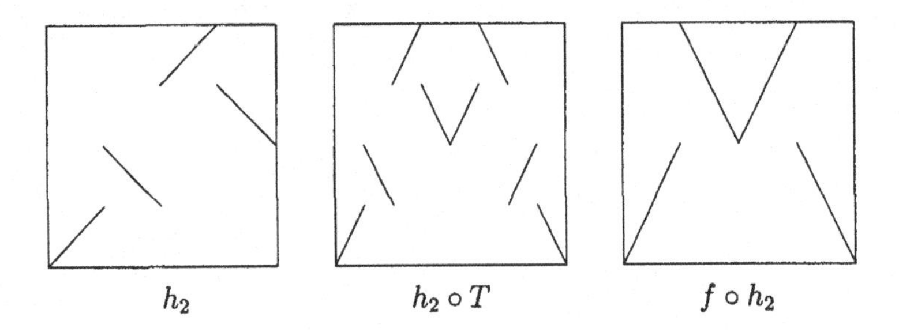

Figure 12.5: Graph of h_2 and its compositions

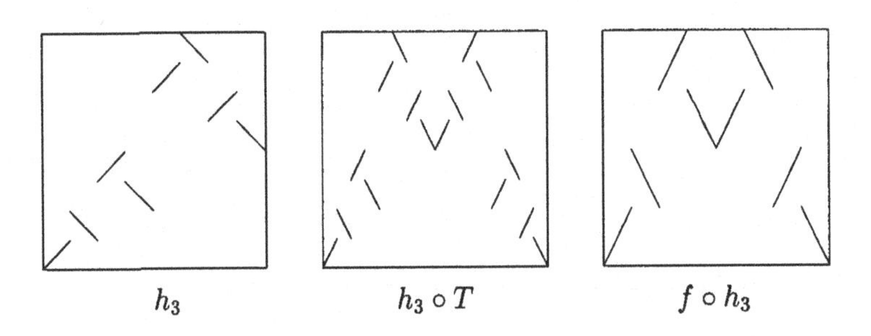

Figure 12.6: Graph of h_3 and its compositions

Notice that, as i increases, the distance between the graphs of $f \circ h_i$ and $h_i \circ T$ decreases. To be precise: For each $i \geq 1$,

$$\sup\{|h_i \circ T(x) - f \circ h_i(x)| \mid x \in [0,1], x \neq n2^{-i}, \quad 0 \leq n \leq 2^i\} \leq 2^{-i},$$

and therefore this supremum goes to 0 as $i \to \infty$. Rather than proving this, notice that $f \circ h_i = h_{i-1} \circ T$. If we can prove this relation, then it follows immediately that for the limit $f \circ h = h \circ T$. Hence, we induct on i to establish that

$$f \circ h_i = h_{i-1} \circ T \text{ for all } i \geq 2. \tag{12.5}$$

Exercise 12.4.8. *Show that* $f \circ h_2 = h_1 \circ T$.

Exercise 12.4.9. *Fix* $i \geq 2$ *and* $x \in J \in P_i$. *Let* J' *be such that* $T(x) \in J' \in P_{i-1}$. *Prove that*

$$I_{i-1}(J,T) = 0 \quad \textit{iff} \quad I_{i-2}(J',T) = 0.$$

Exercise 12.4.10. *Fix* $i \geq 2$ *and* $x \in J \in P_i$. *Let* J' *be such that* $T(x) \in J' \in P_{i-1}$. *Show that*

$$T \circ F_J(x) = F_{J'} \circ T(x).$$

We can now prove formula (12.5). Fix $i \geq 3$ and $x \in J \in P_i$. Let $T(x) \in J' \in P_{i-1}$. We have two cases.

Case 1: Suppose that $I_{i-1}(J,T) = 0$. Then,

$$\begin{aligned} f \circ h_i(x) &= f \circ h_{i-1}(x) && \text{definition of } h_i \\ &= h_{i-2} \circ T(x) && \text{induction hypothesis} \\ &= h_{i-1} \circ T(x) && \text{Exercise 12.4.9 and definition of } h_{i-1}. \end{aligned}$$

Case 2: Suppose that $I_{i-1}(J,T) = 1$. Then,

$$\begin{aligned} f \circ h_i(x) &= f \circ h_{i-1} \circ F_J(x) && \text{definition of } h_i \\ &= h_{i-2} \circ T \circ F_J(x) && \text{induction hypothesis} \\ &= h_{i-2} \circ F_{J'} \circ T(x) && \text{Exercise 12.4.10} \\ &= h_{i-1} \circ T(x) && \text{Exercise 12.4.9 and definition of } h_{i-1}. \end{aligned}$$

We now have formula (12.5).

Exercise 12.4.11.

- *Show that h is one-to-one wherever defined.*
- *Show that m-a.e. point x is recurrent for the map h; see Definition 3.5.1.* **HINT:** *See Exercise 12.2.9. Another strategy would be to use the Poincaré Recurrence Theorem; see [168].*
- ◇ *Is it the case that x is recurrent for h except for at most countably many x's?*

12.5 Ledrappier's Example

Recall the map $\psi : \Sigma \to \Sigma$ from Definition 12.2.5. This map is the symbolic representation of the map h constructed in the previous section (also see Exercises 12.2.7 and 12.4.5). In this set of exercises we study (Σ, ψ) as a dynamical system and point out a connection to *Ledrappier's three-dot example* [106].

Let $e = e_0e_1e_2\ldots$ be any 01 sequence. Write down in a rectangular diagram:

$$E = \{e_{i,j}\}_{i,j\geq 0} = \begin{matrix} e \\ \psi(e) \\ \psi^2(e) \\ \vdots \end{matrix}$$

that is, $e_{i,j} = \psi^i(e)_j$. Figure 12.7 gives an example.

$e =$	1	0	1	1	0	1	0	1	1	1	0	1	1	0	1	0	1	1	1	0
$\psi(e) =$	1	1	0	1	1	0	0	1	0	1	1	0	1	1	0	0	1	0	1	1
$\psi^2(e) =$	1	0	0	1	0	0	0	1	1	0	1	1	0	1	1	1	0	0	1	0
$\psi^3(e) =$	1	1	1	0	0	0	0	1	0	0	1	0	0	1	0	1	1	1	0	0
$\psi^4(e) =$	1	0	1	1	1	1	1	0	0	0	1	1	1	0	0	1	0	1	1	1
$\psi^5(e) =$	1	1	0	1	0	1	0	0	0	0	1	0	1	1	1	0	0	1	0	1
$\psi^6(e) =$	1	0	0	1	1	0	0	0	0	0	1	1	0	1	0	0	0	1	1	0
$\psi^7(e) =$	1	1	1	0	1	1	1	1	1	1	0	1	1	0	0	0	0	1	0	0
$\psi^8(e) =$	1	0	1	1	0	1	0	1	0	1	1	0	1	1	1	1	1	0	0	0
$\psi^9(e) =$	1	1	0	1	1	0	0	1	1	0	1	1	0	1	0	1	0	0	0	0
$\psi^{10}(e) =$	1	0	0	1	0	0	0	1	0	0	1	0	0	1	1	0	0	0	0	0
$\psi^{11}(e) =$	1	1	1	0	0	0	0	1	1	1	0	0	0	1	0	0	0	0	0	0
$\psi^{12}(e) =$	1	0	1	1	1	1	1	0	1	0	0	0	0	1	1	1	1	1	1	1

Figure 12.7: A Ledrappier three-dot pattern with ⌟-shaped patterns

Exercise 12.5.1. *Show that for every $i, j \geq 1$ and $n \geq 0$ we have*

$$e_{i,j} + e_{i-1,j} + e_{i,j-1} \mod 2 = 0. \tag{12.6}$$

HINT: *Use Exercise 12.2.6.*

These patterns were studied by Ledrappier [106] for their unusual measure theoretical properties (namely, a $\mathbb{Z}^2$-shift that is 2-mixing but not 3-mixing). Other research papers on the subject include [27, 97]. We call any pattern $E = \{e_{i,j}\}_{i,j\geq 0}$ (with $e_{i,j} \in \{0, 1\}$) satisfying (12.6) a *three-dot pattern.*

Exercise 12.5.2. *Write down the pattern E starting from $e = 11111\ldots$; see Figure 12.9. Can you discover Pascal's triangle?* **HINT:** *Rewrite Pascal's triangle* mod 2.

Exercise 12.5.3. *Let $E = \{e_{i,j}\}_{i,j\geq 0}$ be a three-dot pattern. Let J be any ⌟-shaped pattern of side length 2^k. (In Figure 12.7 we indicated two ⌟-patterns of side length 1, one ⌟-pattern of side length 2, and one of side length 4. Show that the number of 1's in J is even.* **HINT:** *For $k = 0$, this is Exercise 12.5.1. For $k = 1$, you can cover a ⌟-pattern J of side length 2 by ⌟-patterns of side length 1; see Figure 12.8, left. Notice that each point in J is covered by an odd number of ⌟-patterns of side length 1, while points outside J are covered by an even number of ⌟-patterns of side length 1. For $k > 1$, use induction.*

Exercise 12.5.4. *Show that, for every three-dot pattern E, every $k \geq 0$ and $i, j \geq 2^k$ holds:*

$$e_{i,j} + e_{i-2^k,j} + e_{i,j-2^k} \mod 2 = 0.$$

HINT: *For $k = 0$, this is Exercise 12.5.1. For $k = 1$ and 2, you can cover the three points by ⌟-patterns of side length 1; see Figure 12.8, right. Now use similar arguments as in Exercise 12.5.3. For $k > 2$, use similar convenient covers (or induction).*

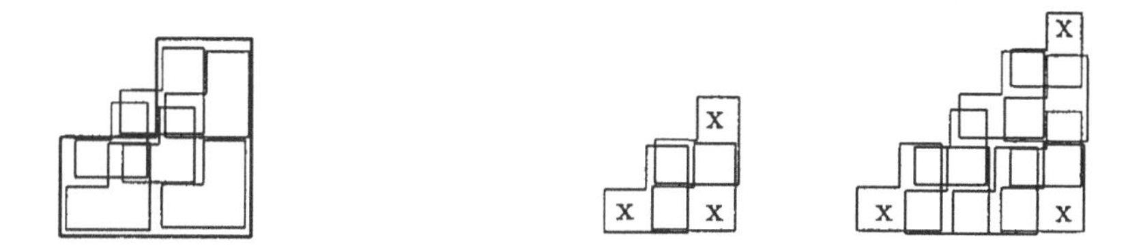

Figure 12.8: Covers with ⌟-shaped patterns

Exercise 12.5.5. *Let $E = \{e_{i,j}\}_{i,j\geq 0}$ be a three-dot pattern. Divide E into 2×2 blocks. Replace each block by a 1 if the number of 1's in the block is odd and by a 0 otherwise. Show that the result is again a three-dot pattern.* **HINT:** *Use (12.6) to show that the parity of a 2×2 block is the same as its left upper corner. Then use Exercise 12.5.4.*

Exercise 12.5.6. *If E and E' are three-dot patterns, show that $E + E'$* mod 2 *is again a three-dot pattern. Hence the collection of three-dot patterns form a group under addition modulo 2. General references for this kind of group are [97, 142].*

In the rest of this subsection, we discuss an interesting relation between three-dot patterns and the dyadic adding machine from Section 5.4. Let us start with the three-dot pattern generated by $e = 1111\dots$; see Figure 12.9. If you look at $e_{1,j}$ for $j \geq 0$, then they switch every time j increases by 1.

$$
\begin{array}{r cccccccccccccccccccc}
e = & 1 & 1 & 1 & 1 & 1 & 1 & 1 & 1 & 1 & 1 & 1 & 1 & 1 & 1 & 1 & 1 & 1 & 1 & 1 & 1 \\
\psi(e) = & 1 & 0 & 1 & 0 & 1 & 0 & 1 & 0 & 1 & 0 & 1 & 0 & 1 & 0 & 1 & 0 & 1 & 0 & 1 & 0 \\
\psi^2(e) = & 1 & 1 & 0 & 0 & 1 & 1 & 0 & 0 & 1 & 1 & 0 & 0 & 1 & 1 & 0 & 0 & 1 & 1 & 0 & 0 \\
\psi^3(e) = & 1 & 0 & 0 & 0 & 1 & 0 & 0 & 0 & 1 & 0 & 0 & 0 & 1 & 0 & 0 & 0 & 1 & 0 & 0 & 0 \\
\psi^4(e) = & 1 & 1 & 1 & 1 & 0 & 0 & 0 & 0 & 1 & 1 & 1 & 1 & 0 & 0 & 0 & 0 & 1 & 1 & 1 & 1 \\
\psi^5(e) = & 1 & 0 & 1 & 0 & 0 & 0 & 0 & 0 & 1 & 0 & 1 & 0 & 0 & 0 & 0 & 0 & 1 & 0 & 1 & 0 \\
\psi^6(e) = & 1 & 1 & 0 & 0 & 0 & 0 & 0 & 0 & 1 & 1 & 0 & 0 & 0 & 0 & 0 & 0 & 1 & 1 & 0 & 0 \\
\psi^7(e) = & 1 & 0 & 0 & 0 & 0 & 0 & 0 & 0 & 1 & 0 & 0 & 0 & 0 & 0 & 0 & 0 & 1 & 0 & 0 & 0 \\
\psi^8(e) = & 1 & 1 & 1 & 1 & 1 & 1 & 1 & 1 & 0 & 0 & 0 & 0 & 0 & 0 & 0 & 0 & 1 & 1 & 1 & 1 \\
\psi^9(e) = & 1 & 0 & 1 & 0 & 1 & 0 & 1 & 0 & 0 & 0 & 0 & 0 & 0 & 0 & 0 & 0 & 1 & 0 & 1 & 0 \\
\psi^{10}(e) = & 1 & 1 & 0 & 0 & 1 & 1 & 0 & 0 & 0 & 0 & 0 & 0 & 0 & 0 & 0 & 0 & 1 & 1 & 0 & 0 \\
\psi^{11}(e) = & 1 & 0 & 0 & 0 & 1 & 0 & 0 & 0 & 0 & 0 & 0 & 0 & 0 & 0 & 0 & 0 & 1 & 0 & 0 & 0 \\
\psi^{12}(e) = & 1 & 1 & 1 & 1 & 0 & 0 & 0 & 0 & 0 & 0 & 0 & 0 & 0 & 0 & 0 & 0 & 1 & 1 & 1 & 1
\end{array}
$$

Figure 12.9: A Ledrappier three-dot pattern for $e = 1111\dots$

The symbols $e_{2,j}$ for $j \geq 0$ switch every time j increases by 2. The symbols $e_{4,j}$ for $j \geq 0$ switch every time j increases by 4, and, in general, the symbols $e_{2^k,j}$ for $j \geq 0$ switch every time j increases by 2^k. Apparently, ψ is recoding a dyadic adding machine, because if we define $k : \text{orb}_\psi(e) \to \{0,1\}^{\mathbb{N}}$ by $k(s_0 s_1 s_2 \dots) = t_1 t_2 t_3 \dots$, where

$$t_n = \lfloor \frac{1}{2^{n-1}} \sum_{i=0}^{n-1} 2^i (1 - s_{2^i}) \rfloor, \tag{12.7}$$

then $k \circ \psi = \tau \circ k$ for the adding machine τ from Section 5.4.

Exercise 12.5.7. *Verify this, that is, that $k \circ \psi = \tau \circ k$ on $orb_\psi(e)$. Also show that k is continuous.*

We conjecture that the space of all three-dots patterns decomposes into uncountably many ψ-invariant components, E_x, and $\psi|E_x$ is isomorphic to the dyadic adding machine.

Exercise 12.5.8. *Assuming the above conjecture is true, show that ψ has no periodic points of period > 1.*

12.6 Maps with Slope < 2

Recall that T_α and f_β are the tent maps resp. β-transformations. In Section 12.4 we saw that T_2 and f_2 are isomorphic. What about other slopes? Are T_α and f_β isomorphic for other values of α and β too? Strictly speaking, this question can be answered only if we know more about measure theory and if we indicate the measures we want to consider for T_α and f_β. One issue is that these measures should be invariant, and Lebesgue measure is not invariant neither for T_α nor for f_β if $\alpha, \beta < 2$. However, assuming that $\alpha, \beta \geq \sqrt{2}$, there are invariant measures, call them μ_α for T_α and ν_β for f_β, which are *equivalent* to Lebesgue measure m. This means that every subset X of the core I_α that has Lebesgue measure $m(X) = 0$ also has $\mu_\alpha(X) = 0$, and vice versa (and similarly for ν_β on $[0,1]$).

Equipped with these measures, T_α and f_β can only be isomorphic if they have the same *measure theoretical entropy*: $h_{\mu_\alpha}(T_\alpha) = h_{\nu_\beta}(f_\beta)$. Measure theoretic entropy was introduced by Kolmogorov [98], and he showed that it is invariant when taking isomorphisms. Ornstein [132] proved that, in the context of Bernoulli shifts (i.e., full two-sided shifts equipped with a stationary product measure), measure theoretic entropy is a complete invariant. In other words, two Bernoulli shifts are isomorphic if and only if they have the same entropy. See [93] for an introduction. Measure theoretical entropy is not the same as the topological entropy h_{top} discussed in Chapter 9. In this particular case, however, the entropies coincide:

$$h_{\mu_\alpha}(T_\alpha) = h_{top}(T_\alpha) = \log\alpha \text{ and } h_{\nu_\beta}(f_\beta) = h_{top}(f_\beta) = \log\beta.$$

So we see that we need $\alpha = \beta$.

But even if $\alpha = \beta$, it is not guaranteed that T_α and f_β are isomorphic. For example, take $\alpha = \beta = \sqrt{2}$, and take the second iterates of the maps.

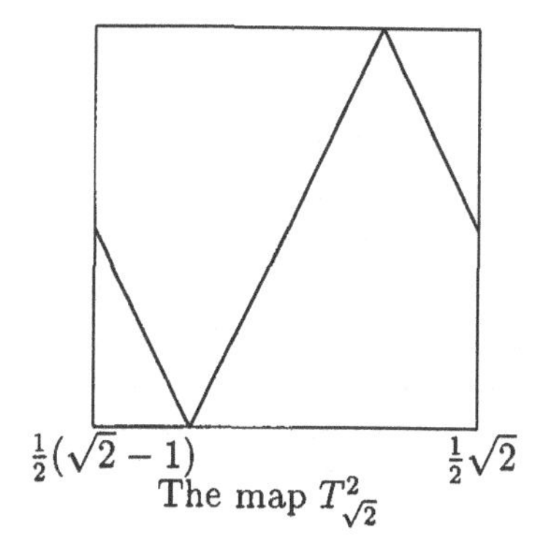

The map $T^2_{\sqrt{2}}$

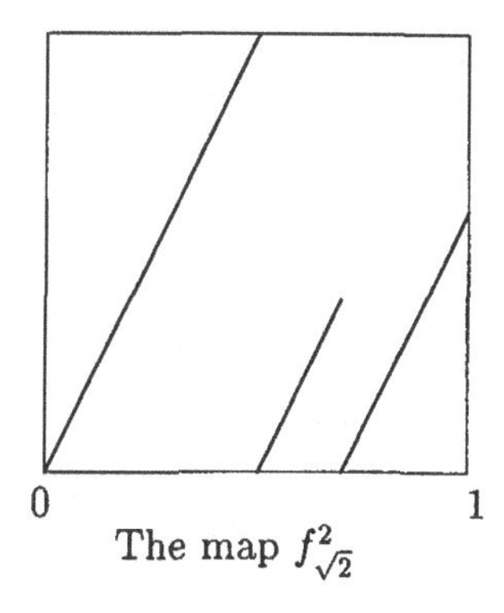

The map $f^2_{\sqrt{2}}$

Both second iterates have the same slope ± 2, but the graphs are of course quite different. The most important difference for our discussion is that, in the left picture, each $x \in I_{\sqrt{2}}$ has exactly two preimages under $T_{\sqrt{2}}^{-2}$ (except for $x = \frac{1}{2}(\sqrt{2}-1)$, $x = \frac{1}{2}$ or $x = \frac{1}{2}\sqrt{2}$). This is not true in the right picture; there is, for example, a whole interval of points with one preimage under $f_{\sqrt{2}}^{-2}$.

Exercise 12.6.1. *Show that the above property prevents T_α and f_α from being isomorphic (cf. Exercise 12.4.3).*

This exercise leads us to believe that T_α and f_α being isomorphic is very unlikely. Still, there are countably many $\alpha \in [\sqrt{2}, 2]$, where T_α and f_α are isomorphic. The remaining exercises and results aim at proving this statement.

Exercise 12.6.2. *Given $n \geq 1$, let α be such that the kneading sequence of T_α is $10^n C$.*

(a) Show that α is the largest real root of $(2-\alpha)\alpha^{n+1} = 1$. (For $n = 1$, α is the golden mean $\frac{1}{2}(\sqrt{5}-1)$.)

(b) Show that α also satisfies $\alpha^{n+1} - \alpha^n - \alpha^{n-1} \cdots - 1 = 0$. Conclude that the f_α-itinerary of 1 is $1^{n+1}0^\infty$.

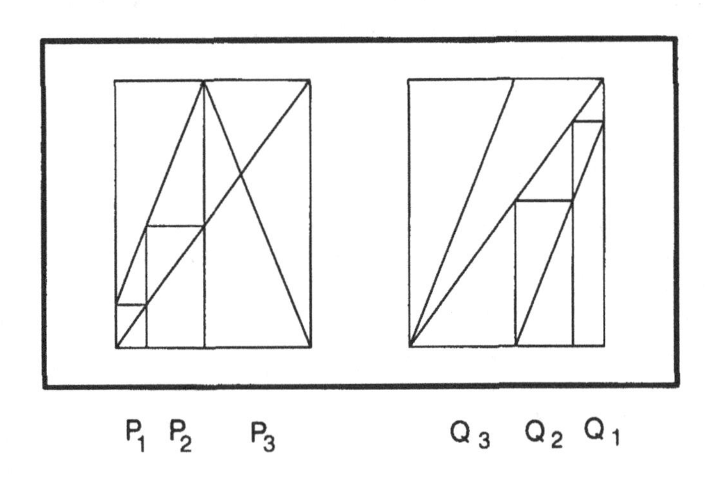

Figure 12.10: Graphs of T_α and f_α with Markov partitions

For each α in Exercise 12.6.2, we define the intervals

$$P_i := \begin{cases} [T^{i+1}(\frac{1}{2}), T^{i+2}(\frac{1}{2})] & \text{if } 1 \leq i \leq n, \\ [\frac{1}{2}, T(\frac{1}{2})] & \text{if } i = n+1. \end{cases}$$

The left part of Figure 12.10 shows these intervals for the case for $n = 2$ (so α is a solution of $\alpha^3 - \alpha^2 - \alpha - 1 = 0$). By construction, $T(P_i)$ is the union of intervals P_j, and $\cup_j P_j$ is the whole core I_α. For this reason is the partition $\{P_j\}$ called a *Markov partition* . One can write down the ways to pass from one partition element to another by means of a matrix. This is the *transition matrix*; it is defined as $A = (a_{i,j})_{i,j=1}^n$ with

$$a_{i,j} = \begin{cases} 0 & \text{if } P_j \not\subset T(P_i), \\ 1 & \text{if } P_j \subset T(P_i). \end{cases}$$

Hence, for the example in Figure 12.10, the transition matrix is

$$\begin{pmatrix} 0 & 1 & 0 \\ 0 & 0 & 1 \\ 1 & 1 & 1 \end{pmatrix}.$$

Exercise 12.6.3. *Compute A explicitly for the other slopes α from Exercise 12.6.2.*

The β-transformation can be treated in the same way; the right side of Figure 12.10 shows the Markov partition $\{Q_1, Q_2, Q_3\}$ for f_α, with the same α as above. To be precise, $Q_1 = [f_\alpha(1), 1]$, $Q_2 = [f_\alpha^2(1), f_\alpha(1)]$ and $Q_3 = [0, f_\alpha^2(1)] = [0, 1/\alpha]$.

Exercise 12.6.4. *a) Show that f_α has a Markov partition $\{Q_i\}$, that is, give an analogous construction for a partition of the interval for f_α and show that the properties of a Markov partition are satisfied.*

b) Show that the Markov partitions $\{P_i\}$ for T_α and $\{Q_i\}$ for f_α give rise to the same transition matrix.

Theorem 12.6.5. *If α is as in Exercise 12.6.2, then (T_α, μ_α) and (f_α, ν_α) are isomorphic.*

Proof. We will not discuss all the measure theoretic aspects of the proof in detail, but at least we indicate what μ_α and ν_α are. First we replace the transition matrix A by the *probability matrix* $B = (b_{i,j})_{i,j=1}^n$, where

$$b_{i,j} = \frac{|P_j|}{|T_\alpha(P_i)|}. \tag{12.8}$$

The numbers $b_{i,j}$ represent the probability that a point from P_i is mapped into P_j.

Exercise 12.6.6. *Verify that $\sum_j b_{i,j} = 1$.*

Exercise 12.6.7. *Show that $(T_\alpha, \{P_i\})$ and $(f_\alpha, \{Q_i\})$ have the same probability matrix (by writing down the analog of (12.8) for $(f_\alpha, \{Q_i\})$).*

The next ingredient is the existence of a left eigenvector $p = (p_1, \ldots, p_n) \neq (0, \ldots, 0)$ with eigenvalue 1. So $pB = p$, and, moreover, all the $p_i \geq 0$. This follows from the general theory of Markov chains, see, for example, [93, Section 2.4] or the linear algebra book [71, Chapter 4.3]. We can multiply p with a positive constant to make $\sum_i p_i = 1$. Now p is the equilibrium probability vector. The entries p_i indicate the frequency that typical points spend in interval P_i:

$$p_i = \lim_{k\to\infty} \frac{1}{k} \#\{0 \leq j < k \mid T^k(x) \in P_i\}. \tag{12.9}$$

It can be shown, since T is piecewise linear, that (12.9) holds for all x with the exception of a set of Lebesgue measure 0.

Define a function $\delta : I_\alpha \to \mathbb{R}$ by

$$\delta(x) := \frac{p_i}{|P_i|} \text{ if } x \in P_i.$$

This is not well defined for the boundary points of the P_i's, but since these are only finitely many points, they play no role in the sequel.

Exercise 12.6.8. *Verify that $\int_{P_i} \delta(x)\, dx = p_i$ and thus $\int_{I_\alpha} \delta(x)\, dx = 1$.*

Now define the measure μ_α by

$$\mu_\alpha(X) = \int_X \delta(x) dx$$

for each set $X \subset I_\alpha$. The function $\delta(x)$ is called the *density* or *Radon-Nykodým derivative* of the measure μ_α. It can be shown that μ_α is invariant, that is, $\mu_\alpha(X) = \mu_\alpha(T^{-1}(A))$, and that

$$\mu_\alpha(X) = \lim_{k\to\infty} \frac{1}{k} \#\{0 \leq j < k; T^j(x) \in X\}.$$

This holds again for all $x \in I_\alpha$ except for a set of Lebesgue measure 0.

Exercise 12.6.9. *Construct in an analogous way the measure ν_α and its density for f_α.*

The proof continues with the observation that each point $x \in I_\alpha$ can be coded using the partition $\{P_i\}$ by $I(x, T) = I_0(x, T)I_1(x, T)I_2(x, T)\ldots$, where

$$I_i(x, T) = k \text{ if } T^i(x) \in P_k.$$

The matrix A with this coding is an example of a *subshift of finite type*; see Section 3.6 and the books of Lind and Marcus [108] and Kitchens [96].

Exercise 12.6.10. *Show that this coding is well defined and unique except for countably many points.*

Exercise 12.6.11. *Construct a similar coding for* f_α *and points* $x \in [0, 1]$. *Show that since* T_α *and* f_α *have the same transition matrix, there is a bijection between the itineraries of* $(T_\alpha, \{P_i\})$ *and* $(f_\alpha, \{Q_i\})$.

The isomorphism can now be defined as $h : I_\alpha \to [0, 1]$,

$$h(x) = I^{-1}(\cdot\,, f_\beta) \circ I(x, T_\alpha).$$

It is now easy to verify that h is indeed a bijection, except for sets of Lebesgue measure 0, and that $h \circ T_\alpha = f_\alpha \circ h$. It is also true that h is measure preserving: $\mu_\alpha(X) = \nu_\alpha(h(X))$ for all measurable sets $X \subset I_\alpha$. □

We conjecture that the slopes α given in this subsection are the only ones such that T_α and f_α are isomorphic.

Chapter 13

Homeomorphic Restrictions in the Unimodal Setting

Given a unimodal map $f : [0,1] \to [0,1]$, we are interested in closed invariant subsets $E \subset [0,1]$ such that the restriction of f to E (denoted $f|E$) is an onto homeomorphism (recall Definition 1.1.37). If such an E were a finite set, it would consist of a finite number of periodic orbits. Hence, we are interested in the case when E is not finite. Chapter 3 and Sections 6.1, 6.2, and 11.1 contain background material for this chapter.

Our interest in this dynamical behavior is motivated by the following example and question. In Chapter 5 we investigated the 2^∞ map, g_*, from the unimodal logistic family $g_a(x) = ax(1-x)$. We found that $\omega(c, g_*)$ was minimal and Cantor, and that $g_*|\omega(c, g_*)$ is an onto homeomorphism (recall Exercises 3.2.5, 5.1.9, and 5.4.7). There are many other maps in the logistic family such that:

1. $\omega(c, g_a)$ is minimal.

2. $\omega(c, g_a)$ is a Cantor set.

3. $g_a|\omega(c, g_a)$ is an onto homeomorphism.

In fact, any infinitely renormalizable map in this family has these three properties; moreover, the Lebesgue measure of such an $\omega(c, g_a)$ is zero [78, 115]. One then asks, are there nonrenormalizable maps with these three properties? If some g_a has an attracting periodic orbit, then $\omega(c, g_a)$ is precisely that orbit. Hence, we are interested in maps from the logistic family that are not renormalizable and which do not have an attracting periodic orbit. Any such map is topologically conjugate to a symmetric tent map (recall Theorem 3.4.27 and see [115, 39]). Thus, one turns to nonrenormalizable symmetric tent maps, that is, T_a's where $a \in (\sqrt{2}, 2]$, and asks when is $T_a|\omega(c, T_a)$ a homeomorphism with $\omega(c, T_a)$ minimal and Cantor? As we are dealing with ω-limit sets, we get onto for free (recall Exercise 3.2.5).

There is no known characterization of when $T_a|\omega(c, T_a)$ is a homeomorphism with $\omega(c, T_a)$ minimal and Cantor. A "combinatoric characterization" would be nice! The construction in Section 12.3 provides examples of unimodal maps f where $Q(k) \leq 1$ for all k, $\omega(c, f)$ is minimal and Cantor but where $f|\omega(c, f)$ is not one-to-one. These examples exist within the logistic and tent families. In [45, Theorem 3] a kneading map Q is constructed such that $\lim_{k\to\infty} Q(k) \neq \infty$ and such that, if f is unimodal (no wandering intervals) with kneading map Q, then $\omega(c, f)$ is Cantor and $f|\omega(c, f)$ is one-to-one. Again, this example exists within the logistic and tent family. Hence, for nonrenormalizable T_a's with $\lim_{k\to\infty} Q_a(k) \neq \infty$, both types of examples exist, that is, $T_a|\omega(c, T_a)$ is (is not) one-to-one with $\omega(c, T_a)$ minimal and Cantor.

In the case $\lim_{k\to\infty} Q_a(k) = \infty$, we have $\omega(c, T_a)$ minimal and Cantor (recall Exercise 6.2.2). However, again there is no known characterization in this case for when $T_a|\omega(c, T_a)$ is one-to-one (and hence an onto homeomorphism). There are examples, in the case $\lim_{k\to\infty} Q_a(k) = \infty$, where $T_a|\omega(c, T_a)$ is one-to-one and examples where it is not one-to-one. Section 13.4 discusses some such examples.

Moving back to the general setting, in this chapter we focus on unimodal maps f for which $Q(k) \to \infty$ and ask when is $f|\omega(c, f)$ one-to-one? If f is has no attracting periodic orbits or wandering intervals, then we have $\omega(c, f)$ being minimal and Cantor (Exercise 6.2.2). Excluding also infinitely renormalizable maps, $f|\omega(c, f)$ can still be one-to-one, and the adding machines $(\Omega, \mathbf{P})$ of Chapter 11 are very useful in finding such examples. Recall from Theorem 11.1.15 that Figure 13.1 is a commuting diagram, that π is onto, and that $\pi^{-1}(c) = \langle 0 \rangle$. If both $\mathbf{P}$ and π were bijections, we would then have $f|\omega(c, f)$ being one-to-one, as desired. In Section 13.3 we provide a characterization for when $\mathbf{P}$ is a bijection, and in Section 13.4 we discuss conditions to guarantee that π is one-to-one (a characterization is not known).

The property $\lim_{k\to\infty} Q(k) = \infty$ is assumed for Theorem 11.1.15. This property guarantees that $\mathbf{P}$ is continuous. In fact, in this case $\mathbf{P}$ is also surjective and minimal. (By minimal we mean the forward orbit of any point in Ω under $\mathbf{P}$ is dense in Ω [75, Theorems 1 and 2].) We discuss $\mathbf{P}$ being continuous and minimal in Section 13.3.

Remark 13.0.12. The adding machines $(\Omega, \mathbf{P})$ that arise here are not the typical dyadic, triadic, and so on. At the time [48] was published, folks in

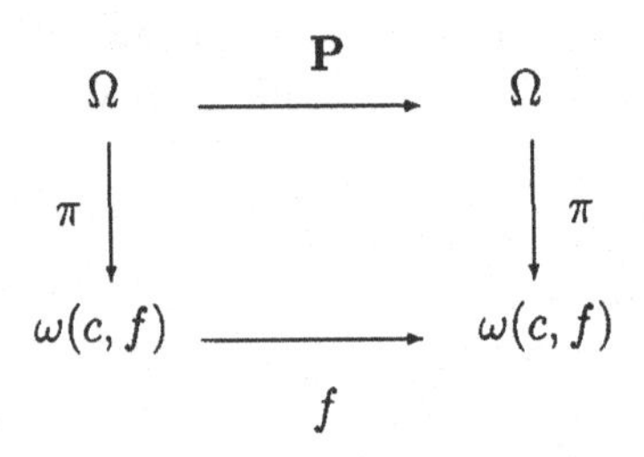

Figure 13.1: Commuting diagram

symbolic dynamics were not aware of these adding machines.

For completeness, in Section 13.2 we discuss a 2^∞ map from a "trapezoidal family," where the map restricted to the ω-limit set of the "turning point" is not one-to-one (contrasting the case of the 2^∞ map from the logistic family). Of course, in both cases we have $Q(k) = \max\{0, k-2\}$ and hence $Q(k) \to \infty$.

Related works include [8, 21, 45, 100].

13.1 First Observations

Theorem 13.1.1 and its proof establish that many examples exist for which $T_a|\omega(c, T_a)$ is a homeomorphism. The proof of Theorem 13.1.1 [45] is constructive, and in the construction it is assumed that $\lim_{k\to\infty} Q(k) = \infty$. Recall from Exercise 6.2.2 that in this case we have $\omega(c, f)$ is minimal and Cantor. Thus, Theorem 13.1.1 shows that for an uncountable and dense set of parameters a we have that $T_a|\omega(c, T_a)$ is an onto homeomorphism with $\omega(c, T_a)$ minimal and Cantor.

Theorem 13.1.1. *[45] There exists a locally uncountable dense set $\mathcal{A} \subset [\sqrt{2}, 2]$ such that for each $a \in \mathcal{A}$ the restricted map $T_a|\omega(c, T_a)$ is a homeomorphism.*

We began our discussion with closed invariant sets E such that $f|E$ is an onto homeomorphism and then moved to the special case where $E = \omega(c, f)$. The remainder of this section shows that indeed this is not a special case, but rather precisely where such behavior lives when one has *locally expanding maps*, such as symmetric tent maps.

Definition 13.1.2. Let (X, ρ) be a compact metric space and $f : X \to X$. We call f *locally expanding* provided there exist $\epsilon > 0$ and $k > 1$ such that $\rho(f(x), f(y)) \geq k\rho(x, y)$ whenever $\rho(x, y) < \epsilon$.

Lemma 13.1.3. *[74] Let (X, ρ) be a compact metric space and $f : X \to X$ a locally expanding onto homeomorphism. Then X is finite.*

Proof. Let ϵ and k be as in the definition of locally expanding. As f^{-1} is continuous, we may choose $\delta > 0$ such that $\rho(x, y) < \epsilon$ implies $\rho(f^{-1}(x), f^{-1}(y)) < \delta$. Use compactness to choose a finite open cover of X, $\{U_1, \dots, U_n\}$ such that $\text{diam}(U_i) < \epsilon$. The definition of δ gives that $\text{diam}(f^{-1}(U)) < \delta$ for each i and hence (by locally expanding) $\text{diam}(f^{-1}(U_i)) < \frac{1}{k}\text{diam}(U_i) < \frac{\epsilon}{k}$. Repeating, we see that for each $j \in \mathbb{N}$ the collection $\{f^{-j}(U_i)\}_{1 \leq i \leq n}$ is an open cover of X with $\text{diam}(f^{-n}(U_i)) < \frac{\epsilon}{k^n}$. Thus, X must be finite. □

As a symmetric tent map T_a with slope $a > 1$ is locally expanding off the turning point c, it follows from Lemma 13.1.3 that if $T_a|E$ is an onto homeomorphism with E an infinite set, then indeed $c \in E$. With our assumption that E is a closed invariant set, we have $\omega(c, T_a) \subset E$. Proposition 13.1.4 shows that indeed $E = \omega(c, T_a)$ modulo a countable set and that $\omega(c, T_a)$ is minimal. We saw in Exercise 3.5.9 that any infinite minimal set is Cantor. Hence, if $T_a|E$ is an onto homeomorphism with E an infinite set, then $\omega(c, T_a) \subset E$, $\omega(c, T_a)$ is minimal and Cantor, $T_a|\omega(c, T_a)$ is an onto homeomorphism, and $E \setminus \omega(c, T_a)$ is at most countable. Thus, searching for where $T_a|E$ is an onto homeomorphism with E infinite leads to searching for where $T_a|\omega(c, T_a)$ is a homeomorphism with $\omega(c, T_a)$ minimal and Cantor. These observations apply to more general unimodal maps as (recall Theorem 3.4.27) any nonrenormalizable unimodal map with no attracting orbits or wandering intervals is topologically conjugate to some symmetric tent map T_a with $a \in (\sqrt{2}, 2]$.

Proposition 13.1.4. *[45] Let T_a be a symmetric tent map with slope $a > 1$. Suppose that $T_a|E$ is an onto homeomorphism with E an infinite set. Then $E = \omega(c, T_a)$ modulo a countable set and $\omega(c, T_a)$ is minimal.*

Exercise 13.1.5. ♣ *Prove Proposition 13.1.4.*

We close this section with an observation, useful here, on symmetric tent maps.

Exercise 13.1.6. *Let T_a be a symmetric tent map with slope $a > 1$. Suppose that $c \in \omega(c, T_a)$ and that $T_a|\omega(c, T_a)$ is one-to-one. Prove $\omega(c, T_a)$ is mimimal.* **HINT:** *Suppose not. Then there exists $x \in \omega(c, T_a)$ such that $c \notin \omega(x, T_a)$. However, $c \notin \omega(x, T_a)$ and $T_a|\omega(x, T_a)$ one-to-one implies (use Lemma 13.1.3) that $\omega(x, T_a)$ is finite and therefore a periodic orbit. This contradicts Remark 3.2.14.*

13.2 A 2^∞ Trapezoidal Map

Let $g_{(a,b,d)}$ be the piecewise linear map given in Figure 13.2; we refer to such maps as *trapezoidal.* In this section we give an example of a 2^∞ (i.e., has periodic points of period 2^n for all $n \geq 0$ and no other periodic points) trapezoidal map $g_{(a,b,d)}$ such that $\omega(c, g_{(a,b,d)})$ is minimal and Cantor but for which $g_{(a,b,d)}|\omega(c, g_{(a,b,d)})$ is not one-to-one. Of course, there is no unique turning point for such maps, and hence we simply choose any $c \in (a, b)$.

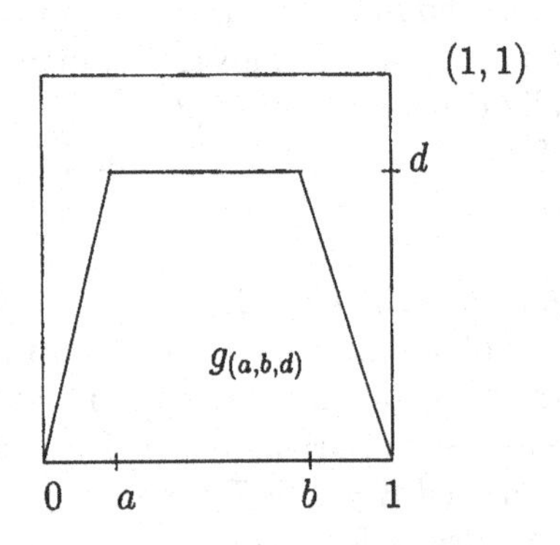

Figure 13.2: Trapezoidal map $g_{(a,b,d)}$

Proposition 13.2.1. *[34] Suppose the slopes of $g := g_{(a,b,d)}$ are at least $\gamma > 1$ (in absolute value). If E is a closed subset of $[0, 1]$ such that $g|E$ is a homeomorphism, then E is a finite set.*

Proof. Fix $\delta > 0$ and let $\{U_i\}_{1\leq i\leq m}$ be an open cover of E with each $U_i = B(x_i, \delta)$ for some $x_i \in E$. As $g|E$ is one-to-one, for each i there is a unique preimage x_{i-1} of x_i in E. For $x_i \neq d$, let $\hat{x}_{i-1}$ be the other preimage of x_i (note that $\hat{x}_{i-1} \notin E$) and if $x_i = d$, let $\hat{x}_{i-1}$ be the point in $[a, b]$ a distance $(b - a)/2$ from x_{i-1}.

For each i, let $U_i^{-1} = B(x_{i-1}, \delta/\gamma) \cap [0, 1]$ and $\hat{U}_i^{-1} = B(\hat{x}_{i-1}, \delta/\gamma) \cap [0, 1]$. Then, $E \subset \cup_i\ (U_i^{-1} \cup \hat{U}_i^{-1})$. Since $g^{-1}|E$ is uniformly continuous and $|x_{i-1} - \hat{x}_{i-1}| \geq (b - a)/2$, we may choose δ small enough to guarantee that $(U_i^{-1} \cup \hat{U}_i^{-1}) \cap E \subset U_i^{-1}$. Thus $\{U_i^{-1}\}_{1\leq i\leq m}$ is again a finite cover of E (with cardinality m); however, the diameter of each $U_i^{-1} < \delta/\gamma$. Iterating, we can cover E with m open sets of arbitrarily small diameter. Hence E is a finite set. □

Remark 13.2.2. It follows from Proposition 13.2.1 that we cannot have $g_{a,b,d}|E$ an onto homeomorphism with E infinite if the slopes of $g_{a,b,d}$ are larger than 1 in absolute value. The proofs of Lemma 13.1.3 and Proposition 13.2.1 are essentially the same.

For simplicity, in the next definition we move to *symmetric* trapezoidal maps.

Definition 13.2.3. For $e \in (0, 1/2)$ let g_e be the piecewise linear trapezoidal map shown in Figure 13.3. Then map g_e is referred to as a *symmetric* trapezoidal map.

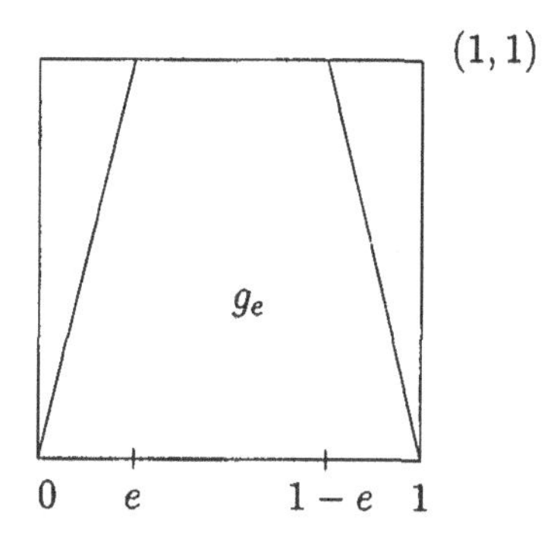

Figure 13.3: Symmetric trapezoidal map g_e

Remark 13.2.4. Given $e \in (0, 1/2)$, we can form a one-parameter family of trapezoidal maps given by $\{ag_e\}_{a \in [0,1]}$. Note that for $a > e$ we have the slopes (in absolute value) of ag_e greater than 1. Hence Proposition 13.2.1 applies.

Exercise 13.2.5. *Fix $e \in (0, 1/2)$. Prove there exists $a > e$ such that the map ag_e is a 2^∞ map (i.e., has periodic points of period 2^n for all $n \geq 0$ and no other periodic points).* **HINT:** *The kneading sequence for this 2^∞ trapezoidal map is the same as the kneading sequence for the 2^∞ logistic map. (In fact, there is a unique such parameter value a; can you prove uniqueness?)*

Exercise 13.2.6. *Fix $e \in (0, 1/2)$ and $a > e$ such that ag_e is a 2^∞ map. Let $c \in (e, 1-e)$ and set $E = \omega(c, ag_e)$. Prove that*

1. *E is a minimal set.*

2. *E is a Cantor set.*

3. *$ag_e|E$ is not one-to-one.*

Exercise 13.2.7. *Fix $e \in (0, 1/2)$, $c \in (e, 1-e)$, and $a > e$ such that ag_e is a 2^∞ map. Let $E = \omega(c, ag_e)$. Prove that e and $1-e$ are in E. Is ag_e one-to-one on $E \setminus \{e, 1-e\}$?*

As remarked in Section 13.1, for the 2^∞ logistic map g_*, the Lebesgue measure of $\omega(c, g_*)$ is 0. The same is true for the 2^∞ trapezoidal maps. This fact follows from the Theorem 13.2.8, which is interesting in its own right. For further discussion on trapezoidal maps see [10, 11, 169]. In Theorem 13.2.8, λ denotes Lebesgue measure.

Theorem 13.2.8. *[36] Fix $0 < a < b < 1$ and let $f : [0,1] \to [0,1]$ be a map such that on $[0,a]$ and $[b,1]$, f is class C^2, and*

$$\inf \{|f'(x)| \mid x \in [0,a] \cup [b,1]\} > 1.$$

Then

$$\lambda\left(\{x \in [0,1] \mid f^n(x) \notin (a,b) \ \text{ for } \ n \geq 0\}\right) = 0.$$

Proof. Fix $a < c < d < b$ and a map $g : [0,c] \cup [d,1] \to [0,1]$ such that

1. $g = f$ on $[0,a] \cup [b,1]$.
2. g is class C^2 on each of $[0,c]$ and $[d,1]$.
3. g maps each of $[0,c]$ and $[d,1]$ onto $[0,1]$.
4. $\inf\{|g'(x)| \mid x \in [0,c] \cup [d,1]\} = \alpha > 1$.

See Figure 13.4.

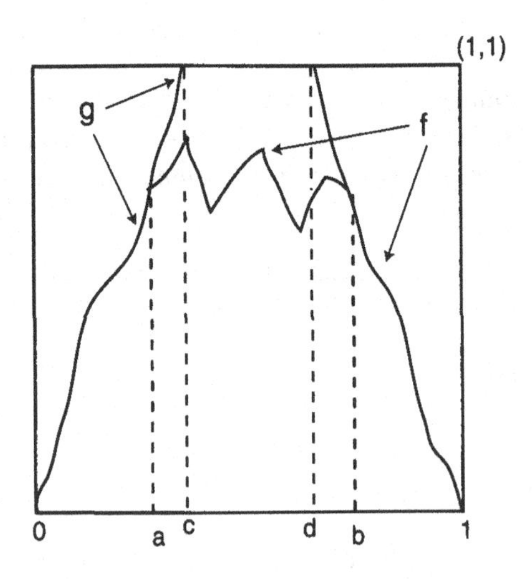

Figure 13.4: Graph of g

Let $D_0 = E_0 = [0,1]$ and for $n \geq 1$ set

$$\begin{aligned} D_n &= \{x \in [0,1] \mid f^i(x) \notin (a,b) \text{ for } 0 \leq i \leq n-1\} \\ &= \{x \in [0,1] \mid g^i(x) \notin (a,b) \text{ for } 0 \leq i \leq n-1\} \\ E_n &= \{x \in [0,1] \mid g^i(x) \in [0,c] \cup [d,1] \text{ for } 0 \leq i \leq n-1\}. \end{aligned}$$

Then $D_n \subset E_n$ for all n. We want $\lambda\left(\cap_{n=0}^{\infty} D_n\right) = 0$; we prove the stronger statement that $\lambda\left(\cap_{n=0}^{\infty} E_n\right) = 0$. Thus, we investigate the sets E_n.

Although $[0,c] \cup [d,1]$ is the domain of the map g, the domain of g^2 is the union of four disjoint subintervals whose union sits properly in $[0,c] \cup [d,1]$; moreover, E_2 is precisely the domain of g^2. Inducting on n we have that E_n is a union of 2^n disjoint intervals $\Delta_{k,n}$, $0 \leq k \leq 2^n - 1$, and each is mapped onto $[0,1]$ by g^n. Indeed, E_{n+1} is obtained from E_n by taking each $\Delta_{k,n}$ and dividing it into three subintervals:

$$\left(g^n|_{\Delta_{k,n}}\right)^{-1}([0,c]), \quad \left(g^n|_{\Delta_{k,n}}\right)^{-1}([d,1]), \quad \left(g^n|_{\Delta_{k,n}}\right)^{-1}([c,d]).$$

The first two of these subintervals are components of E_{n+1} contained in E_n, whereas the third is disjoint from E_{n+1}. Next, we need to compare the Lebesgue measure of E_{n+1} to that of E_n.

We show there exists a constant γ with $d - c < \gamma$ and such that for each n we have:

$$\lambda(E_n) \leq \left(1 - \frac{d-c}{\gamma}\right)^n. \tag{13.1}$$

It follows from (13.1) that

$$\lambda\left(\cap_{n=0}^{\infty} E_n\right) \leq \lim_{n\to\infty} \left(1 - \frac{d-c}{\gamma}\right)^n = 0,$$

and therefore $\lambda\left(\cap_{n=0}^{\infty} E_n\right)$. We now work to find the constant γ.

Set

$$\beta = \sup |(\log|g'(z)|)'| = \sup \left|\frac{g''(z)}{g'(z)}\right|.$$

As g is C^2, we have that β is finite (we have g'' bounded as it is continuous and has a compact domain; also, $|g'| > 1$). For $x, y \in \Delta_{k,n}$ we have:

$$\begin{aligned} |\log|(g^n)'(x)| - \log|(g^n)'(y)|| &= \left|\sum_{i=0}^{n-1} \log|g'(g^i(x))| - \sum_{i=0}^{n-1} \log|g'(g^i(y))|\right| \\ &\leq \sum_{i+0}^{n-1} |\log|g'(g^i(x))| - \log|g'(g^i(y))|| \\ &\leq \sum_{i=0}^{n-1} \beta|g^i(x) - g^i(y)|, \end{aligned} \tag{13.2}$$

as $g^i(x)$ and $g^i(y)$ lie in the same component of $[0,c] \cup [d,1]$. We also have

$$1 \geq |g^n(x) - g^n(y)| \geq \alpha^{n-i}|g^i(x) - g^i(y)|. \tag{13.3}$$

Combining (13.2) and (13.3) we have:

$$|\log|(g^n)'(x)| - \log|(g^n)'(y)|| \leq \beta \sum_{i=0}^{n-1} \alpha^{i-n} < \beta \sum_{j=1}^{\infty} \alpha^{-j}.$$

Set

$$\gamma = exp\left(\beta \sum_{j=1}^{\infty} \alpha^{-j}\right).$$

The constant γ is finite, since $\alpha > 1$. Thus,

$$\sup_{x,y \in \Delta_{k,n}} \left|\frac{(g^n)'(x)}{(g^n)'(y)}\right| \leq \gamma,$$

and therefore

$$\frac{\inf_{x \in \Delta_{k,n}} |(g^n)'(x)|}{\sup_{x \in \Delta_{k,n}} |(g^n)'(x)|} \geq \frac{1}{\gamma}. \tag{13.4}$$

Setting $G_{k,n} = (g^n|\Delta_{k,n})^{-1}((c,d))$, we have (from simple integration)

$$1 = \int_{\Delta_{k,n}} |(g^n)'(x)|\, dx \geq \lambda(\Delta_{k,n}) \inf_{x \in \Delta_{k,n}} |(g^n)'(x)| \tag{13.5}$$

and

$$d - c = \int_{G_{k,n}} |(g^n)'(x)|\, dx \leq \lambda(G_{k,n}) \sup_{x \in G_{k,n}} |(g^n)'(x)|. \tag{13.6}$$

Combining (13.4), (13.5), and (13.6) we obtain

$$\frac{\lambda(G_{k,n})}{\lambda(\Delta_{k,n})} \geq \frac{d-c}{\gamma}. \tag{13.7}$$

As (13.7) holds for all k, we have

$$\lambda(E_{n+1}) \leq \lambda(E_n)\left(1 - \frac{d-c}{\gamma}\right),$$

and therefore (by induction) we have (13.1). □

Remark 13.2.9. Theorem 13.2.8 is known to hold in $C^{1+\epsilon}$ and to be false in C^1 [26].

13.3 The Adding Machine $(\Omega, \mathbf{P})$

In this section we further analyze the adding machine $(\Omega, \mathbf{P})$ from Section 11.1. To avoid repetition, see Section 11.1 for first definitions and theorems regarding $(\Omega, \mathbf{P})$. Throughout this section, it is assumed that f is a unimodal map with kneading map Q and cutting times $S_0, S_1, S_2, \ldots$. *Unless specifically stated, it is not assumed that* $Q(k) \to \infty$. As discussed at the beginning of this chapter, we are interested in when $\mathbf{P}$ is a bijection.

Before discussing when $\mathbf{P}$ is a bijection, we need to look more closely at the system $(\Omega, \mathbf{P})$.

For $\omega = \omega_0, \omega_1, \ldots \in \Omega$, recall that $\omega(j)$ is defined as

$$\omega(j) = \sum_{i \geq 0}^{j} \omega_j S_j.$$

Also recall,

$$\begin{aligned}
\Omega &= \{\omega = \omega_0, \omega_1, \ldots \mid j \geq 0, \omega_j = 1, Q(j+1) \leq i \leq j-1 \Rightarrow \omega_i = 0\} \\
&= \{\omega = \omega_0, \omega_1, \ldots \mid \sum_{i=0}^{j} \omega_i S_i < S_{j+1} \text{ for all} j \geq 0\} \\
\Omega_0 &= \{\omega \in \Omega \mid \exists\, M_\omega \in \mathbb{N} \text{ such that for all } j \geq M_\omega, \omega(j) < S_{j+1} - 1\}.
\end{aligned}$$

First some beginning observations about elements of Ω.

Lemma 13.3.1. *Fix $\omega \in \Omega$ and suppose that $\omega(j) = S_{j+1} - 1$ for some j. Then, $\omega_j = 1$.*

Proof. Suppose to the contrary that $\omega_j = 0$. If $j = 0$, then $\omega(0) = S_1 - 1 + 1$ implies that $\omega_0 = 1$, a contradiction. Hence, $j \neq 0$.

Suppose $j > 0$. Then $\omega_j = 0$ implies $\omega(j-1) = \omega(j) = S_{j+1} - 1$. However, $\omega \in \Omega$ gives $\omega(j-1) \leq S_j - 1$. Thus, $S_{j+1} - 1 = \omega(j) = \omega(j-1) \leq S_j - 1$, a contradiction. Hence, $\omega_j = 1$. □

Exercise 13.3.2. *[48, Lemma 1] Let $\omega \in \Omega$. If $\omega(j) = S_{j+1} - 1$ and $Q(j+1) > 0$ for some $j \geq 0$, then $\omega(Q(j+1) - 1) = S_{Q(j+1)} - 1$.*

Lemma 13.3.3. *Fix $\omega \in \Omega$ and suppose $\{l_i\}_{i \geq 0}$ are consecutive nonnegative integers with $\omega(l_i) = S_{l_i+1} - 1$ for all i. Then,*

1. $Q(l_0 + 1) = 0$,

2. $l_0 > 0$ *implies* $\omega(l_0 - 1) = 0$,

3. $l_0 = 1$ *implies* $\omega_0 = 1$,

4. $Q(l_j + 1) > 0$ *for all* $j > 0$, *and*

5. $Q(l_j + 1) = l_{j-1} + 1$ *for all* $j > 0$.

Proof. Suppose to the contrary that $Q(l_0 + 1) > 0$. Then, Exercise 13.3.2 and $\omega(l_0) = S_{l_0+1} - 1$ give that $\omega(Q(l_0+1) - 1) = S_{Q(l_0+1)} - 1$, contradicting l_0 being minimal such that $\omega(j) = S_{j+1} - 1$. Hence, item (1) holds.

Suppose $l_0 > 0$. Then

$$\begin{aligned} \omega(l_0) &= S_{l_0+1} - 1 \\ &= \omega(l_0 - 1) + \omega_{l_0} S_{l_0} \\ &= \omega(l_0 - 1) + S_{l_0} && \text{(by Exercise 13.3.2)} \\ &= \omega(l_0 - 1) + S_{l_0+1} - 1 && \text{(by item (1))}. \end{aligned}$$

Hence, $\omega(l_0 - 1) = 0$. We now have item (2). Item (3) is immediate from Lemma 13.3.1.

Suppose that $Q(l_j + 1) = 0$ for some $j > 0$. Then,

$$\begin{aligned} S_{l_{j+1}} - 1 &= \omega(l_j) \\ &= \omega(l_j - 1) + S_{l_j} && \text{(by Exercise 13.3.2)} \\ &= \omega(l_j - 1) + S_{l_j+1} - 1 && \text{(by } Q(l_j + 1) = 0). \end{aligned}$$

Hence, $\omega(l_j - 1) = 0$, contradicting $\omega_{l_0} = 1$. Thus, item (4) holds.

Fix $j > 0$. Then $\omega(l_j) = S_{l_j+1} - 1$, $Q(l_j + 1) > 0$, and Exercise 13.3.2 give

$$\omega(Q(l_j + 1) - 1) = S_{Q(l_j+1)}. \tag{13.8}$$

From (13.8), the definition of the sequence $\{l_i\}$, and $Q(k) < k$, we have

$$Q(l_j + 1) - 1 \in \{l_0, l_1, \cdots, l_{j-1}\}.$$

Suppose to the contrary (of item (5)) that $Q(l_j + 1) - 1 = l_m$ for $m < j - 1$. Then, using Lemma 13.3.1, we have

$$\omega(l_j - 1) = S_{l_j+1} - S_{l_j} - 1 = S_{Q(l+j+1)} - 1 = S_{l_m+1} - 1,$$

and therefore

$$S_{l_{j-1}+1} - 1 = \omega(l_{j-1}) \leq \omega(l_j - 1) = S_{l_m+1} - 1 \leq S_{l_{j-1}+1} - 1.$$

Hence, $S_{l_{j-1}+1} \leq S_{l_{j-1}}$, a contradiction. ∎

Remark 13.3.4. Let $\omega \in \Omega$ and let $\{q_i\}$ be such that $\omega_j = 1 \iff j \in \{q_i\}$. From Theorem 11.1.6, we have $\omega_i = 1$ implies $\omega_j = 0$ for $Q(i+1) \leq j < i$. Hence it is always the case that

$$Q(q_{i+1} + 1) \geq q_i + 1. \tag{13.9}$$

Compare (13.9) to item (v) of Lemma 13.3.3. Under the hypothesis of Lemma 13.3.3, we have equality in (13.9).

Proposition 13.3.5. *Fix $\omega \in \Omega$. Suppose the sequence $\{l_i\}_{i \geq 0}$ are consecutive nonnegative integers such that $\omega(l_i) = S_{l_i+1} - 1$ for all i. Then,*

$$\omega_j = 1 \iff j \in \{l_i\}. \tag{13.10}$$

If $\{l_i\}_{i \geq 0}$ is a finite sequence, say $\{l_i\}_{i=0}^{i=n}$, then (13.10) holds for $j \leq l_n$ and $\omega_{l_n+1} = 0$.

Proof. The "only if" direction is immediate from Lemma 13.3.1. We have:

$$S_{l_{i+1}+1} - 1 = \omega(l_{i+1}) = \omega(l_i) + \sum_{p=l_i+1}^{l_{i+1}-1} \omega_p S_p + S_{l_{i+1}}$$

implies

$$\begin{aligned} S_{l_i+1} - 1 + \sum_{p=l_i+1}^{l_{i+1}-1} \omega_p S_p &= \omega(l_i) + \sum_{p=l_i+1}^{l_{i+1}-1} \omega_p S_p \\ &= S_{l_{i+1}+1} - S_{l_{i+1}} - 1 \\ &= S_{l_i+1} - 1 \qquad \text{by Lemma 13.3.3(5).} \end{aligned}$$

Thus, $\sum_{p=l_i+1}^{l_{i+1}-1} \omega_p S_p = 0$.

Lastly, suppose that $\{l_i\}_{i \geq 0}$ is indeed a finite sequence, say $\{l_i\}_{i=0}^{i=n}$; thus, $\omega(l_n + 1) < S_{l_n+2} - 1$. Suppose to the contrary that $\omega_{l_n+1} = 1$. Then

$$\begin{aligned} \omega(l_n + 1) &= \omega(l_n) + S_{l_n+1} \\ &= S_{l_n+1} - 1 + S_{l_n+1} \\ &< S_{l_n+2} - 1, \end{aligned}$$

and therefore $2S_{l_n+1} < S_{l_n+2}$ or $S_{l_n+1} < S_{l_n+2} - S_{l_n+1} = S_{Q(l_n+2)}$. Thus

$$S_{l_n+1} < S_{Q(l_n+2)},$$

contradicting $Q(k) < k$ for all k. □

Remark 13.3.6. Let $\omega \in \Omega_0$. It follows from Proposition 13.3.5 that exactly one of the following hold.

1. For all $j \geq 0$ we have $\omega(j) < S_{j+1} - 1$.

2. There exists $n \geq 0$ and $0 \leq d_0 < d_1 < \cdots < d_n$ such that

   - $i \leq d_n$ and $\omega_i = 1$ implies that $i \in \{d_0, \ldots, d_n\}$ and $\omega(i) = S_{i+1} - 1$, and
   - $i > d_n$ implies that $\omega(i) < S_{i+1} - 1$.

Remark 13.3.7. Let $\omega = \omega_0, \omega_1, \omega_2, \ldots \in \Omega_0$. If item (1) of Remark 13.3.6 holds, then $\omega_0 = 0$ and the add and carry operator $\mathbf{P}$ acts on ω by

$$\mathbf{P}(\omega) = 1, \omega_1, \omega_2, \ldots .$$

If item (2) of Remark 13.3.6 holds, then

$$\begin{aligned} \mathbf{P}(\omega) &= 0^{d_n+1}, \omega_{d_n+1} + 1, \omega_{d_n+2}, \omega_{d_n+3} \\ &= 0^{d_n+1}, 1, \omega_{d_n+2}, \omega_{d_n+3}, \omega_{d_n+4}, \ldots \quad \text{(by Proposition 13.3.5).} \end{aligned}$$

Remark 13.3.8. Let $\omega = \omega_0, \omega_1, \ldots \in \Omega$ with $\omega_j = 1 \iff j \in \{q_i\}_{i \geq 0}$ (we do not assume this sequence is infinite). It follows from Lemma 13.3.3(5), Proposition 13.3.5, and Remark 13.3.7 that $\mathbf{P}(\omega) = \langle 0 \rangle$ if and only if the sequence $\{q_i\}$ is infinite, $Q(q_0 + 1) = 0$, and $Q(q_j + 1) = q_{j-1} + 1$ for all $j \geq 1$.

Definition 13.3.9. Let $\mathcal{P}_0$ be those elements $\omega = \omega_0, \omega_1, \ldots \in \Omega$ for which there exists an infinite strictly increasing sequence of positive integers $\{l_i\}$ with $\omega(l_i) = S_{l_i+1} - 1$ for all i and where $\omega(j) = S_{j+1} - 1 \iff j \in \{l_i\}$.

Remark 13.3.10. It follows from Lemma 13.3.3, Proposition 13.3.5, and Remark 13.3.8 that $\mathbf{P}(\omega) = \langle 0 \rangle \iff \omega \in \mathcal{P}_0$.

Exercise 13.3.11. *Assuming that $\mathcal{P}_0 \neq \emptyset$, prove that $\mathbf{P}$ is surjective.*

Remark 13.3.12. It follows from Remark 13.3.7 that $\mathbf{P}$ is injective off $\mathcal{P}_0$. Thus, (recall Exercise 13.3.11) to show that $\mathbf{P}$ is a bijection on Ω it suffices to show that $\mathcal{P}_0$ consists of exactly one element.

We now work to provide a characterization of when $\mathbf{P}$ is a bijection, that is, when $\mathcal{P}_0$ consists of exactly one element. Still we have some fixed unimodal map f with kneading map Q and cutting times $S_0, S_1, S_2, \ldots$. Of course, Ω is determined by the sequence of cutting times.

Lemma 13.3.13. *[45] Suppose there is an infinite sequence $\{k_i\}$ such that for each i and all $k > k_i$ we have:*

- *either $Q(k) \geq k_i$,*

- *or* $Q(k) < k_i$ *and there are only finitely many* $l > k$ *with* $Q^n(l) = k$ *for some* $n \in \mathbb{N}$.

Then $\mathbf{P}$ *is one-to-one, that is,* $\mathcal{P}_0$ *consists of at most one element.*

Proof. Let $\{q_i\}$ be such that $Q(q_{i+1}+1) = q_i+1$ for $i \geq 1$ and $Q(q_0+1) = 0$. We will show there is at most one such sequence, and hence that $\mathcal{P}_0$ has at most one element.

Claim 1: The sequence $\{k_j - 1\}$ is a subsequence of $\{q_i\}$.

As $Q(q_{i+1} + 1) = q_i + 1$ for all $i \geq 1$, we have that q_j determines $q_{j'}$ for $j' < j$. Hence, using Claim 1, there can be only one sequence $\{q_i\}$ with $Q(q_{i+1} + 1) = q_i + 1$ for $i \geq 1$ and $Q(q_0 + 1) = 0$. Thus, $\mathcal{P}_0$ has at most one element. □

Exercise 13.3.14. *Prove Claim 1 from the proof of Lemma 13.3.13.*

Remark 13.3.15. Lemma 13.3.13 provides conditions to guarantee that $\mathbf{P}$ is one-to-one. The characterizations given in Propositions 13.3.21 and 13.3.23 require that $Q(k) \to \infty$.

Proposition 13.3.16. *Assume that* $Q(k) \to \infty$. *Suppose there exist* $s_0 \in \mathbb{N}$ *and sequences of positive integers* $\{l_i\}_{i\geq 1}$, $\{n_i\}_{i\geq 1}$ *such that*

- $l_1 < l_2 < l_3 < \cdots$, *and*
- $Q^{n_i}(l_i) = s_0$ *for all* i.

Then there exist $\{q_i\}_{i\geq 0}$ *with* $Q(q_0 + 1) = 0$ *and* $Q(q_i + 1) = q_{i-1} + 1$ *for all* $i \geq 1$.

Proof. Let s_0, $\{l_i\}$, and $\{n_i\}$ be as above. Since $\lim_{k\to\infty} Q(k) = \infty$, we have that $\{n_i\}$ is not bounded. Thus, passing to a subsequence if needed, we may assume $n_1 < n_2 < n_3 \cdots$. We construct $\{q_i\}$.

Step one: Since $\lim_{k\to\infty} Q(k) = \infty$, the set $\{m \mid Q(m) = s_0\}$ is a finite set. Hence, $\{Q^{n_i-1}(l_i) \mid n_i - 1 \geq 1\}$ is a finite set. Choose $\gamma_1 \in \{Q^{n_i-1}(l_i) \mid n_i - 1 \geq 1\}$ such that for infinitely many j we have $Q^{n_j-1}(l_j) = \gamma_1$. Passing to a subsequence if needed, assume that $Q^{n_i-1}(l_i) = \gamma_1$ and $n_i - 1 \geq 1$ for all i. Notice that $Q(\gamma_1) = s_0$.

Step two: Again, since $\lim_{k\to\infty} Q(k) = \infty$, the set $\{m \mid Q(m) = \gamma_1\}$ is a finite set. As in step one, choose $\gamma_2 \in \{Q^{n_i-2}(l_i) \mid n_i - 2 \geq 1\}$ such that for infinitely many j we have $Q^{n_i-2}(l_i) = \gamma_2$. Passing to a subsequence if needed, assume that $Q^{n_i-2}(l_i) = \gamma_2$ and $n_i - 2 \geq 1$ for all i. Notice that $\gamma_2 > \gamma_1$, $Q(\gamma_2) = \gamma_1$, and $Q(\gamma_1) = s_0$.

Continuing the above steps, we generate $\gamma_1 < \gamma_2 < \gamma_3 \cdots$ such that

$$Q(\gamma_1) = s_0 \text{ and } Q(\gamma_{i+1}) = \gamma_i \text{ for } i \geq 1.$$

Since $Q(k) < k$, we may choose a minimal $m_1 \in \mathbb{N}$ such that $Q^{m_1}(s_0) = 0$. (Notation: $Q^0(k) \equiv k$.) Define

- $q_i = Q^{m_1-(i+1)}(s_0) - 1$ for $0 \leq i \leq m_1 - 1$, and
- $q_{m_1+j} = \gamma_{j+1} - 1$ for $j \geq 0$.

□

Remark 13.3.17. In Proposition 13.3.16, notice that the construction of the sequence $\{q_i\}$ is such that $s_0 = q_i + 1$ for some i.

Corollary 13.3.18. *In the case $Q(k) \to \infty$ we have $\mathcal{P}_0 \neq \emptyset$.*

Proof. Suppose that $s_0 = 0$ in Proposition 13.3.16 (which is disallowed in the lemma as $s_0 \in \mathbb{N}$). Then the construction for the sequence $\{\gamma_i\}$ would still work and setting $q_i = \gamma_{i+1} - 1$ for $i \geq 0$ would still generate $e \in \mathcal{P}_0$. Hence, $\mathcal{P}_0 \neq \emptyset$. □

Remark 13.3.19. If Ω is formed from some general increasing sequence of positive integers, not necessarily from cutting times, it may be the case that $\mathcal{P}_0 = \emptyset$. However, in the event Ω is formed from a sequence of cutting times and $Q(k) \to \infty$, then it is the case that $\mathbf{P}$ is a continuous minimal surjection. Continuity and minimal are discussed at the end of the section. First we characterize (for the case $Q(k) \to \infty$) when $\mathcal{P}_0$ consists of exactly one element, that is, when $\mathbf{P}$ is a bijection.

Lemma 13.3.20. *Assume $Q(k) \to \infty$ and $\mathbf{P}$ is a bijection. Suppose $\omega \in \mathcal{P}_0$ with $\{q_j\}_{j\geq 0}$ the index sequence of the nonzero entries of ω. For each $i \geq 0$ set*

$$k_i = q_i + 1.$$

Then, for each i and all $k > k_i$ we have:

- *either $Q(k) \geq k_i$,*
- *or $Q(k) < k_i$ and there are only finitely many $l > k$ with $Q^n(l) = k$ for some $n \in \mathbb{N}$.*

Proof. Fix i and $k > k_i$. Suppose that $Q(k) < k_i$. Then, $k \neq k_i$ for all i. Suppose further that there are infinitely many $l > k$ such with $Q^n(l) = k$ for some $n \in N$. Then Proposition 13.3.16 and Remark 13.3.17 give another element $\omega' \neq \omega$ with $\omega' \in \mathcal{P}_0$, contradicting $\mathbf{P}$ being a bijection. □

Proposition 13.3.21. *Assume $Q(k) \to \infty$. The map $\mathbf{P}$ is a bijection if and only if there exists a sequence $\{k_i\}$ such that for each i and all $k > k_i$ we have:*

- *either $Q(k) \geq k_i$,*
- *or $Q(k) < k_i$ and there are only finitely many $l > k$ with $Q^n(l) = k$ for some $n \in \mathbb{N}$.*

Proof. If such a sequence $\{k_i\}$ exists, then Lemma 13.3.13 gives that $\mathcal{P}_0$ consists of at most one point. However, we have seen that $\mathcal{P}_0 \neq \emptyset$ (Corollary 13.3.18), and hence $\mathbf{P}$ is a bijection.

Suppose $\mathbf{P}$ is a bijection. We need to show that such a sequence $\{k_i\}$ exists. By Corollary 13.3.18 we have that $\mathcal{P}_0 \neq \emptyset$, and hence we obtain the sequence $\{k_i\}$ by Lemma 13.3.20. □

Proposition 13.3.21 provides one characterization when $\mathbf{P}$ is a bijection, in the setting that $Q(k) \to \infty$. (We see in Proposition 13.3.31 that $\mathbf{P}$ is continuous if and only if $Q(k) \to \infty$.) Proposition 13.3.23 gives a second characterization.

Definition 13.3.22. Let $s \in \mathbb{N}$. We say s satisfies property *drop* provided there exist sequences of positive integers $l_1 < l_2 < l_3 < \cdots$ and $n_1 < n_2 < n_3 < \cdots$ such that

$$Q^{n_i}(l_i) = s$$

for all i.

Proposition 13.3.23. *Assume $Q(k) \to \infty$. Set*

$$\mathcal{M} = \{s \in \mathbb{N} \mid s \ \text{ satisfies property drop}\}.$$

Then $\mathbf{P}$ is a bijection $\iff$ $s_1 < s_2 \in \mathcal{M}$ implies there exists $t \geq 1$ such that $Q^t(s_2) = s_1$.

Proof. Suppose that $\mathbf{P}$ is a bijection and $s_1 < s_2 \in \mathcal{M}$. The construction in the proof of Proposition 13.3.16 gives the sequences $\{q_i\}_{i\geq 0}$ and $\{\tilde{q}_i\}_{i\geq 0}$ with $Q(q_0+1) = Q(\tilde{q}_0+1) = 0$, $Q(q_i+1) = q_{i-1}+1$, and $Q(\tilde{q}_i+1) = \tilde{q}_{i-1}+1$ for all $i \geq 1$. Moreover, $s_1 = q_i + 1$ for some i and $s_2 = \tilde{q}_j + 1$ for some j.

As P is a bijection, we have $q_i = \tilde{q}_i$ for i. Hence, there is some t such that $Q^t(s_2) = s_1$.

Suppose that P is not a bijection (we want a contradiction). Then, there are distinct sequences $\{q_i\}_{i\geq 0}$ and $\{\tilde{q}_i\}_{i\geq 0}$ with $Q(q_0+1) = Q(\tilde{q}_0+1) = 0$, $Q(q_i+1) = q_{i-1}+1$, and $Q(\tilde{q}_i+1) = \tilde{q}_{i-1}+1$ for $i \geq 1$. Choose a minimal j such that $q_i = \tilde{q}_i$ for $i < j$ and (without loss of generality) $q_j < \tilde{q}_j$. Then, also, $q_l \neq \tilde{q}_l$ for $l > j$. Set $s_1 = q_j + 1$ and $s_2 = \tilde{q}_j + 1$. Thus, $s_1 < s_2 \in \mathcal{M}$. If $j > 0$, then there is no t with $Q^t(s_2) = s_1$ since otherwise some $\tilde{q}_i = q_j$ for some $i < j$, a contradiction to $\tilde{q}_i = q_i < q_j$. If $j = 0$, then (without loss of generality) $q_0 + 1 < \tilde{q}_0 + 1 \in \mathcal{M}$. But for all $t \geq 1$, $Q^t(\tilde{q}_0+1) = 0 < q_0+1$, a contradiction to the assumption that such a t exists. □

Exercise 13.3.24. *Let $Q(k) = k-1$ for $k \geq 1$ and $Q(0) = 0$. Form $(\Omega, \mathbf{P})$. Prove $\mathbf{P}$ is a bijection.*

Exercise 13.3.25. *Fix $d \in \mathbb{N}$ with $d \geq 2$. Let $Q(k) = \max\{0, k-d\}$ for $k \geq 0$. Form $(\Omega, \mathbf{P})$. Prove $\mathbf{P}$ is not a bijection; indeed, prove $|\mathbf{P}^{-1}(\langle 0 \rangle)| = d$.*

We now discuss the continuity of $\mathbf{P}$.

Definition 13.3.26. From Remark 13.3.6 we have for $\omega = \omega_0, \omega_1, \ldots \in \Omega$ either $\omega(j) < S_{j+1} - 1$ for all $j \geq 0$ *or* there is a finite or infinite sequence of indices $(d_0, d_1, \ldots)$, depending on ω, such that $\omega(k) = S_{k+1} - 1 \iff k \in (d_0, d_1, \ldots)$. In the latter case, if the sequence of indices is infinite, then $\omega_k = 1 \iff k \in (d_0, d_1, \ldots)$ and $\mathbf{P}(\omega) = 0$. For a finite sequence of indices $(d_0, \ldots, d_n)$, we have for $k \leq d_n$ that $\omega_k = 1 \iff k \in (d_0, d_1, \ldots, d_n)$. In the event that the sequence of indices is finite, define $\gamma(\omega) = (d_0, d_1, \ldots, d_n)$ and set

$$\Delta = \{(d_0, d_1, \ldots,) \mid \text{ there exists } \omega \in \Omega \text{ with } \gamma(\omega) = (d_0, d_1, \ldots, d_n)\}.$$

Lemma 13.3.27. *For $k \geq 2$, express $\langle S_k - 1 \rangle$ as*

$$\langle S_k - 1 \rangle_i = 1 \text{ if and only if } i \in \{n_{(k,1)}, n_{(k,2)}, \ldots, n_{(k,j(k))}\}.$$

Then

$$(n_{(k,1)}, n_{(k,2)}, \ldots, n_{(k,j(k))}) \in \Delta;$$

that is, letting $\omega = \langle S_k - 1 \rangle$, we have $\omega(n_{(k,i)}) = S_{n_{(k,i)}+1} - 1$ for each $1 \leq i \leq j(k)$ and $\omega(n_{(k,j(k))} + 1) < S_{n_{(k,j(k))}+2} - 1$.

Proof. Let $\omega = \langle S_k - 1 \rangle$. We induct on $j(k)$. Suppose $j(k) = 1$ for some k. Then $S_k - 1 = S_{n_{(1,1)}}$ and hence $n_{(1,1)} = k - 1$. Thus,

$$\omega(n_{(1,1)}) = S_{n_{(1,1)}} = S_{k-1} = S_k - 1,$$

and therefore $(n_{(1,1)}) \in \Delta$.

Assume the result holds for $j(k) \leq m$. Consider $j(k) = m + 1 > 1$. Since $j(k) > 1$, we have that $S_k - 1 > S_{k-1}$ and hence that $n_{(k,j(k))} = k - 1$. Thus,

$$\omega(n_{(k,j(k))}) = \omega(n_{(k,j(k)-1)}) + S_{k-1} = S_k - 1, \tag{13.11}$$

and hence $\omega(n_{(k,j(k)-1)}) = S_{Q(k)} - 1$. As the representation $\langle S_{Q(k)} - 1 \rangle$ is unique, we have by induction that

$$\omega(n_{(k,i)}) = S_{n_{(k,i)}+1} - 1 \text{ for } 1 \leq i \leq j(k) - 1. \tag{13.12}$$

Putting (13.11) and (13.12) together, the proof is complete. □

Exercise 13.3.28. *Let $\omega = \langle S_k - 1 \rangle$ for some $k \geq 2$ and let $\gamma(\omega) = (d_0, \ldots, d_n)$. By Lemma 13.3.27, we have $\gamma(\omega) \in \Delta$. Suppose there exist positive integers $d_n < e_1 < e_2 < \cdots$ such that $(d_0, \ldots, d_n, e_i) \in \Delta$ for all i. Prove $\mathbf{P}$ is not continuous at ω.*

Exercise 13.3.29. *Prove $\mathbf{P}$ is continuous on $\Omega \setminus \langle \mathbb{N} \rangle$.* **HINT:** *Use Remarks 13.3.6, 13.3.7, and 13.3.8.*

Proposition 13.3.30 ([75, Theorem 1]). *The map $\mathbf{P}$ is continuous if and only if for all $(d_0, d_1, \ldots, d_k) \in \Delta$ the set $\{d > d_k \mid (d_0, d_1, \ldots, d_k, d) \in \Delta\}$ is finite.*

Proof. Assume $\mathbf{P}$ is continuous. Suppose $(d_0, d_1, \ldots, d_n) \in \Delta$ is such that $(d_0, d_1, \ldots, d_n, d) \in \Delta$ for infinitely many d's. For each $k \in \mathbb{N}$, let $y^k \in \Omega$ and $e_k \in \mathbb{N}$ be such that $\gamma(y^k) = (d_0, d_1, \ldots, d_n, e_k)$. Passing to a subsequence if needed, assume $e_1 < e_2 < \cdots$ and $\lim_{k\to\infty} y^k = y$ (recall Definition 1.1.26 and that Ω is compact; Remark 11.1.4). From $\lim_{k\to\infty} y^k = y$ and $\lim_{k\to\infty} e_k = \infty$, we have $\gamma(y) = (d_0, d_1, \ldots, d_n)$ and hence $\mathbf{P}(y) \neq 0$. It follows from Remark 13.3.7 that $\lim_{k\to\infty} \mathbf{P}(y^k) = 0$, contradicting our assumption that $\mathbf{P}$ is continuous.

Recall (Definition 11.1.2) that the metric on $(\Omega, \mathbf{P})$ is such that points in Ω are "close" provided they have long initial segments that agree. Hence, using the characterization of $\mathcal{P}_0$ given in Remark 13.3.8 and item (2) of Remark 13.3.7, we have that $\mathbf{P}$ is continuous on $\mathcal{P}_0$. Thus $\mathbf{P}$ is continuous on $\mathcal{P}_0$ independent of any conditions on Δ.

Assume for each $(d_0, \ldots, d_n) \in \Delta$ that we have $(d_0, \ldots, d_n, d) \in \Delta$ for at most finitely many $d \in \mathbb{N}$. We need to show $\mathbf{P}$ is continuous. By the above paragraph it suffices to show $\mathbf{P}$ is continuous on $\Omega \setminus \mathcal{P}_0$. Fix $\omega \in \Omega \setminus \mathcal{P}_0$. If $\omega(j) < S_{j+1} - 1$ for all j, continuity follows from Remarks 13.3.6 and 13.3.7. Lastly, suppose $\gamma(\omega) = (d_0, \ldots, d_n)$. For y close to ω we have that $\gamma(y)$ begins with $\gamma(\omega)$. If $\omega_j = 1$ for infinitely many j, then continuity follows from Remarks 13.3.6 and 13.3.7. Suppose $\omega_j = 1$ for only finitely many j's. Again, using Remarks 13.3.6 and 13.3.7, the only way to break continuity at ω would be for $\omega = \langle m \rangle$ for some m with $\omega_j = 0$ for $j > d_n$, and to have positive integers $d_n < e_1 < e_2 < \cdots$ and $\{y^k \in \Omega\}$ with $\gamma(y^k) = (d_0, \ldots, d_n, e_k)$. Then $\lim_{k\to\infty} y^k = \omega$, and $\lim_{k\to\infty} \mathbf{P}(y^k) = 0 \neq \mathbf{P}(\omega)$. However, our assumption on Δ gives that such a sequence $e_1 < e_2 < \cdots$ does not exist. □

Proposition 13.3.31. *The map* $\mathbf{P}$ *is continuous if and only if* $Q(k) \to \infty$.

Proof. We prove that $Q(k) \to \infty$ if and only if, for all $(d_0, d_1, \ldots, d_k) \in \Delta$, the set $\{d > d_k \mid (d_0, d_1, \ldots, d_k, d) \in \Delta\}$ is finite. The result then follows by Proposition 13.3.30.

Assume that $Q(k) \to \infty$. We argue by contradiction. Suppose there exists $(d_0, d_1, \ldots, d_n) \in \Delta$ and positive integers $e_1 < e_2 < e_3 < \cdots$ such that $(d_0, d_1, \ldots, d_n, e_i) \in \Delta$ for all i. Then

$$\omega(e_i) = S_{d_0} + S_{d_1} + \cdots + S_{d_n} + S_{e_i} = S_{e_i+1} - 1,$$

and therefore (using $S_{k+1} - S_k = S_{Q(k+1)}$)

$$S_{Q(e_i+1)} = S_{d_0} + S_{d_1} + \cdots + S_{d_n} + 1 \tag{13.13}$$

for all i. As the right-hand side of (13.13) is fixed and independent of i, we have $\lim_{k\to\infty} Q(k) \neq \infty$, a contradiction.

Next, assume that for all $(d_0, d_1, \ldots, d_k) \in \Delta$ the set

$$\{d > d_k \mid (d_0, d_1, \ldots, d_k, d) \in \Delta\}$$

is finite. We argue by contradiction. Suppose $Q(k)$ does not go to infinity with k. Then there exist positive intergers $e_0 < e_1 < \cdots$ and $\beta \in \mathbb{N}$ such that $Q(e_i + 1) = \beta$ for all i.

Let $\omega = \langle S_\beta - 1 \rangle$ and say

$$\langle S_\beta - 1 \rangle_j = 1 \text{ if and only if } j \in \{d_0, d_1, \ldots, d_n\}.$$

By Lemma 13.3.27 we have $(d_0, d_1, \ldots, d_n) \in \Delta$. Since $\omega(d_n) = S_\beta - 1 = S_{Q(e_i+1)} - 1$ and $S_{e_i+1} - S_{e_i} = S_{Q(e_i+1)}$, we have

$$\omega(d_n) + S_{e_i} = S_{e_i+1} - 1. \tag{13.14}$$

It follows from (13.14) that $(d_0, d_1, \ldots, d_n, e_i) \in \Delta$ for each $e_i > d_n$, a contradiction. □

We next provide a second proof of Proposition 13.3.31. This proof does not make use of the machinery $\gamma(\omega)$. Although this next proof is shorter than the former, we include the former due to its connection to $\gamma(\omega)$.

Alternate Proof of Proposition 13.3.31

Proof. Assume $\mathbf{P}$ is continuous. We show $\lim_{k\to\infty} Q(k) = \infty$. Suppose the contrary, that is, choose M such that $Q(k) = M$ for infinitely many k. Set $\omega = \langle S_M - 1\rangle$ and note that $\mathbf{P}(\omega) = \langle S_M\rangle$. For k large such that $Q(k) = M$, set $\omega^k = \langle S_M - 1 + S_k\rangle$. The strings ω and ω^k coincide except that ω^k has an extra 1 in position $k-1$. Note that $\mathbf{P}(\omega^k) = \langle S_M + S_k\rangle = \langle S_{Q(k)} + S_{k-1}\rangle = \langle S_k\rangle$.

Choose $k_1 < k_2 < k_3 < \cdots$ such that $Q(k_i) = M$ for all i. Then $\lim_{i\to\infty} \omega^{k_i} = \omega$ and $\lim_{i\to\infty} \mathbf{P}(\omega^{k_i}) = \langle 0\rangle \neq \mathbf{P}(\omega)$, contradicting the continuity of $\mathbf{P}$.

Next, assume $\lim_{k\to\infty} Q(k) = \infty$. We show $\mathbf{P}$ is continuous. Let M be arbitrary and $L > M$ such that $Q(k) > M$ for all $k \geq L$. Let $\omega, \tilde{\omega} \in \Omega$ be such that $\omega_i = \tilde{\omega}_i$ for all $i \leq L$. Then $\mathbf{P}(\omega)_j = \mathbf{P}(\tilde{\omega})_j$ for $j \leq M$. Continuity of $\mathbf{P}$ follows as M was arbitrary. □

We now discuss the minimality of the map $\mathbf{P}$. From Definition 13.3.32 we see that, to prove $\mathbf{P}$ is minimal, we need to show that the forward orbit of any $\omega \in \Omega$, under $\mathbf{P}$, is dense in Ω.

Definition 13.3.32. Let $g : E \to E$. We say g is *minimal* provided for all $x \in E$ we have $\overline{\{g^i(x)\}_{i\geq 0}} = E$.

Lemma 13.3.33. *Let $\omega \in \Omega$ with $\omega_j = 1 \iff j \in \{q_i\}_{i\geq 0}$. The sequence $\{q_i\}$ may be finite or infinite. Suppose*

$$Q(q_{i_0+1} + 1) > q_{i_0} + 1 \tag{13.15}$$

for some $i_0 \geq 0$. Set $l = S_{q_{i_0}+1} - \omega(q_{i_0})$. Define $\hat{\omega}$ by $\hat{\omega}_j = 1 \iff j \in \{\hat{q}_i\}_{i\geq 0}$, where $\hat{q}_0 = q_{i_0} + 1$ and $\hat{q}_j = q_{i_0+j}$ for $j \geq 1$. Then $\hat{\omega} \in \Omega$ and $\mathbf{P}^l(\omega) = \hat{\omega}$.

Proof. First note that $q_{i_0+1} > q_{i_0} + 1$, since otherwise (13.15) gives $Q(q_{i_0} + 2) = Q(q_{i_0+1} + 1) > q_{i_0} + 1$, contradicting the general fact that $Q(k) < k$ with $k = q_{i_0} + 2$. Hence $\hat{q}_1 = q_{i_0+1} > q_{i_0} + 1 = \hat{q}_0$.

Recalling the characterization of Ω given by

$$\Omega = \{\omega = \omega_0, \omega_1, \ldots \mid j \geq 0,\ \omega_j = 1, Q(j+1) \leq i \leq j-1 \Rightarrow \omega_i = 0\}, \tag{13.16}$$

it is easy to see that $\hat{\omega} \in \Omega$. Lastly, (13.15) gives:

$$l = S_{q_{i_0}+1} - \omega(q_{i_0}) < S_{Q(q_{i_0+1}+1)} - \omega(q_{i_0}) = S_{q_{i_0+1}+1} - \omega(q_{i_0+1}).$$

Thus, inductively (recall 13.9) we obtain

$$S_{q_{i_0+j}+1} - \omega(q_{i_0+j}) > l = S_{q_{i_0}+1} - \omega(q_{i_0})$$

for $j \geq 1$, and therefore $\mathbf{P}^l(\omega) = \hat{\omega}$. □

Exercise 13.3.34. *Let $\omega \in \Omega$ with $\omega_j = 1 \iff j \in \{q_i\}_{i \geq 0}$. Prove*

$$\langle \omega(q_j) \rangle = \omega_0, \omega_1, \ldots, \omega_{q_j}, 0^\infty$$

for $j \geq 0$. **HINT:** *First do the case where $\omega = \langle n \rangle$ for $n \in \mathbb{N}$. Then use that $\Omega = \overline{\langle \mathbb{N} \rangle}$ and recall that points in Ω are close provided they agree for longer initial strings.*

Lemma 13.3.35. *Let $\omega \in \Omega$ with $\omega_j = 1 \iff j \in \{q_i\}_{i \geq 0}$. Assume the sequence $\{q_i\}$ is infinite. Suppose $i_0 = \min\{i \mid Q(q_{j+1} + 1) = q_j + 1$ for $j \geq i\}$ exists (note $i_0 \geq 0$). Set $m = \min\{l \mid Q^l(q_{i_0} + 1) = 0\}$ (note $m \geq 1$). Define $\hat{\omega}$ by $\hat{\omega}_j = 1 \iff j \in \{\hat{q}_i\}_{i \geq 0}$, where*

$$\begin{aligned}
\hat{q}_{m-1} &= q_{i_0}, \\
\hat{q}_{m-2} &= Q(q_{i_0} + 1) - 1, \\
&\vdots \\
\hat{q}_1 &= Q^{m-2}(q_{i_0} + 1) - 1, \\
\hat{q}_0 &= Q^{m-1}(q_{i_0} + 1) - 1, \quad \textit{and} \\
\hat{q}_{m+j} &= q_{i_0+(j+1)} \quad \textit{for } j \geq 0.
\end{aligned}$$

Set $l = \hat{\omega}(\hat{q}_{m-1}) - \omega(q_{i_0})$. Then $\hat{\omega} \in \Omega$, $\mathbf{P}(\hat{\omega}) = \langle 0 \rangle$, $l \geq 0$, and $\mathbf{P}^l(\omega) = \hat{\omega}$.

Proof. The construction of $\hat{\omega}$ gives that $Q(\hat{q}_{i+1} + 1) = \hat{q}_i + 1$ for $i \geq 0$, and hence, using (13.16), we have $\hat{\omega} \in \Omega$. From $Q(\hat{q}_0 + 1) = Q^m(q_{i_0} + 1) = 0$ and Remark 13.3.8, we see that $\mathbf{P}(\hat{\omega}) = \langle 0 \rangle$.

If $m = 1$, then $\hat{q}_0 = \hat{q}_{m-1} = q_{i_0}$ and $Q(q_{i_0} + 1) = Q(\hat{q}_0 + 1) = 0$. Hence, $i_0 = 0$ and therefore $\omega = \hat{\omega}$. Thus the lemma holds.

Suppose $m > 1$. We have $\hat{q}_{m-1} = q_{i_0}$ from the defintion of $\hat{\omega}$ and want $l = \hat{\omega}(\hat{q}_{m-1}) - \omega(q_{i_0}) \geq 0$ with $\mathbf{P}^l(\omega) = \hat{\omega}$.

First, notice that

$$\begin{aligned}
\hat{q}_{m-2} &= Q(q_{i_0} + 1) - 1 \\
&\geq q_{i_0 - 1} \quad \text{by (13.9).}
\end{aligned}$$

Suppose that $\hat{q}_{m-2} > q_{i_0-1}$. In this case, we show $\hat{\omega}(\hat{q}_{m-2}) > \omega(q_{i_0-1})$ (and therefore $l > 0$) and $\mathbf{P}^l(\omega) = \hat{\omega}$. If, to the contrary, $\hat{\omega}(\hat{q}_{m-2}) \leq \omega(q_{i_0-1})$, then $\omega(q_{i_0-1}) \geq S_{\hat{q}_{m-2}}$ and hence there exists some $j \geq \hat{q}_{m-2}$ such that $\langle \omega(q_{i_0-1}) \rangle_j = 1$, contradicting Exercise 13.3.34. We leave it to the reader to check that $\mathbf{P}^l(\omega) = \hat{\omega}$.

Suppose $\hat{q}_{m-2} = q_{i_0-1}$. If $m = 2$, then again $\omega = \hat{\omega}$ and we are done. Hence assume $m > 2$. Then

$$\begin{aligned} \hat{q}_{m-3} &= Q^2(q_{i_0} + 1) - 1 \\ &= Q(\hat{q}_{m-2} + 1) - 1 \quad \text{definition of } \hat{q}_{m-2} \\ &= Q(q_{i_0-1} + 1) - 1 \quad \text{by } \hat{q}_{m-2} = q_{i_0-1} \\ &\geq q_{i_0-2} \quad \text{by (13.9).} \end{aligned}$$

If $\hat{q}_{m-3} > q_{i_0-2}$, then as above $\hat{\omega}(\hat{q}_{m-3}) > \omega(q_{i_0-2})$ and thus $l > 0$ with $\mathbf{P}^l(\omega) = \hat{\omega}$. If $\hat{q}_{m-3} = q_{i_0-2}$ and $m = 3$, then $\omega = \hat{\omega}$ and we are done. Otherwise, $m > 3$ and continue as above using $\hat{q}_{m-4} \geq q_{i_0-3}$. □

Lemmas 13.3.33 and 13.3.35 give Proposition 13.3.36.

Proposition 13.3.36. *Let $\omega = \omega_0, \omega_1, \ldots \in \Omega$ with $\omega_j = 1 \iff j \in \{q_i\}_{i \geq 0}$. Then $\mathbf{P}^l(\omega) = \langle 0 \rangle$ for some $l > 0$ if and only if the following hold.*

1. *The sequence $\{q_i\}$ is infinite.*

2. *There exists $i_0 = \min\{i \mid Q(q_{j+1} + 1) = q_j + 1 \text{ for } j \geq i\}$.*

Theorem 13.3.37. *If $\lim_{k \to \infty} Q(k) = \infty$, then the map $\mathbf{P}$ is minimal.*

Proof. Let $\omega = \omega_0, \omega_1, \ldots \in \Omega$ with $\omega_j = 1 \iff j \in \{q_i\}_{i \geq 0}$. We need to show that $\overline{\{\mathbf{P}^i(\omega)\}}_{i \geq 0} = \Omega$. As $\Omega = \overline{\langle \mathbb{N} \rangle}$, we only need consider the case where the sequence $\{q_i\}$ is infinite.

If $i_0 = \min\{i \mid Q(q_{j+1} + 1) = q_j + 1 \text{ for } j \geq i\}$ exists, then by Proposition 13.3.36 we have $\mathbf{P}^l(\omega) = \langle 0 \rangle$ for some l and hence $\overline{\{\mathbf{P}^i(\omega)\}}_{i \geq 0} = \Omega$.

If such an i_0 does not exist, then there exists $j_0 < j_1 < \cdots$ such that for $k \geq 0$ we have $Q(q_{j_k+1} + 1) > q_{j_k} + 1$. For $k \geq 0$, set $l_k = S_{q_{j_k}+1} - \omega(q_{j_k})$. Then, by Lemma 13.3.33,

$$\lim_{k \to \infty} \mathbf{P}^{l_k}(\omega) = \langle 0 \rangle,$$

and therefore $\overline{\{\mathbf{P}^i(\omega)\}}_{i \geq 0} = \Omega$. Note that it is here that we use the continuity of $\mathbf{P}$ and hence that $Q(k) \to \infty$. □

13.4 The Case $Q(k) \to \infty$

Recall that we are looking for unimodal maps f with the turning point c such that $f|\omega(c, f)$ is an onto homeomorphism and such that $\omega(c)$ is minimal and Cantor. As previously remarked, since we are dealing with ω-limit sets, we get onto for free and as $\lim_{k\to\infty} Q(k) = \infty$ is assumed we have that $\omega(c, f)$ is minimal and Cantor. Hence, at issue is whether $f|\omega(c, f)$ is one-to-one. In this section we discuss conditions on $Q(k)$ (in addition to the assumption that $Q(k) \to \infty$) that are sufficient but not necessary to obtain one-to-one. For further discussion see [45]. We begin by identifying three conditions on $Q(k)$.

Throughout this section we assume f is unimodal with kneading map Q. Moreover, f is not renormalizable, has no periodic attractors, and has no wandering intervals.

Definition 13.4.1. We say a kneading map Q satisfies *Condition A* provided there is some $K_A \in \mathbb{N}$ such that for $k \geq K_A$ we have

$$Q(k+1) > Q(Q^2(k) + 1). \tag{13.17}$$

Observe that (13.17) says that one of $z_{Q(Q^2(k)+1)}, \hat{z}_{Q(Q^2(k)+1)}$ is in $\langle c_{S_k}; c_{S_{Q^2(k)}} \rangle$ (recall Exercise 6.1.7). See Figure 13.5.

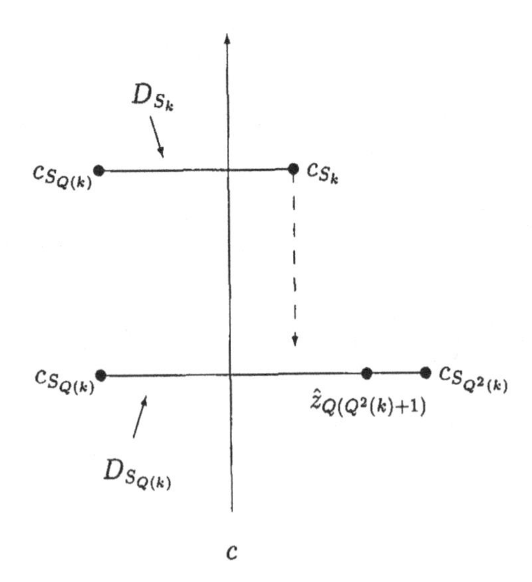

Figure 13.5: Geometry of Condition A

Definition 13.4.2. [45] We say a kneading map Q satisfies *Condition B* provided there is some $K_B \in \mathbb{N}$ such that for $k \geq K_B$ we have

$$Q(k+1) > Q(Q^2(k)+1)+1. \tag{13.18}$$

Condition B is slightly stronger than Condition A. More precisely, observe that (13.18) says there are at least two z_l's or two $\hat{z}_l$'s sitting in $\langle c_{S_k}; c_{S_{Q^2(k)}} \rangle$ (recall Exercise 6.1.7 and see Figure 13.5).

Definition 13.4.3. [45] We say a kneading map Q satisfies *Condition C* provided there is some $K_C \in \mathbb{N}$ such that for $s, \tilde{s} > K_C$ we have

$$Q(s+1) = Q(\tilde{s}+1) \text{ for } s \neq \tilde{s} \text{ implies } Q^{n+1}(s) \neq Q^{\tilde{n}+1}(\tilde{s})$$

for any $n, \tilde{n} \geq 0$ such that $Q^n(s) \neq Q^{\tilde{n}}(\tilde{s})$.

Example 13.4.4. *[45, Example 1] Let $\{k_i\}$ be a sequence of positive intergers such that $k_{i+1} > k_i + 10$ for all $i \geq 1$. Construct a kneading map Q such that*

$$Q(k) = \begin{cases} k_i & \text{for } k = k_i + 4 \\ k_i + 2 & \text{for } k = k_i + 5 \\ k - 2 & \text{otherwise.} \end{cases}$$

Then $\lim_{k\to\infty} Q(k) = \infty$, and Conditions B and C hold.

First some discussion about these three conditions and what one can conclude from them.

Lemma 13.4.5. *Assume that $\lim_{k\to\infty} Q(k) = \infty$ and that Condition C holds. Then,*

$$\left.\begin{array}{lcl} K_C < p < q & & \\ Q(p+1) & = & Q(q+1) \\ Q(p), Q(q) & \neq & 0 \end{array}\right\} \Rightarrow Q^j(q) = p, \quad \text{for some } j > 0.$$

Proof. Let $\mathcal{Z} = \{k \in \mathbb{N} \mid Q(k) = 0\}$. As $Q(k) \to \infty$, we have that $\mathcal{Z}$ is a finite set. Express $\mathcal{Z}$ as

$$\mathcal{Z} = \{i_0, i_1, \dots, i_d\}.$$

For each $k \in \mathbb{N} \setminus \mathcal{Z}$, let $m_k \in \mathbb{N}$ be minimal such that $Q^{m_k}(k) = 0$. Then $m_k \geq 2$ and $Q^{m_k-1}(k) \neq 0$ for all $k \in \mathbb{N} \setminus \mathcal{Z}$. For $0 \leq j \leq d$ set

$$E_{i_j} = \{k \in \mathbb{N} \mid Q^{m_k-1}(k) = i_j\}.$$

Notice that some, but not all, of the E_{i_j}'s may be empty. We have:

$$\cup_{j=0}^{j=d} E_{i_j} = \mathbb{N} \setminus \mathcal{Z} \text{ and } E_{i_j} \cap E_{i_{j'}} = \emptyset \text{ for } j \neq j'.$$

Let $p, q > K_C$.

Claim 1: If $p \in E_{i_j}$ and $q \in E_{i_{j'}}$ for $j \neq j'$, then $Q(p+1) \neq Q(q+1)$.

Proof of Claim 1. Suppose to the contrary that $Q(p+1) = Q(q+1)$. Then

$$Q^{m_p-1}(p) = i_j \neq i_{j'} = Q^{m_q-1}(q)$$

and Condition C imply that

$$Q^{m_p}(p) = 0 \neq 0 = Q^{m_q}(q),$$

a contradiction. This completes the proof of Claim 1.

Hence, the only way $Q(p+1) = Q(q+1)$ for $q \neq p \in \mathbb{N} \setminus \mathcal{Z}$ is for p, q to be in the same E_{i_j} (recall we have assumed that $p, q > K_C$).

Claim 2: Let $p \neq q \in E_{i_j}$ for some j. Then $Q(p+1) = Q(q+1)$ implies that $m_p \neq m_q$.

Proof of Claim 2. Suppose to the contrary that $m_p = m_q$. Then, Condition C, $p \neq q$, and $Q(p+1) = Q(q+1)$ imply

$$i_j = Q^{m_p-1}(p) \neq Q^{m_p-1}(q) = Q^{m_q-1}(q) = i_j,$$

a contradiction. This completes the proof of Claim 2.

Claim 3: Suppose $p \neq q \in E_{i_j}$ for some j, and $Q(p+1) = Q(q+1)$. Then,

$$\begin{aligned} m_q > m_p \quad &\Rightarrow \quad Q^{m_q-m_p}(q) = p \\ m_p > m_q \quad &\Rightarrow \quad Q^{m_p-m_q}(p) = q. \end{aligned}$$

Proof of Claim 3. We do the case $m_q > m_p$ (note that from Claim 2 we know that $m_p \neq m_q$). Then Condition C, $Q(p+1) = Q(q+1)$, and $p \neq Q^{m_q-m_p}(q)$ imply

$$Q^{m_p-1}(p) = i_j \neq Q^{m_p-1}(Q^{m_q-m_p}(q)) = Q^{m_q-1}(q) = i_j,$$

a contradiction. This completes the proof of Claim 3.

Claims 1–3 prove the lemma. □

We include Remarks 13.4.6, 13.4.9, and 13.4.10 to help illuminate the consequences of Conditions A (or B) and C holding. Remark 13.4.10 is used in the proof of Theorem 13.4.26.

Remark 13.4.6. Assume $Q(k) \to \infty$ and Condition C holds. Then the only way we can have $Q(p+1) = Q(q+1)$ for $K_C < p < q$ with $Q(p), Q(q) \neq 0$ is for $Q^j(q) = p$ for some j. If, in addition, Condition A holds, then in fact j must be odd (see Exercise 13.4.7).

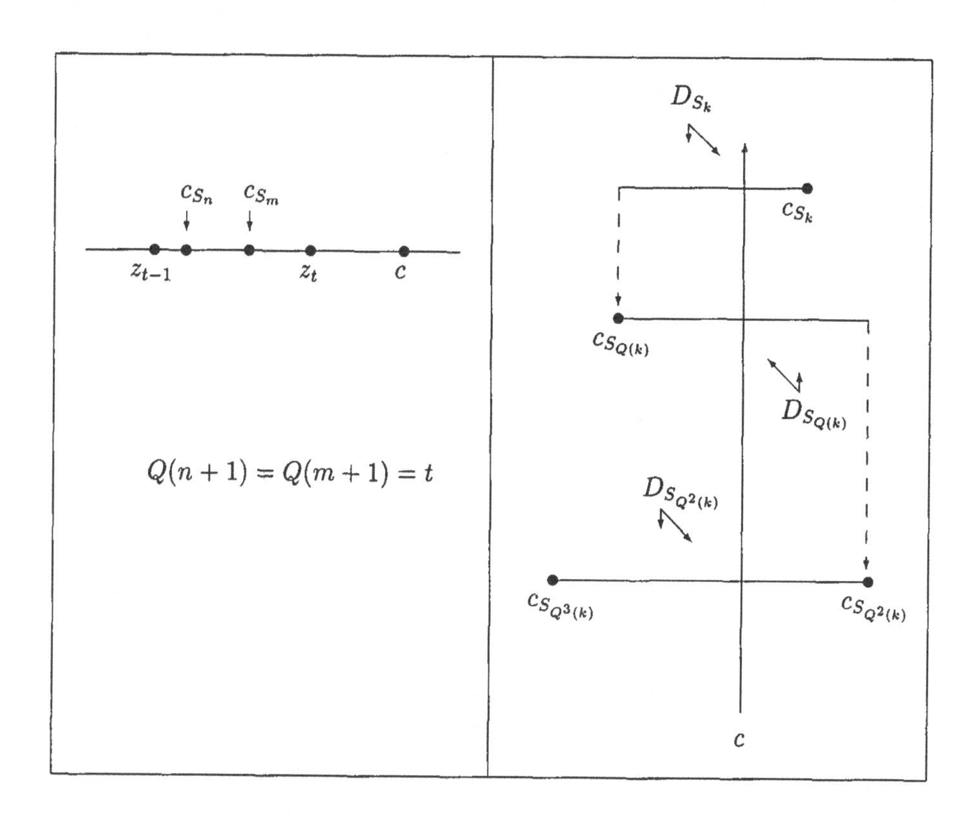

Figure 13.6: Placement via kneading map

Exercise 13.4.7. *Suppose $Q(k) \to \infty$ and that Conditions A and C hold. Let $K_C < p < q$ be such that $Q(p), Q(q) \neq 0$ and $Q(p+1) = Q(q+1)$. Prove $Q^j(q) = p$ for some odd j. Hence c_{S_p} and c_{S_q} lie on opposite sides of c.* **HINT:** *See Figure 13.6 and Exercise 6.1.7.*

Remark 13.4.8. The cutting times S_k, $S_{Q(k)}$, and $S_{Q^2(k)}$ shown in the Hofbauer tower of Figure 13.6 are not necessarily consecutive cutting times.

Remark 13.4.9. Suppose $Q(k) \to \infty$ and that Conditions A and C hold. Let M be such that $k > M$ implies $Q(k) \neq 0$. Set $K = \max\{K_C, M\}$. Let $K + 1 < p < q < r$. Then it is **not** the case that

$$Q(p) = Q(q) = Q(r).$$

For otherwise there exist j_1, j_2, j_3 odd such that $Q^{j_1}(r-1) = q-1$, $Q^{j_2}(q-1) = p-1$ and $Q^{j_3}(r-1) = p-1$. However, then $j_3 = j_1 + j_2$, contradicting each j_i being odd.

Remark 13.4.10. Suppose $Q(k) \to \infty$ and that Conditions A and C hold. It follows from Remark 13.4.9 that there exists $L \in \mathbb{N}$ such that $t > L$ implies that each of (z_{t-1}, z_t), $(\hat{z}_t, \hat{z}_{t-1})$ contains at most one c_{S_k}.

Lemma 13.4.11. *[45] Assume Condition A holds, that is, for $k \geq K_A$ we have $Q(k+1) > Q(Q^2(k)+1)$. Suppose further that $S_t < n = S_t + S_r < S_{t+1}$ and $t \geq K_A$. Then D_n contains at most one closest precritical point (i.e., at most one z_l ($\hat{z}_l$)).*

Proof. Suppose D_n contains a closest precritical point. Without loss of generality, assume D_n lies to the left of c. Then, since one endpoint of D_n is precisely c_{S_r}, either $z_{Q(r+1)}$ or $z_{Q(r+1)-1}$ is in D_n. Again, without loss of generality, assume $z_{Q(r+1)} \in D_n$. Then, $S_t + S_r + S_{Q(r+1)} = S_{t+1}$, and therefore

$$S_{r+1} = S_r + S_{Q(r+1)} = S_{t+1} - S_t = S_{Q(t+1)}.$$

Thus,

$$Q(t+1) = r + 1. \tag{13.19}$$

Both $z_{Q(r+1)}, z_{Q(r+1)+1} \in D_n$ imply

$$S_{t+2} = S_{t+1} + S_{Q(r+1)+1} - S_{Q(r+1)},$$

and therefore $Q(t+2) = Q(Q(r+1)+1)$. Thus, using (13.19), $Q(t+2) = Q(Q^2(t+1)+1)$, contradicting Condition A, that is, (13.17). Hence D_n contains at most one closest precritical point. □

Exercise 13.4.12. *Assume Condition B holds and that q_0, q_1, and $\tilde{q}_0$ are such that the following hold:*

$$\begin{aligned} \tilde{q}_0 &> K_B \\ q_1 &> q_0 + 1 \\ Q(q_1 + 1) &= q_0 + 1 \\ Q(\tilde{q}_0 + 1) &> Q(q_0 + 1). \end{aligned}$$

Prove that

$$c_{S_{\tilde{q}_0}} \notin D_{S_{q_1} + S_{q_0}}. \tag{13.20}$$

HINT: *Show that if (13.20) fails, then $D_{S_{\tilde{q}_0} + S_{Q(q_0+1)}}$ contains more than one z_l ($\hat{z}_l$), contradicting Lemma 13.4.11.*

Exercise 13.4.13. *[45] Assume Condition B holds. Let $S_t < n = S_t + S_r < S_{t+1}$ with $t \geq K_B$. Then D_n does not contain both c_{S_k} and a point from $\{z_{Q(k+1)-1}, \hat{z}_{Q(k+1)-1}\}$ in its interior for any $k \geq K_B$.* **HINT:** *Suppose the contrary. Show $D_{S_k + S_{Q(r+1)}}$ contains two closest precritical points.*

This concludes our discussion of Conditions A, B, and C. We now discuss the *nesting* of the map π from Theorem 11.1.15.

Remark 13.4.14. Assume that $\lim_{k\to\infty} Q(k) = \infty$. Review Theorem 11.1.15, Remark 11.1.17, and Figure 11.1. Fix $\omega = \omega_0, \omega_1, \ldots \in \Omega$ with $\omega_j = 1 \iff j \in \{q_i\}_{i\geq 0}$. Assume $\{q_i\}$ is an infinite sequence. We discuss the *nesting* shown in Figure 11.1. We have

$$\begin{aligned} \omega(q_i) &= \sum_{j\leq i} S_{q_j} \\ \pi(\omega) &= \cap_{i\geq 0} D_{\omega(q_i)}. \end{aligned}$$

Moreover (review Figure 11.1),

$$\begin{aligned} \omega(q_0) &= S_{q_0} \\ \omega(q_1) &= S_{q_1} + S_{q_0} = S_{q_1} + \omega(q_0) \\ \omega(q_2) &= S_{q_2} + S_{q_1} + S_{q_0} = S_{q_2} + \omega(q_1) \\ \omega(q_3) &= S_{q_3} + S_{q_2} + S_{q_1} + S_{q_0} = S_{q_3} + \omega(q_2) \\ &\vdots \end{aligned}$$

and

$$\begin{aligned} \omega(q_0) = S_{q_0} &< S_{q_1} \\ S_{q_1} &< \omega(q_1) < S_{q_1+1} \\ S_{q_2} &< \omega(q_2) < S_{q_2+1} \\ S_{q_3} &< \omega(q_3) < S_{q_3+1} \\ &\vdots \end{aligned}$$

In general we have:

$$S_{q_i} < \omega(q_i) < S_{q_i+1} \tag{13.21}$$

for $i \geq 1$.

Exercise 13.4.15. *Prove equation (13.21).* **HINT:** *Recall that for $i \geq 1$ we have $S_{q_i+1} - S_{q_i} = S_{Q(q_i+1)}$ and $Q(q_i + 1) \geq q_{i-1} + 1$. (see equations (13.9) and (6.2)).*

Thus, setting $x = \pi(\omega)$, we have that:

- the intervals $\{D_{\omega(q_i)}\}_{i\geq 0}$ are nested and their intersection is x (see Figure 11.1), and
- $c \in D_{\omega(q_i)}$ if and only if $i = 0$.

Hence each $\omega \neq \langle 0\rangle \in \Omega \setminus \langle \mathbb{N}\rangle$ has associated with it a nested sequence of closed intervals $\{D_{\omega(q_i)}\}_{i\geq 0}$ such that $\pi(\omega) = \bigcap_{i\geq 0} D_{\omega(q_i)}$. Recall that $\pi(\langle n\rangle) = c_n$ for $n \geq 1$, $\pi(\langle 0\rangle) = c$, and π is onto. If both $\mathbf{P}$ and π are bijections, then $f|\omega(c, f)$ is one-to-one. We have already characterized when $\mathbf{P}$ is a bijection, and hence we are interested in when π is one-to-one. Before investigating when π is one-to-one, we make some further observations about the nestings.

Using equation (13.9) we have:

$$\begin{aligned}
S_{q_0+1} - \omega(q_0) &= S_{q_0+1} - S_{q_0} \quad = \quad S_{Q(q_0+1)} \\
S_{q_1+1} - \omega(q_1) &= S_{q_1+1} - S_{q_1} - \omega(q_0) \quad = \quad S_{Q(q_1+1)} - \omega(q_0) \\
&\geq S_{q_0+1} - \omega(q_0) \\
S_{q_2+1} - \omega(q_2) &= S_{q_2+1} - S_{q_2} - \omega(q_1) \quad = \quad S_{Q(q_2+1)} - \omega(q_1) \\
&\geq S_{q_1+1} - \omega(q_1) \\
&\vdots
\end{aligned}$$

Thus,

$$S_{q_i+1} - \omega(q_i) \geq S_{q_{i-1}+1} - \omega(q_{i-1}) \quad \text{for} \quad i \geq 1, \tag{13.22}$$

and therefore the sequence $\{S_{q_i+1} - \omega(q_i)\}_{i\geq 0}$ is a monotone increasing sequence (not necessarily strictly increasing). See Figure 13.7, where the levels for $D_{S_{q_1}}$, $D_{S_{q_1+1}}$, $D_{S_{q_i}}$, and $D_{S_{q_i+1}}$ are shown in the Hofbauer tower; however, the actual intervals are not shown.

Exercise 13.4.16. *Suppose there exists $j_1 < j_2 < j_3 < \ldots$ such that $Q(q_{j_i} + 1) > q_{j_i-1} + 1$ for all i. Prove that*

$$\lim_{i\to\infty} S_{q_i+1} - \omega(q_i) = \infty.$$

HINT: *Use (13.22).*

It follows from Theorem 11.1.15 that

$$\pi(\mathbf{P}^n(\omega)) = f^n(x) \quad \text{for all} \quad n \geq 1.$$

Let $\{\tilde{q}_i\}_{i\geq 0}$ be such that $(\mathbf{P}^n(\omega))_j = 1 \iff j \in \{\tilde{q}_i\}$. We have seen that $\mathbf{P}$ is a minimal map and hence there are no periodic points for $\mathbf{P}$. Therefore $\mathbf{P}^n(\omega) \neq \omega$ and hence $\{q_i\} \neq \{\tilde{q}_i\}$. However, it follows from Remark 13.3.7 that these two sequences have a common tail, that is, there exists $m, K \in \mathbb{N}$ such that $q_i = \tilde{q}_{i+m}$ for $i \geq K$.

This concludes Remark 13.4.14.

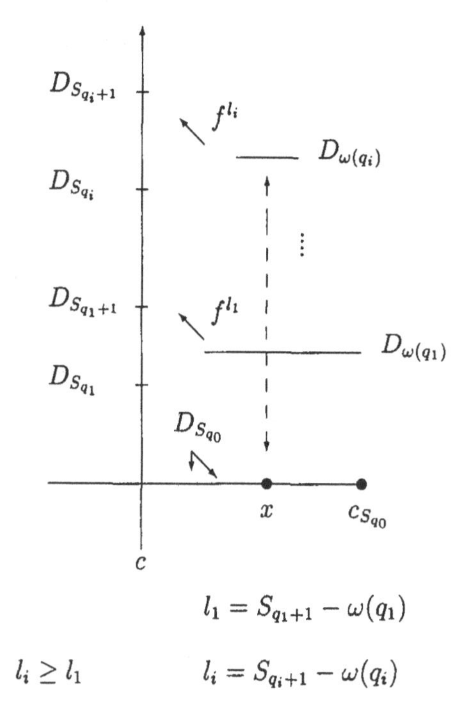

Figure 13.7: Nesting for $x = \pi(\omega)$

The remaining definitions, remarks, and lemmas are used to prove Theorem 13.4.26.

Definition 13.4.17. Given $(\Omega, \mathbf{P})$, set

$$\mathcal{C} = \cup_{n \geq 1} \mathbf{P}^{-n}(\langle 0 \rangle).$$

Remark 13.4.18. Assume $Q(k) \to \infty$ and that $\mathbf{P}$ is a bijection. To show π is one-to-one, it suffices to show π is one-to-one on $\Omega \setminus \mathcal{C}$. To see this, recall Theorem 11.1.15, in particular that $\pi^{-1}(c) = \langle 0 \rangle$.

Definition 13.4.19. Assume $Q(k) \to \infty$. For $\langle 0 \rangle \neq \omega = \omega_0, \omega_1, \omega_2, \ldots \in \Omega$ define

$$\begin{aligned} q_0(\omega) &= \min\{j \mid \omega_j = 1\} \\ q_1(\omega) &= \min\{j \mid j > q_0(\omega) \text{ and } \omega_j = 1\}. \end{aligned}$$

Note that when ω has only one nonzero entry, we have that $q_1(\omega)$ is not defined.

Remark 13.4.20. Let $\omega \in \Omega$ with $\omega_j = 1 \iff j \in \{q_i\}_{i \geq 0}$. Assume $\{q_i\}$ is an infinite sequence. It follows from Proposition 13.3.36 that $\omega \in \Omega \setminus \mathcal{C}$ if and only if there exists $j_1 < j_2 < j_3 < \cdots$ such that $Q(q_{j_i+1}+1) > q_{j_i}+1$ for all i.

Lemma 13.4.21. *Assume $Q(k) \to \infty$ and let $\omega, \tilde{\omega} \in \Omega \setminus C$. Suppose that $\pi(\omega) = \pi(\tilde{\omega})$ and $n_1 < n_2 < n_3 < \cdots$ are such that $\lim_{i\to\infty} q_0(\mathbf{P}^{n_i}(\omega)) = \infty$. Then,*

$$\lim_{i\to\infty} q_0(\mathbf{P}^{n_i}(\tilde{\omega})) = \infty. \tag{13.23}$$

Proof. Suppose to the contrary that (13.23) fails. Let $M \in \mathbb{N}$ and $k_1 < k_2 < k_3 < \cdots$ be such that $q_0(\mathbf{P}^{n_{k_i}}(\tilde{\omega})) \leq M$ for all $k \geq 1$. Observe that

$$x_i \equiv \pi(\mathbf{P}^{n_{k_i}}(\omega)) = \pi(\mathbf{P}^{n_{k_i}}(\tilde{\omega}))$$

for all i. Since $|D_n| \to 0$ and both c and x_i are in $D_{q_0(\mathbf{P}^{n_{k_i}}(\omega))}$ with $q_0(\mathbf{P}^{n_i}(\omega)) \to \infty$, we have $\lim_{i\to\infty} x_i = c$.

For ease of notation set $q_0(i) = q_0(\mathbf{P}^{n_{k_i}}(\tilde{\omega}))$ and $q_1(i) = q_1(\mathbf{P}^{n_{k_i}}(\tilde{\omega}))$. For each i we have

$$x_i \in D_{q_0(i)} \cap D_{q_1(i)}. \tag{13.24}$$

The sequence $\{q_1(i)\}$ is either bounded or not. However, either one along with $|D_n| \to 0$, $q_0(i) \leq M$ for all i, and (13.24) contradict $\lim_{i\to\infty} x_i = c$. □

Remark 13.4.22. Assume $Q(k) \to \infty$ and let $\omega \in \Omega \setminus C$. Let $\{q_i\}_{i\geq 0}$ be such that $\omega_j = 1 \iff j \in \{q_i\}$. Assume $\{q_i\}$ is an infinite sequence. From Remark 13.4.20 we may choose $j_1 < j_2 < j_3 < \cdots$ such that $Q(q_{j_i+1}+1) > q_{j_i}+1$ for all i. For each i, set $n_i = S_{q_{j_i}+1} - \omega(q_{j_i})$. Using Lemma 13.3.33 we have that

$$q_0(\mathbf{P}^{n_i}(\omega)) = q_{j_i}+1$$

for each i. Thus,

$$\lim_{i\to\infty} q_0(\mathbf{P}^{n_i}(\omega)) = \infty.$$

Lemma 13.4.23. *[45] Assume $Q(k) \to \infty$. Suppose $\omega \neq \tilde{\omega} \in \Omega \setminus C$ are such that $\pi(\omega) = \pi(\tilde{\omega})$. Suppose $q_0 = \tilde{q}_0$, where $\omega_i = 1 \iff i \in \{q_j\}_{j\geq 0}$ and $\tilde{\omega}_i = 1 \iff i \in \{\tilde{q}_j\}_{j\geq 0}$. Then there exists $l \in \mathbb{N}$ such that $q_0(\mathbf{P}^l(\omega)) \neq q_0(\mathbf{P}^l(\tilde{\omega}))$.*

Proof. Notice we have not assumed that either $\{q_i\}$ or $\{\tilde{q}_i\}$ is an infinite sequence. However, as $\pi(\langle m\rangle) = c_m$ for all $m \in \mathbb{N}$ and $\pi(\omega) = \pi(\tilde{\omega})$, it must be that one of the sequences is infinite. Without loss of generality assume $\{q_i\}$ is infinite.

Fix $i_0 = \min\{i \mid \omega(q_i) \neq \tilde{\omega}(\tilde{q}_i)\}$. (Such an i_0 exists, for otherwise the sequence $\{\tilde{q}_i\}$ is finite, say $\tilde{q}_0, \ldots \tilde{q}_n$, and $q_j = \tilde{q}_j$ for $0 \leq j \leq n$. However, then $\pi(\tilde{\omega}) = cs_{\tilde{\omega}(\tilde{q}_n)}$ and $cs_{\tilde{\omega}(\tilde{q}_n)} = cs_{\omega(q_n)} \notin Ds_{\omega(q_{i+1})}$, contradicting $\pi(\omega) = \pi(\tilde{\omega}) = cs_{\tilde{\omega}(\tilde{q}_n)}$.) Without loss of generality assume $\omega(q_{i_0}) < \tilde{\omega}(\tilde{q}_{i_0})$. Then $q_{i_0} < \tilde{q}_{i_0}$. Set $l = S_{\tilde{q}_{i_0}+1} - \tilde{\omega}(\tilde{q}_{i_0})$. Then $(\mathbf{P}^l(\tilde{\omega}))_j = 0$ for $j \leq \tilde{q}_{i_0}$, and therefore $q_0(\mathbf{P}^l(\tilde{\omega})) > \tilde{q}_{i_0} > q_{i_0}$. On the other hand, $\omega(q_{i_0}) < \tilde{\omega}(\tilde{q}_{i_0})$ and $l + \omega(q_{i_0}) < S_{\tilde{q}_{i_0}+1}$ imply there is some $j \leq \tilde{q}_{i_0}$ with $(\mathbf{P}^l(\omega))_j = 1$. Thus,

$$q_0(\mathbf{P}^l(\tilde{\omega})) > \tilde{q}_{i_0} > q_{i_0} > q_0(\mathbf{P}^l(\omega)).$$

□

Exercise 13.4.24. *Assume $Q(k) \to \infty$. Suppose $\omega \neq \tilde{\omega} \in \Omega \setminus C$ are such that $\pi(\omega) = \pi(\tilde{\omega})$. Fix $M \in \mathbb{N}$. Prove there exists $l \in \mathbb{N}$ such that $q_0(\mathbf{P}^l(\omega)) \neq q_0(\mathbf{P}^l(\tilde{\omega}))$ and $M < q_0(\mathbf{P}^l(\omega)), q_0(\mathbf{P}^l(\tilde{\omega}))$.* **HINT:** *Use Remark 13.4.22 and Lemma 13.4.23.*

Lemma 13.4.25. *Let $\omega, \tilde{\omega} \in \Omega \setminus C$ with $\omega_j = 1 \iff j \in \{q_i\}_{i \geq 0}$ and $\tilde{\omega}_j = 1 \iff j \in \{\tilde{q}_i\}_{i \geq 0}$. Suppose that $\pi(\omega) = \pi(\tilde{\omega})$ and $Q(q_0 + 1) < Q(\tilde{q}_0 + 1)$. Then*

$$Q(q_1 + 1) = q_0 + 1.$$

Proof. Again, we have not assumed that either sequence $\{q_i\}$ or $\{\tilde{q}_i\}$ is infinite. However, $Q(q_0 + 1) < Q(\tilde{q}_0 + 1)$ and $x \in [c, cs_{\tilde{q}_0}]$ guarantee that q_1 exists. Set $x = \pi(\omega) = \pi(\tilde{\omega})$. Since $x \in [c, cs_{\tilde{q}_0}]$, $Q(q_0 + 1) < Q(\tilde{q}_0 + 1)$, and $x \in D_{\omega(q_1)}$, we have (without loss of generality) that $z_{Q(q_0+1)} \in D_{\omega(q_1)}$. Hence, $S_{q_1+1} - \omega(q_1) = S_{Q(q_0+1)}$, and therefore $S_{q_1+1} - S_{q_1} - S_{q_0} = S_{Q(q_0+1)}$. Thus, $S_{Q(q_1+1)} = S_{q_0} + S_{Q(q_0+1)} = S_{Q(q_0+1)}$ and hence $Q(q_1 + 1) = q_0 + 1$. □

Theorem 13.4.26 ([45, Theorem 2]). *Assume that $Q(k) \to \infty$, $\mathbf{P}$ is a bijection and Conditions B and C hold. Then $f|\omega(c, f)$ is one-to-one.*

Proof. We show that $\pi|\Omega \setminus C$ is one-to-one. Suppose to the contrary that $\omega \neq \tilde{\omega} \in \Omega \setminus C$ with $\pi(\omega) = \pi(\tilde{\omega})$.

Let $L \in \mathbb{N}$ be as in Remark 13.4.10. From Exercise 13.4.24 (passing to $\mathbf{P}^l(\omega) = \mathbf{P}^l(\tilde{\omega})$ and relabeling $\mathbf{P}^l(\omega)$ as ω and the same for $\tilde{\omega}$) we may assume $q_0 \neq \tilde{q}_0 > K_B$ and $Q(q_0+1), Q(\tilde{q}_0+1) > L$ (note that Remark 13.4.10 uses only the weaker Condition A). Thus (using Remark 13.4.10) we have $Q(q_0+1) \neq Q(\tilde{q}_0+1)$; without loss of generality assume $Q(\tilde{q}_0+1) > Q(q_0+1)$.

From Lemma 13.4.25 we have that $Q(q_1 + 1) = q_0 + 1$. However, now we meet the assumptions of Exercise 13.4.12 (it is here that we use the stronger Condition B) and thus $cs_{\tilde{q}_0} \notin Ds_{q_1+S_{q_0}} = D_{\omega(q_1)}$, a contradiction; see Figure 13.8. □

Remark 13.4.27. For $a \in [\sqrt{2}, 2]$, let $Q_a(k)$ be the kneading map for the symmetric tent map T_a. Let $\mathcal{I} = \{a \mid Q_a(k) \to \infty\}$. For $a \in \mathcal{I}$, let $(\Omega_a, \mathbf{P}_a)$ be the adding machine obtained from the cutting times generated by $Q_a(k)$. Set

$$\mathcal{E} = \{a \in \mathcal{I} \mid \mathbf{P}_a \text{ is a bijection and Conditions B and C hold}\}.$$

It is shown in [45, Example 1] that $\mathcal{E}$ is a locally uncountable and dense subset of $[\sqrt{2}, 2]$.

Question 13.4.28. Do there exist $a \in \mathcal{E}$ and $\rho \in [0,1] \setminus \mathbb{Q}$ such that Π_ρ is well defined (recall Definition 11.2.2)? Keep in mind that a conjugacy is not possible.

Exercise 13.4.29. ♣ *Prove the set $\mathcal{E}$ from Remark 13.4.27 is indeed a locally uncountable and dense subset of $[\sqrt{2}, 2]$.*

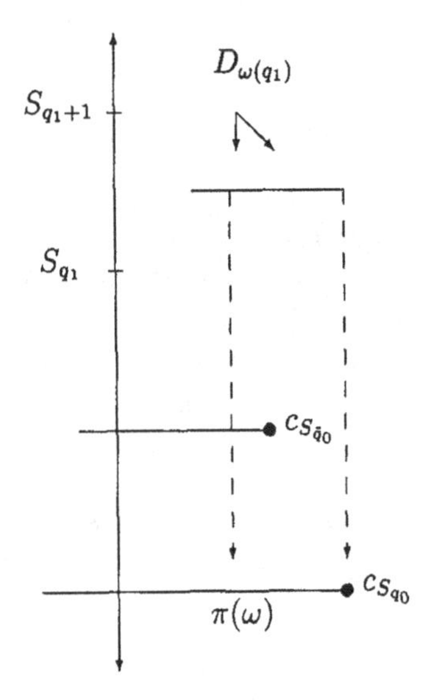

Figure 13.8: $c_{S_{\tilde{q}_0}} \in D_{\omega(q_1)}$

Remark 13.4.30. The proof of Theorem 13.4.26 consists of the following three main pieces:

1. $q_0 \neq \tilde{q}_0$ can be taken arbitrarily large.
2. q_0, $\tilde{q}_0$ can be taken with $Q(q_0 + 1) \neq Q(\tilde{q}_0 + 1)$.
3. Condition B and items (1) and (2) result in a contradiction to $\pi(\omega) = \pi(\tilde{\omega})$.

We obtain item (1) with no added conditions other than $Q(k) \to \infty$. We used Remark 13.4.10 to obtain item (2). Remark 13.4.10 assumes Conditions A and C. One can obtain item (2) with Condition C only; see Exercise 13.4.31. However, Condition B is needed for item (3), and hence nothing is lost by using Conditions A and C to obtain item (2). (Recall that Condition B implies Condition A.)

Exercise 13.4.31. *[45, Lemma 4 - proof] Suppose Condition C holds. Assume $\pi(\omega) = \pi(\tilde{\omega})$, $K_C < q_0 < \tilde{q}_0$, and $Q(q_0+1) = Q(\tilde{q}_0+1)$. The sequences $\{q_i\}$ and $\{\tilde{q}_i\}$ are defined in the proof of Theorem 13.4.26. Set $l = Q(q_0+1)$, $s = q_0(\mathbf{P}^l(\omega))$, and $\tilde{s} = q_0(\mathbf{P}^l(\tilde{\omega}))$. Prove:*

1. *$K_C < s, \tilde{s}$ and $s \neq \tilde{s}$.*
2. *$Q(s+1) \neq Q(\tilde{s}+1)$.*

HINT: *Prove (1) directly and use Condition C for (2).*

Definition 13.4.32. We say a kneading map Q satisfies *Condition D* provided there exists $L \in \mathbb{N}$ such that $t \geq L$ implies that each of (z_{t-1}, z_t), $(\hat{z}_t, \hat{z}_{t-1})$ contains at most one c_{S_k}.

Remark 13.4.33. Assume $Q(k) \to \infty$. It follows from Remark 13.4.10 that Conditions A and C imply Condition D. Hence one could restate Theorem 13.4.26 with the assumption that Conditions B and D hold.

Question 13.4.34. Can one prove Theorem 13.4.26 with Conditions A and C? What are the necessary conditions on Q for $f|\omega(c, f)$ being one-to-one under the assumption that $\mathbf{P}$ is a bijection?

Exercise 13.4.35. ♣ *The following construction is given in [45, Section 5]. Let k_1 be arbitrary, $k_2 = k_1 + 1$, and $k_i = 2k_{i-1} - k_{i-2} + 1$ for $i \geq 3$. Define a kneading map by*

$$\begin{aligned} Q(k_i) &= k_i - 1 \quad \text{for} \quad i \geq 3 \\ Q(k_i + j) &= Q(k_{i-1} + j - 1) \quad \text{for} \quad i \geq 2 \quad \text{and} \quad 1 \leq j < k_{i+1} - k_i, \end{aligned}$$

and for $k \leq k_2$ let $Q(k)$ be arbitrary (although chosen so that Q is admissible; recall Definition 6.1.12). Let f be a unimodal map with kneading sequence Q. It is shown in [45] that $f|\omega(c, f)$ is one-to-one. Note that $\lim_{k\to\infty} Q(k) \neq \infty$. Prove or disprove that $\omega(c, f)$ is minimal. **HINT:** *Recall that $\omega(c, f)$ is minimal if and only if c is uniformly recurrent.*

Chapter 14

Complex Quadratic Dynamics

Complex dynamics (i.e., the theory of dynamical systems on the complex plane) is a rich area in which powerful techniques from complex analysis are available. In our presentation, we try to avoid the more involved techniques as much as possible. Therefore, we work in settings that allow simplified definitions and theorems yet still allow for challenging results. For example, our definition of Julia sets (Section 14.1) is only valid for polynomials, not for arbitrary analytic functions on $\mathbb{C}$. Comprehensive introductions can be found in [9, 50, 53, 64, 118, 121].

We only discuss quadratic polynomials on $\mathbb{C}$ because they are the straightforward, and the most frequently studied, complexifications of unimodal maps. Our goal is to introduce and study symbolic dynamics of them. Just as in the real case, symbolic approaches give a lot of information on the various dynamical behaviors that the system can exhibit. In principle, itineraries can be defined for complex quadratic maps just the same as for real unimodal maps. The difficulty is that the Julia set, the set on which interesting dynamics take place, is much more complicated than the interval. Instead of just "left of the critical point" and "right of critical point," we need to be much more inventive to decide on which side of the critical point points lie. To make this decision, we will introduce the notion of external rays and external angles. At that point, we will want to determine the external angle of a given point in the Julia set. This is the contents of Section 14.1.

We will also see cutting times again; these will get a new name *internal address*, because they serve as an "address book" of the parameter space, the *Mandelbrot set* $\mathcal{M}$. We give the precise definitions and some of the properties in Section 14.2.

In Section 14.3 we finally define itineraries and kneading sequences for complex quadratic maps. We will also introduce the notion of the *Hubbard tree*, which plays the role of dynamical core in the Julia set. The Hubbard tree is a central notion in connecting many of the combinatoric and symbolic aspects of Julia sets. It will finally be used to determine which $0-1$ sequences can actually appear as a kneading invariant of a complex quadratic polynomial.

We assume the reader is familiar with the complex plane $\mathbb{C}$. The *norm* of $z \in \mathbb{C}$ is denoted $|z|$. Most often we express $z \in \mathbb{C}$ by $z = re^{i\theta}$ and refer to θ as the *polar angle* for z. An introduction to $\mathbb{C}$ and complex analysis can be found in [54, 102]. Chapter 3 contains background material for this chapter.

14.1 Julia Sets and External Rays

In this section we investigate Julia sets and external rays for quadratic maps on the complex plane $\mathbb{C}$. It is convenient in this setting to write the map as $f_c : z \mapsto z^2 + c$. The critical point is now 0, being the only point z where $f_c'(z) = 0$, and $c = f_c(0)$ is the *critical value*. This use of c as critical value may be somewhat confusing compared to previous chapters but is standard in this area.

Exercise 14.1.1. *Prove that every quadratic map $g(z) = \alpha z^2 + \beta z + \gamma$ with $\alpha \neq 0$ is conjugate to $f_c(z) = z^2 + c$ for some c.* **HINT:** *Let the conjugacy be given by $h(z) = kz + l$, set $h(g(z)) = f_c(h(z))$, and solve for k, l.*

Exercise 14.1.2. *Show that f_c is topologically conjugate to $z \mapsto \mu z(1-z)$ for $c = \frac{1}{2}\mu - \frac{1}{4}\mu^2$.*

If $|z|$ is sufficiently large, say $|z| > 3(|c|+1)$, then $|f_c(z)| = |z^2 + c| > |z|^2 - |c| > 2|z|$. Therefore $|f^n(z)| > 2^n|z|$ and $f^n(z) \to \infty$. Thus, we say that ∞ is an attracting fixed point for the map f_c. To make this rigorous, we would like to include ∞ in our space, and hence one defines the *Riemann sphere*.

Definition 14.1.3. The *Riemann sphere* is the set $\overline{\mathbb{C}} = \mathbb{C} \cup \{\infty\}$ with the topology such that $U_K = \{z \in \mathbb{C} \mid |z| > K\}$ are open neighborhoods of ∞.

Every polynomial f on $\mathbb{C}$ can be extended to $\overline{\mathbb{C}}$ by putting $f(\infty) = \infty$. It is customary to give $\overline{\mathbb{C}}$ a new metric in which ∞ is no longer infinitely remote from 0. Think of the complex plane as xy-plane in $\mathbb{R}^3$, and let S be the unit sphere in $\mathbb{R}^3$. For each point $z = x + iy \in \mathbb{C}$, you can connect $(x, y, 0)$ with the north pole $(0, 0, 1)$ of S with a straight line. This line intersects S in a second point (other than $(0,0,1)$), call it z'. For example, if $z = 0$, then $z' = (0,0,-1)$ is the south pole.

For each $z \in \mathbb{C}$, z' is different from the north pole; however the larger $|z|$, the closer z' is to the north pole. It is plausible to define $\infty' = (0,0,1)$. The map $z \mapsto z'$ is called the *stereographic projection*. See Figure 14.1.

The new metric measures the distance over the sphere of the projected points: So $\rho(z,w) = d(z',w')$, where d is the shortest distance over the surface of S. For example, $\rho(0,\infty) = \pi$, because $0'$ is the south pole, ∞' is the north pole, and the shortest line between the two poles is a half-circle of radius 1.

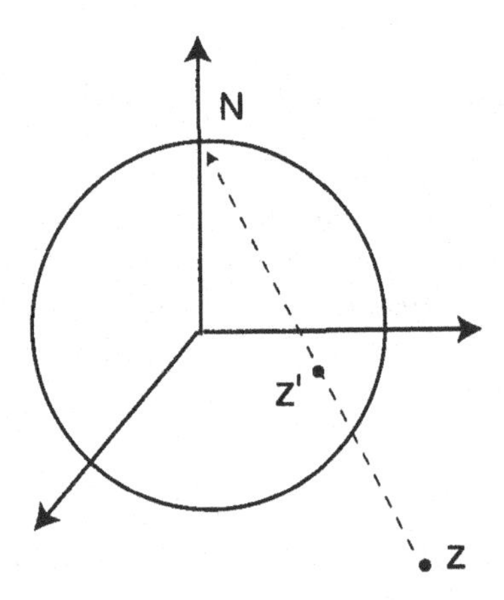

Figure 14.1: Stereographic projection

Exercise 14.1.4. *Show that ρ defines a topology such that the sets U_K in Definition 14.1.3 are indeed open sets.*

Exercise 14.1.5. *[9] The metric ρ can be computed by the integral*

$$\rho(z_1, z_2) = \int_{z_1}^{z_2} \frac{2dz}{1+|z|^2}.$$

Verify that $\rho(0,\infty) = \pi$.

Exercise 14.1.6. *Let f be any polynomial on $\overline{\mathbb{C}}$. Show that ∞ is an attracting (recall Definition 3.1.6) fixed point.*

Definition 14.1.7. The *basin of attraction* of ∞ for the map f_c is defined as

$$A_c(\infty) = \{z \in \mathbb{C} \mid f_c^n(z) \to \infty \text{ as } n \to \infty\}.$$

More simply, we call $A_c(\infty)$ the *basin of* ∞.

Definition 14.1.8. The *Julia set* of f_c is the boundary of the basin of ∞: $J_c := \partial A_c(\infty)$. The *filled-in Julia set* K_c is the set of all points that do not converge to infinity: $K_c = \mathbb{C} \setminus A_c(\infty)$. The *Fatou set* is the complement of the Julia set: $F_c := \overline{\mathbb{C}} \setminus J_c$.

The Julia and Fatou sets were named after Gaston Julia and Pierre Fatou, who in the 1920s did the first comprehensive research of complex dynamics.

Exercise 14.1.9. *Show that J_c and K_c are closed sets and that F_c is open. Show that both F_c and J_c are completely invariant, that is, $f_c(J_c) = f_c^{-1}(J_c) = J_c$ and likewise for F_c.*

Definition 14.1.10. We call $E \subset \mathbb{C}$ *locally connected* provided every point $z \in E$ has arbitrarily small neighborhoods U such that $E \cap U$ is connected.

We list without proof a few beginning facts on J_c, sending the interested reader to [9, 50, 53, 64, 118] for details. These facts hold for the Julia set of any polynomial of degree at least 2.

- J_c is infinite.
- J_c is perfect (recall Definition 1.1.22).
- J_c has an empty interior (for nonpolynomial functions, it can happen that the Julia set is precisely $\mathbb{C}$).
- J_c is locally connected when 0 is either periodic or preperiodic for f_c.
- $A_c(\infty)$ is connected.

Example 14.1.11. *Let $c = 0$; thus $f_c(z) = f_0(z) = z^2$. It is easy to check that $f_0^n(z) \to \infty$ whenever $|z| > 1$. However, if $|z| = 1$, then $|f_0^n(z)| = 1$ for all n. Thus J_0 is the unit circle in $\mathbb{C}$ and F_0 consists of two disks in $\overline{\mathbb{C}}$. If $z \in F_0$, then either $f_0^n(z) \to \infty$ or $f_0^n(z) \to 0$. See Figure 14.2.*

Exercise 14.1.12. *Prove J_0 and K_0 are as stated in Example 14.1.11. Thus $J_0 \neq K_0$ and K_0 has interior points. More generally, prove that $K_c \neq J_c$ whenever f_c has a bounded attracting periodic orbit. Is the converse true?* **HINT:** *For the converse consider an n-periodic point p with $|(f^n)'(p)| = 1$.*

Exercise 14.1.13. *Show that $f_0 : J_0 \to J_0$ is topologically conjugate to the angle doubling map $d : \mathbb{R}/\mathbb{Z} \to \mathbb{R}/\mathbb{Z}$ defined as $d(\vartheta) = 2\vartheta \mod 1$. (Recall definition of $\mathbb{R}/\mathbb{Z}$ from Section 8.1.)*

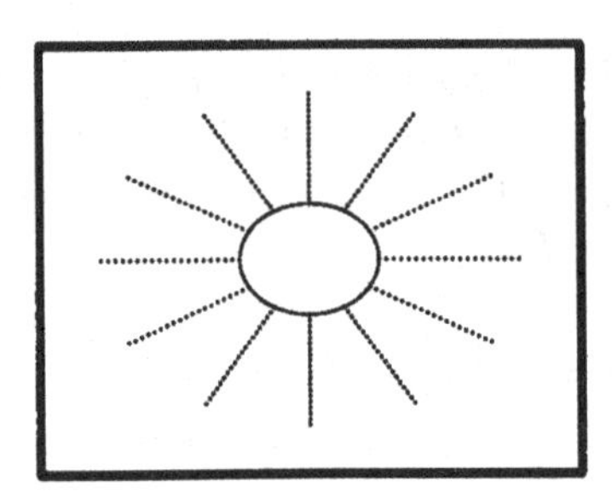

Figure 14.2: Julia set J_0 and external rays

We wish to extend the conjugacy of Example 14.1.13 to $A_0(\infty)$. To do this, we define a *foliation of external rays* of $A_0(\infty)$ by setting $R_\vartheta = \{re^{2\pi i\vartheta} \mid r > 1\}$ for $\vartheta \in [0, 1)$. We call R_ϑ an *external ray* with *external angle* ϑ. The importance of this foliation is that it provides an extension of the conjugacy via

$$f_0(R_\vartheta) = R_{d(\vartheta)}. \tag{14.1}$$

This looks a bit trivial, but the point we try to make in this section is that for all quadratic maps f_c one can find a foliation of $A_c(\infty)$ into external rays such that (14.1) holds (with f_c replacing f_0). More generally, R_ϑ takes the form

$$R_\vartheta = h_c(\{re^{2\pi i\vartheta} \mid r > 1\}),$$

where $h_c : \{z \mid |z| > 1\} \to A_c(\infty)$ is a *conformal mapping*; the existence of such an h_c follows from the Riemann Mapping Theorem; see [150] for details. For $c = 0$, we have that h_c is the identity map.

Remark 14.1.14. In our discussion of external rays, we focus mainly on the setting where 0 is preperiodic or periodic for f_c. More generally, the discussion holds when J_c is locally connected (or connected). (Recall that J_c is locally connected when 0 is preperiodic or periodic.)

Example 14.1.15. *Set $c = -2$. In this case the Julia set is the real interval $[-2, 2]$. Figure 14.3 shows J_{-2} along with some external rays.*

Exercise 14.1.16. *Recall, from Exercise 14.1.2, that f_{-2} is conjugate to the full unimodal map given by $z \mapsto 4z(1-z)$. Show that $f_{-2}^n(x) \to \infty$ if $x \in \mathbb{R}$ with $|x| > 2$. Compare this result to Example 14.1.15.*

Exercise 14.1.17. *Show that f_{-2} and f_0 are semiconjugate (recall Remark 3.3.2) on their Julia sets via $h(z) = z + \frac{1}{z}$. More precisely: $h : \{|z| = 1\} \to [-2, 2]$ is $2-1$ onto and $h \circ f_0 = f_{-2} \circ h$.*

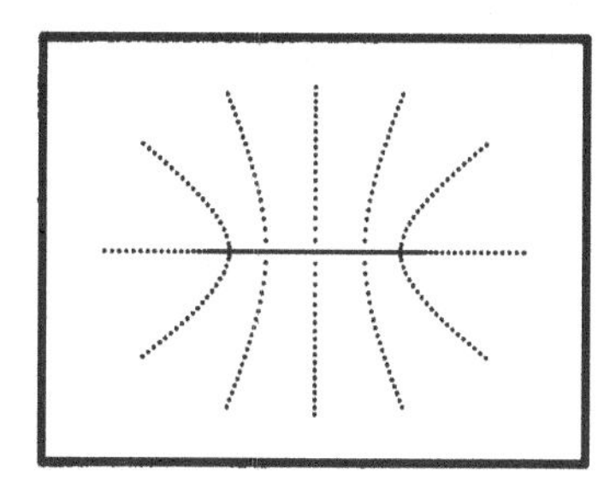

Figure 14.3: Julia set J_{-2} and external rays

We use h to define external rays for f_{-2}:

$$R_\vartheta = h(\{re^{2\pi i\vartheta} \mid r > 1\}).$$

Now, if $z \in R_\vartheta$, say $z = h(re^{2\pi i\vartheta}) = re^{2\pi i\vartheta} + \frac{1}{r}e^{-2\pi i\vartheta}$, then

$$f_{-2}(z) = z^2 - 2 = (r^2 e^{4\pi\vartheta} + \frac{1}{r^2}e^{-4\pi\vartheta}) = h(r^2 e^{2\pi d(\vartheta)}).$$

This proves (14.1) for f_{-2}, that is, $f_{-2}(R_\vartheta) = R_{d(\vartheta)}$.

Exercise 14.1.18. *Use external rays to prove that indeed* $A_{-2}(\infty) = \overline{\mathbb{C}} \setminus [-2, 2]$.

Example 14.1.19. *Set* $c = i$. *In this case, the Julia set is already too complicated to describe. See Figure 14.4. The critical point* 0 *has a finite orbit:*

$$0 \to i \to i-1 \to -i \to i-1 \to -i \to \cdots .$$

Hence 0 *is strictly preperiodic of period* 2.

Definition 14.1.20. Given f_c, h_c, and $\vartheta \in [0,1)$, we say the external ray $R_\vartheta = h_c(\{re^{2\pi i\vartheta} \mid r > 1\})$ *lands* at a point z if the limit $\lim_{r\searrow 1} h_c(re^{2\pi i\vartheta})$ exists and is equal to z.

Definition 14.1.21. Given f_c and $z \in J_c$, we call ϑ an *external angle of* z provided the external ray $R_\vartheta = h_c(\{re^{2\pi i\vartheta} \mid r > 1\})$ lands at z.

For the case $c = i$, it is next to impossible to give explicit formulas for the external rays. Moreover, due to the ragged structure of J_i (see Figure 14.4), it is not so clear that external rays actually land. Is it possible for a ray to oscillate more and more as $r \searrow 1$? We provide an answer in Theorem 14.1.39.

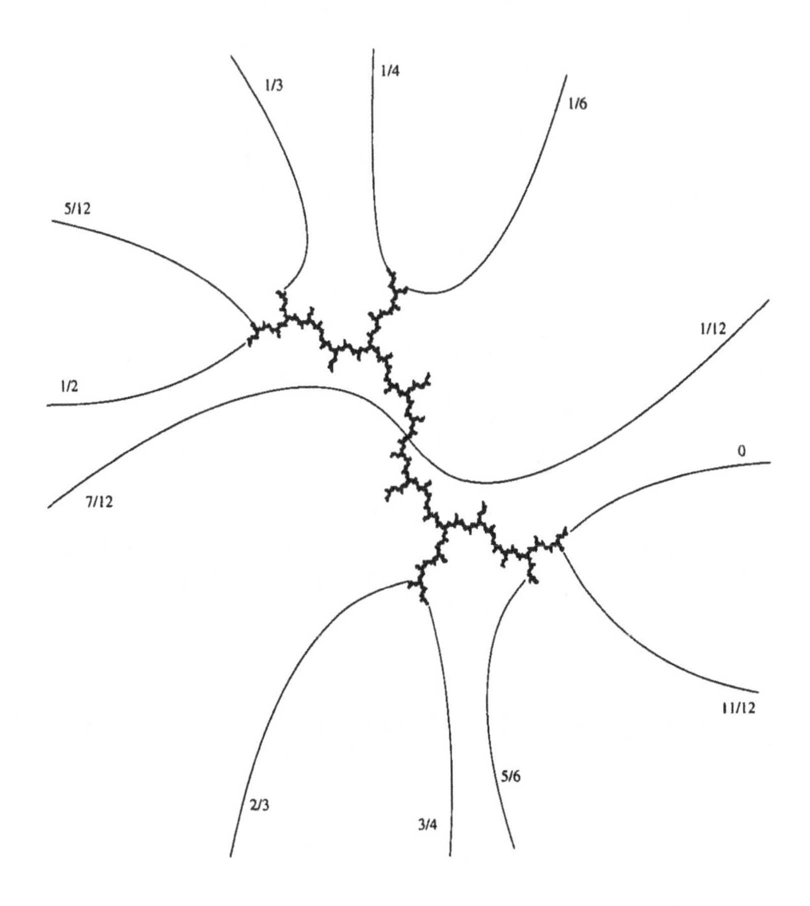

Figure 14.4: Julia set J_i

Exercise 14.1.22. *The point $-i$ is periodic of period 2 for f_i. Compute the derivative of f_i^2 at $-i$ to verify that $-i$ is repelling. Obviously $-i \notin A_i(\infty)$. Use the following proposition to show that $-i$ belongs to the Julia set J_i.*

Proposition 14.1.23. *If $\{(f_c^n)'(z)\}_{n\geq 1}$ is unbounded, then either $f_c^n(z) \to \infty$ or $z \in J_c$.*

The proof of Proposition 14.1.23 relies on a result from complex analysis called Montel's Theorem; see [53, 64, 50]. We remark that $\{(f_c^n)'(z)\}_{n\geq 1}$ is bounded for $z \in K_c^\circ$; again, one needs Montel's Theorem.

Exercise 14.1.24. *Set $c = i$. Assume (as is the case) that there is indeed a (unique) external ray R_ϑ landing at $-i$. Show that $f_i^2(R_\vartheta) = R_\vartheta$. Argue* *that* $\vartheta = \frac{2}{3}$. **HINT:** *Recall formula (14.1). Apparently we need an external angle such that* $d^2(\vartheta) = \vartheta$.

Exercise 14.1.24 is a special case of the question: How do we determine the external angle(s) of $z \in J_c$?

Observe that $z \in J_c$ need not have a single external angle. For example, when $c = -2$ (as in Example 14.1.11), we have that $J_c = [-2, 2]$ and each $z \in (-2, 2)$ has precisely two external angles. Such points are called *biaccessible*. More complicated Julia sets may contain triple, quadruple, ... accessible points.

Definition 14.1.25. A point $z \in J_c$ is called n-fold accessible if n external rays land at z or, equivalently, if $J_c \setminus \{z\}$ consists of n components (i.e., consists of n pairwise disjoint connected sets).

Be warned that the equivalence in Definition 14.1.25 only holds if J_c is locally connected. The equivalence is a corollary of Caratheodory's Theorem; see [53]. For a general discussion of when Julia sets are locally connected see [53, Chapter V.4]. As an example: If p is a polynomial with connected Julia set J such that each critical point of p belonging to J is preperiodic, then J is locally connected (z is a critical point of p provided $p'(z) = 0$).

Definition 14.1.26. The Mandelbrot set $\mathcal{M}$ is defined as the set of those parameters $c \in \mathbb{C}$ for which $\{f_c^n(0) \mid n \geq 0\}$ is bounded.

We discuss $\mathcal{M}$ in the next section; however it is helpful to have it for our discussion below.

In the case that J_c is locally connected (which is the predominant case), there is an elegant algorithm for finding external angles. This algorithm is easiest explained for $c \in [-2, \frac{1}{4}]$, and hence our discussion focuses on such c. We remark that $\mathcal{M} \cap \mathbb{R} = [-2, \frac{1}{4}]$. For such c, let $\beta_c = \frac{1+\sqrt{1-4c}}{2}$ be the largest (real) fixed point of f_c. See [50, Sections 4.2 and 6.3] and [64] for a description of J_c when $c \in [-2, \frac{1}{4}]$.

Exercise 14.1.27. *Show that if* $c \in [-2, \frac{1}{4}]$, *then* $K_c \cap \mathbb{R} = [-\beta_c, \beta_c]$, $R_0 = (\beta_c, \infty)$ *and* $R_{\frac{1}{2}} = (-\infty, -\beta_c)$. **HINT:** *The ray* R_0 *has the properties that it maps into itself and points on it go off to infinity under iteration of* f_c.

Remark 14.1.28. For $c \in \mathcal{M} \setminus \mathbb{R}$ with J_c connected and locally connected, one can again find a fixed point β_c as landing point of the ray R_0 and connect, within K_c, β_c and $-\beta_c$ by an arc. If K_c has empty interior, then this arc is unique and contains 0. Otherwise, the arc is not unique, but we can simply take the shortest arc connecting β_c and $-\beta_c$ that contains 0. We call this arc the *spine* of K_c.

For $c \in [-2, \frac{1}{4}]$, set $R = R_0 \cup [-\beta_c, \beta_c] \cup R_{\frac{1}{2}}$. Note that the spine is precisely $[-\beta_c, \beta_c]$.

Definition 14.1.29. For $c \in [-2, \frac{1}{4}]$, we code each $z \in A_c(\infty)$ by an infinite sequence $b(z, A_c(\infty)) = b_0 b_1 b_2 \cdots \in \{0,1\}^{\mathbb{N}}$ as follows:

$$b_i = \begin{cases} 0 & \text{if the polar angle of } f_c^i(z) \text{ is in } [0, \pi), \\ 1 & \text{if the polar angle of } f_c^i(z) \text{ is in } [\pi, 2\pi). \end{cases}$$

Exercise 14.1.30. *Take $c = 0$ and $|z| > 1$. Show that $b(z)$ gives the binary expansion of the polar angle of z, that is, $z = re^{2\pi i \vartheta}$ for $\vartheta = \sum_{i=0}^{\infty} b_i 2^{-i-1}$.*

Remark 14.1.31. Exercise 14.1.30 holds only for $c = 0$. It is only for $c = 0$ that the external angle coincides with the polar angle for $z \in J_c$.

Definition 14.1.32. Let $z \in J_c$ and U be an open disk containing z. Then $U \setminus K_c$ consists of components, at least one of which contains z in its boundary. Choose such a component and call it U_0. Define $b(z, J_c)$ by

$$b_i = \begin{cases} 0 & \text{if } f_c^i(z) = \beta_c \\ 1 & \text{if } f_c^i(z) = -\beta_c \\ 0 & \text{if } f_c^i(U_0) \text{ lies above } R \text{ for } U \text{ sufficiently small} \\ 1 & \text{if } f_c^i(U_0) \text{ lies below } R \text{ for } U \text{ sufficiently small.} \end{cases}$$

Remark 14.1.33. Take $c = -2$ and $z = -1$. Here, $J_c = [-2, 2]$ and $f_c(-1) = -1$. Thus, $z \in J_c$ and $z \notin A_c(\infty)$. We would have $b(z, A_c(\infty)) = 111111\ldots$, which is different from $b(z, J_c) = 0101010101\ldots$. Hence one needs to use the appropriate definition of $b(z)$ depending on whether $z \in A_c(\infty)$ or $z \in J_c$.

Remark 14.1.34. The coding $b(z, J_c)$ depends on the component U_0 chosen. For example, when $c = -2$ we have $J_c = [-2, 2]$. One has a choice of components as shown in Figure 14.5.

Proposition 14.1.35. *If $c \in [-2, \frac{1}{4}]$, $z \in J_c$, and J_c is locally connected, then $b(z, J_c)$ is the binary expansion of (one of) the external angle(s) of z; that is, $\vartheta = \sum_{i=0}^{\infty} b_i 2^{-i-1}$ and the external ray R_ϑ lands at z.*

Remark 14.1.36. The binary expansion of the external angle should not be confused with the itinerary of a point (discussed in Section 14.3). There is an algorithm to translate the external angle into an itinerary, but they are not the same.

Exercise 14.1.37. *Let $c \in [-2, \frac{1}{4}]$ and $z \in J_c$ such that J_c is locally connected. Show that all points on an external ray R_ϑ that land at z have the same code b. Prove Proposition 14.1.35.*

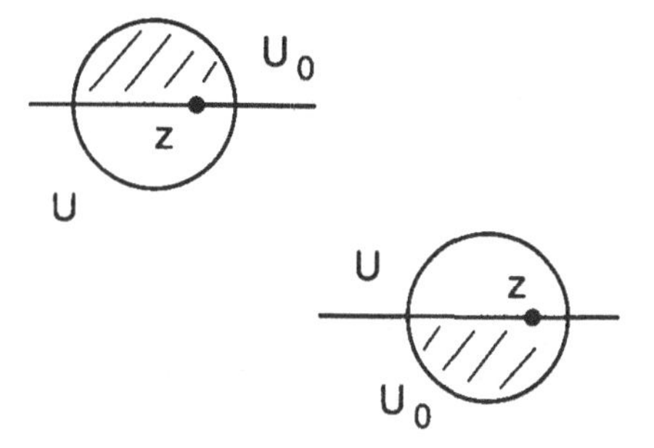

Figure 14.5: Choice of component U_0

Exercise 14.1.38. *Assume that* $c \in [-2, \frac{1}{4}]$ *and* $z \in J_c \cap \mathbb{R}$. *Show that* $1 - \vartheta = \sum_{i=0}^{\infty}(1 - b_i)2^{-i-1}$ *is also an external angle of* z.

For $c \in [-2, \frac{1}{4}]$, we have shown how to compute external rays/angles for $z \in J_c$ with J_c locally connected. The next theorem tells us that the foliation and its properties described above hold for $c \in \mathbb{C}$ with 0 either periodic or preperiodic. The proof of Theorem 14.1.39 is beyond the scope of the text.

Theorem 14.1.39. *Let* $f_c(z) = z^2 + c$ *be a quadratic map with a periodic or preperiodic critical point. Then:*

1. J_c *is connected and locally connected.*

2. $A(\infty)$ *admits a foliation of external rays* R_ϑ *such that* $f_c(R_\vartheta) = R_{d(\vartheta)}$.

3. *Every external ray lands at a point of* J_c.

4. *The landing point of* R_ϑ *depends continuously on* ϑ.

It follows that, for f_c satisfying the hypothesis of Theorem 14.1.39, the map $L : \vartheta \mapsto \lim_{r \searrow 1} h_c(re^{2\pi i\vartheta})$ is a semiconjugacy between the angle doubling map and the action on the Julia set. In other words, $L : \mathbb{R}/\mathbb{Z} \to J_c$ is continuous and the diagram

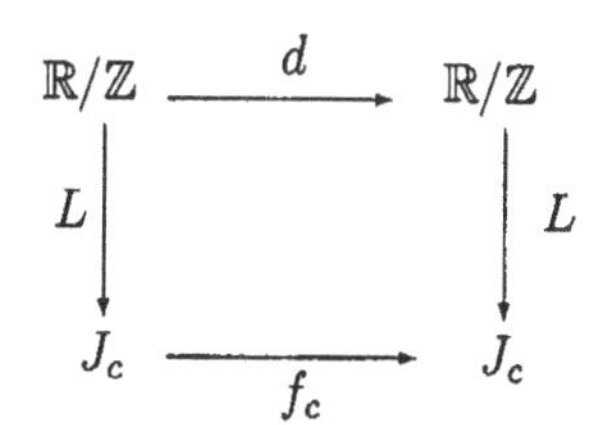

commutes. The map L is not a homeomorphism, except for the case $c = 0$. In most cases, L maps pairs, triples, and so on, of points to the same point in J_c (multiply accessible points).

Remark 14.1.40. Theorem 14.1.39 holds whenever c is attracted to a periodic orbit. For most (in the sense of harmonic measure, see [53]) $c \in \partial\mathcal{M}$, Theorem 14.1.39 holds; however, there are $c \in \partial\mathcal{M}$ where the theorem fails, namely, where not all external rays land [50, Section 5.1]. If J_c is locally connected, then all rays land.

A few general comments are in order.

1. The set of $c \in [-2, \frac{1}{4}]$ for which 0 is periodic or preperiodic is countable and hence of Lebesgue measure 0.

2. If $|c| < \frac{1}{4}$, then J_c is a simple closed curve [64, Section 3.6]. Here c lies in the *main cardioid* of $\mathcal{M}$, which is discussed in the next section (see also Exercise 14.2.6).

3. For $|c|$ sufficiently large, J_c is homeomorphic to the Cantor set K and the action of f_c on J_c is conjugate to the shift map σ on K (represent K as $\{0, 1\}^{\mathbb{N}}$) [50, Section 2]. See Example 14.1.41.

4. If 0 is preperiodic under f_c, but not periodic, then J_c is a dendrite (connected, simply-connected, with empty interior). An example is $c = i$.

In the case that J_c is a Cantor set, one can still, to some extent, define external rays. Some of these rays, however, are no longer rays in the original sense, but more complicated sets, called *bouncing rays* or *branched rays*. We refer to [7] for more details.

We close with an example where J_c is indeed a Cantor set.

Example 14.1.41. *Take c outside the Mandelbrot set, for example, $c = i + 0.5$; see Figure 14.6. Let us argue that J_c is a Cantor set.*

We first show that J_c is bounded and 0 belongs to the Fatou set. To see this, let D be a closed disk such that

- *J_c lies in the interior of D;*
- *c lies outside D;*
- *$f_c(D) \supset D$.*

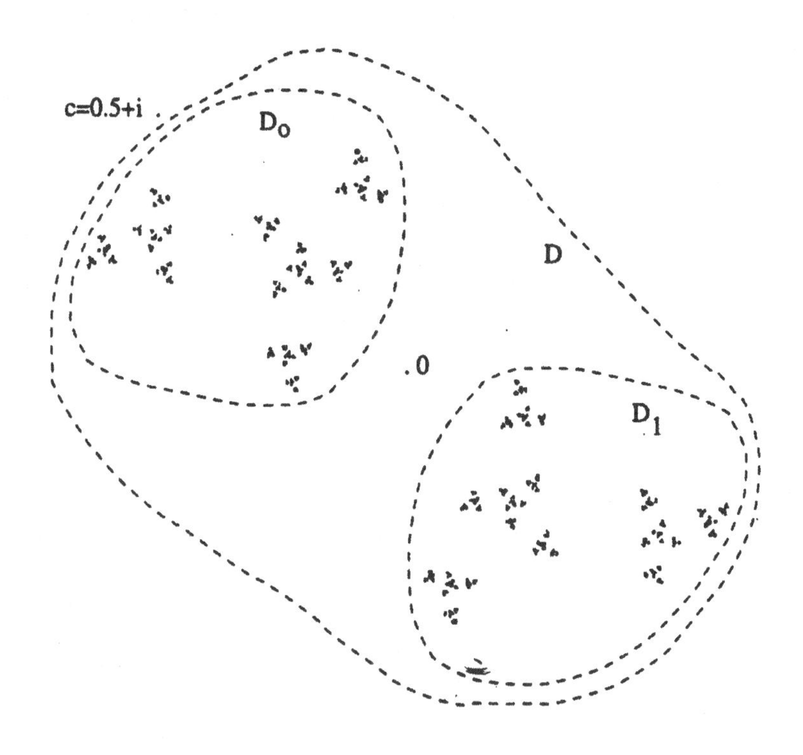

Figure 14.6: Julia set $J_{i+0.5}$ with disks D, D_0, and D_1

Exercise 14.1.42. *Find such a disk.* **HINT:** *Take a large round disk U such that $f_c(U) \supset U$, and then take the inverse image under an appropriate iterate of f_c.*

Exercise 14.1.43. *Show that each $z \in \mathbb{C} \setminus \{c\}$ has two separate preimages under f_c.*

According to Exercise 14.1.43, $f_c^{-1}(D)$ consists of two closed disks, say D_0 and D_1. Check that D_0 and D_1 are indeed disjoint and that they lie in the interior of D. Next take the preimage of D_0; it consists of two disks, say $D_{00} \subset D_0$ and $D_{01} \subset D_1$. Also, $f_c^{-1}(D_1)$ consists of two disks, say $D_{10} \subset D_0$ and $D_{11} \subset D_1$. Continue this way: At step N there are 2^N disks $D_{s_1 \ldots s_N}$, $s_i \in \{0,1\}$.

Exercise 14.1.44. *For each $N \in \mathbb{N}$ show that $J_c \subset \cup_{s_1 \ldots s_N} D_{s_1 \ldots s_N}$ and that $J \cap D_{s_1 \ldots s_N} \neq \emptyset$ for each sequence $s_1 \ldots s_N \in \{0,1\}^N$.*

Exercise 14.1.44 shows that J_c has at least 2^N components, and as N is arbitrary, J_c must have infinitely many components.

Exercise 14.1.45. *Show that J_c has no isolated points.*

One can use Montel's Theorem to prove that the diameters of the disks $D_{s_1 \ldots s_N}$ tend to 0 as $N \to \infty$ [53]. It follows that J_c is totally disconnected.

Exercise 14.1.46. *Use the coding outlined above to provide a homeomorphism $h : J_c \to \{0,1\}^{\mathbb{N}}$. Moreover, the action of f_c on J_c is conjugate to the shift map σ on $\{0,1\}^{\mathbb{N}}$; the conjugacy is precisely h.*

14.2 The Mandelbrot Set

The Mandelbrot set is one of the most intricate sets that we know in mathematics; see Figure 14.7. We will only have a glimpse of its beauty. As before, $f_c(z) = z^2 + c$ for $c \in \mathbb{C}$. It is very important to realize that $\mathcal{M}$ lies in parameter space $\mathbb{C}$, and we do not iterate in parameter space itself. For distinct $c \in \mathcal{M}$, the corresponding mappings f_c behave differently under iteration. For a discussion of connections between the Mandelbrot set and the Farey tree see [65].

Definition 14.2.1. The Mandelbrot set $\mathcal{M}$ is defined as the set of those parameters c for which $\{f_c^n(0) \mid n \geq 0\}$ is bounded.

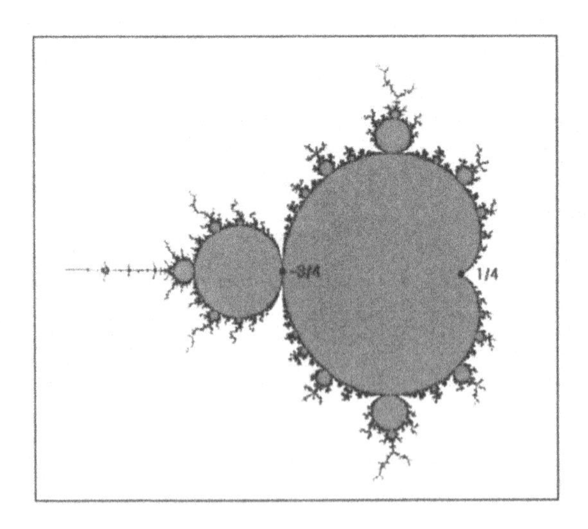

Figure 14.7: Mandelbrot set

Exercise 14.2.2. *Show that the Mandelbrot set is closed and contained in* $\{c \in \mathbb{C} \mid |c| \leq 4\}$. *Figure 14.7 suggests that* $\mathcal{M} \subset \{c \in \mathbb{C} \mid |c| \leq 2\}$. *Can you prove that too?*

Exercise 14.2.3. *Use the results of the previous section to show that* $\mathcal{M} = \{c \in \mathbb{C} \mid J_c \text{ is connected}\}$.

Proposition 14.2.4. *The Mandelbrot set is connected and contains no holes (i.e.,* $\mathbb{C} \setminus \mathcal{M}$ *consists of only one component).*

See [53] for a proof of Proposition 14.2.4. As a matter of fact, Benoit Mandelbrot, who lent his name to the set, originally conjectured that $\mathcal{M}$ has infinitely many connected components.

Exercise 14.2.5. *Show that* $\mathcal{M} \cap \mathbb{R} = [-2, \frac{1}{4}]$.

Exercise 14.2.6. *The large heart-shaped region of* $\mathcal{M}$ *is called the* main cardioid *(from the Greek word* $\kappa\alpha\rho\delta\iota\alpha$*, which means heart). It is the set of parameters for which* f_c *has an attracting fixed point. Show that the boundary of the main cardioid can be given by the formula* $\{\frac{1}{2}e^{\rho i} - \frac{1}{4}e^{2\rho i} \mid \rho \in [0, 2\pi)\}$. **HINT:** *Use Exercise 14.1.2 and show that the intersection of the main cardioid with the real line is* $(-\frac{3}{4}, \frac{1}{4})$.

Let c be in the interior of $\mathcal{M}$. Then $c' \in \mathcal{M}$ for c' sufficiently close to c. This suggests that $0 \notin J_c$. Indeed, if $0 \in J_c$, then $0 \in \partial A_c(\infty)$ and it seems plausible that, by an appropriate small perturbation in the parameter (c' instead of c), we can push 0 into $A_{c'}(\infty)$. But then $c' \notin \mathcal{M}$, so c could not have been an interior point of $\mathcal{M}$. Hence we conclude that if $c \in \mathcal{M}^\circ$, then $0 \notin J_c$. When $c \in \mathcal{M}$ and $0 \notin J_c$, we say 0 belongs to a *bounded Fatou component* of f_c. We call $W \subset \mathcal{M}^\circ$ a *component* of $\mathcal{M}^\circ$ provided W is connected and there does not exist $W' \subset \mathcal{M}^\circ$ connected such that W sits properly in W'.

Exercise 14.2.7. *Let* $c \in \mathcal{M}^\circ$. *Assume* $\{f_c^n(0)\}$ *converges to an attracting periodic orbit, say of period* k. *Prove that if* c' *is sufficiently close to* c*, then* $\{f_{c'}^n(0)\}$ *also converges to an attracting periodic orbit of period* k.

Set

$$\mathcal{H} = \{c \in \mathbb{C} \mid f_c \text{ has an attracting periodic orbit }\}.$$

One can show that $\mathcal{H} \subset \mathcal{M}$ and that $\mathcal{H}$ is open. It is conjectured that $\mathcal{H} = \mathcal{M}^\circ$. A component W of $\mathcal{M}^\circ$ is called *hyperbolic* provided $W \subset \mathcal{H}$. Let W be a hyperbolic component of $\mathcal{M}^\circ$. It is known that W is conformally equivalent to an open disk [16, Theorem 10.8] (i.e., there is a conformal bijection between W and the open unit disk) and that $\{f_c^n(0)\}_{n\geq 0}$ converges to an attracting periodic orbit for $c \in W$. Moreover, as in Exercise 14.2.7, the period of the orbit is constant over W. If the period is k, we call W a hyperbolic component of period k. In fact, W contains a unique c such that $f_c^k(0) = 0$. This c is called the *center of the hyperbolic component* W. For example, the main cardioid is the hyperbolic component of period 1 and 0 is its center. For some periods (actually every period $n \geq 3$) there exist more than one hyperbolic component. For example, there are three hyperbolic components of period 3: the largest bulb on top of the main cardioid, the largest bulb below the main cardioid, and a cardioid-shaped region to the left of the main cardioid. A component of $\mathcal{M}^\circ$ that is not hyperbolic is called a *queer component.* It is conjectured that there are no queer components.

Exercise 14.2.8. *Prove or disprove: If $c \in \partial\mathcal{M}$, then $0 \in J_c$ and hence $c \in J_c$.* **HINT:** *Consider $c = \frac{1}{4}$. Establish that $c \in \partial\mathcal{M}$ and $0 \notin J_{\frac{1}{4}}$.*

Exercise 14.2.9. ◇ *Show that the disk $\{c \mid |c - \frac{1}{2}| < \frac{1}{4}\}$ is a hyperbolic component of period 2. In fact, it is the only hyperbolic component of period 2. What is the center of this hyperbolic component?*

We can use the hyperbolic components and their periods to code parameters in the Mandelbrot set. This idea is worked out in detail in [105]; we describe it here.

Let $c_0 \in \mathcal{M}$. By Proposition 14.2.4 there is a path γ in $\mathcal{M}$ connecting 0 and c_0. The path is not allowed to have self-intersections. In general, γ passes through several hyperbolic components W; let us stipulate that whenever it passes through W, then $\gamma \cap W$ is the shortest curve passing through the center of W. In this fashion, γ is uniquely determined.

The lowest period of the hyperbolic components that γ passes through is 1. Indeed, γ starts in the main cardioid. We write $S_0 = 1$. Next define S_k to be the minimal period $> S_{k-1}$ of all centers[1] that γ passes through from the center of period S_{k-1} to c_0. The convention is made such that, if c_0 is the center of a hyperbolic component of period n, then the process is finite and stops with n, that is, we obtain a finite sequence $S_0 = 1, \dots, S_k = n$.

Definition 14.2.10. The sequence $S_0 \to S_1 \to \dots$ (or $S_0 \to S_1 \to \cdots \to S_k$ if finite) is called the *internal address* of c_0.

Exercise 14.2.11. *Show that $c_0 = -2$ has internal address $1 \to 2 \to 3 \to 4 \to \dots$* **HINT:** *Show that there is a sequence of real parameters $a_i \searrow -2$*

[1] For brevity we wrote center of period n instead of center of the hyperbolic component with period n.

such that f_{a_i} has an i-periodic critical point and the kneading sequence is $10^{i-2}C$.

Exercise 14.2.12. *The parameter $c = -1$ is the center of the hyperbolic component of period 2. Show that the internal address of -1 is $1 \to 2$ but that the internal address of $c = -0.999$ is just 1. Thus the internal address is not constant within a hyperbolic component. Nevertheless, we define the* internal address of a hyperbolic component *to be the address of its center.*

Exercise 14.2.13. *Show that there are two hyperbolic components with internal address $1 \to 3$ and one with internal address $1 \to 2 \to 3$. Prove there are four solutions c to $f_c^3(0) = 0$. So where is the fourth internal address?* **HINT:** *Think of the main cardioid.*

Exercise 14.2.14. *For $c \in \{-1, 0\}$, prove that the internal address of c is exactly the set of cutting times of f_c. Note that for these parameters f_c has periodic points only of period 1 or 2 in its core [50, Lemma 4.5]. (Recall that $f_c(z) = z^2 + c$ is topologically conjugate to $g_\mu(z) = \mu z(1-z)$, where $c = \frac{1}{2}\mu - \frac{1}{4}\mu^2$. Cutting times for g_μ were discussed in Chapter 6 and are similarly defined for f_c. For $c = \frac{1}{2}\mu - \frac{1}{4}\mu^2$, the sequence of cutting times is the same for both g_μ and f_c.)*

More generally we have the following.

Theorem 14.2.15. *Let $c \in [-2, \frac{1}{4}]$, and if c belongs to a hyperbolic component W, assume c is its center. Then the internal address of c consists of precisely the cutting times of f_c.*

Proof. We use the kneading sequence of f_c. Recall $\nu(c) = \nu = \nu_1\nu_2\nu_3\ldots$, where

$$\nu_i = \begin{cases} 0 & \text{if } f_c^i(0) > 0, \\ * & \text{if } f_c^i(0) = 0, \\ 1 & \text{if } f_c^i(0) < 0. \end{cases}$$

Given ν, let us also define a function $\mathcal{P} : \mathbb{N} \to \mathbb{N}$ by

$$\mathcal{P}(n) = \min\{i > n \mid \nu_i \neq \nu_{i-n}\}. \tag{14.2}$$

(This resembles the function $\mathcal{R}$ given in Definition 10.5.3. More precisely, $\mathcal{P}(n) = n + \mathcal{R}(n)$.)

Exercise 14.2.16. *Verify that the cutting times are the orbit of 1 under $\mathcal{P}$, that is, the k-th cutting time $S_k = \mathcal{P}^k(1)$.*

Due to Exercise 14.2.14, we may assume $c_0 = c \leq -\frac{5}{4}$. The path γ is the real interval $[c_0, 0]$. As c decreases from 0 to c_0, the kneading sequence will change from $*^\infty = ***\dots$ to $\nu(c_0)$. Just to the left of 0, the kneading sequence is $1^\infty = 111....$, so $\mathcal{P}(1) = \infty$ and 1 is the only cutting time.

The kneading sequence depends continuously on c, except when it switches at some entry, say from $\nu_i = 0$ to $\nu_i = 1$, or vice versa. But this only happens when $f_c^i(0)$ passes through 0, and at this point $\nu_i = *$, so $\nu = (\nu_1 \dots \nu_{i-1}*)^\infty$.

Since $c_0 \leq -\frac{5}{4}$, we have $\nu_2(c_0) = 0$. For ν to change from 1^∞ to $\nu(c_0)$, the second symbol must change. This can only happen via the symbol $*$, when c is the center of the hyperbolic component of period 2. At this point ν changes from 1^∞ via $(1*)^\infty$ to $(10)^\infty$, and the cutting times change from just 1 to $1 \to \mathcal{P}(1) = 2$.

Next, for $(10)^\infty$ to change into $\nu(c_0)$, the first entry at which they differ, say ν_k, must switch, say from 1 to 0. This switch can only go through the symbol $*$ when $f_c^k(0) = 0$ and c is the center of a hyperbolic component of period k. Here $(\nu_1 \dots \nu_{k-1}1)^\infty$ changes into $(\nu_1 \dots \nu_{k-1}0)^\infty$ via $(\nu_1 \dots \nu_{k-1}*)^\infty$, and the orbit of $\mathcal{P}$ changes from $1 \to \mathcal{P}(1)$ onto $1 \to \mathcal{P}(1) \to k = \mathcal{P}(1)$. Repeating this argument we arrive at the assertion. □

14.3 Itineraries and Hubbard Trees

Most of the material in this subsection cannot be found in published literature. Related references include [90, 95, 120] We start with a definition of itinerary for complex quadratic maps. Let J_c be the Julia set for some $c \in \partial\mathcal{M}$ and assume that J_c is locally connected. Then there exists (at least) one external ray R_ϑ landing on the critical value $c \in J_c$. This ray has two preimage rays $R_{\vartheta/2}$ and $R_{(1+\vartheta)/2}$ that both land on 0. Therefore $R_{\vartheta/2} \cup \{0\} \cup R_{(1+\vartheta)/2}$ separates the complex plane into two parts, say $\mathbb{C}_0$ and $\mathbb{C}_1$, where by convention $c \in \mathbb{C}_1$.

Definition 14.3.1. The *itinerary* of $z \in J_c$ is the sequence

$$I(z, f_c) = I_0(z, f_c)I_1(z, f_c)I_2(z, f_c)\dots,$$

where

$$I_i(z, f_c) = \begin{cases} 0 & \text{if } f_c^i(z) \in \mathbb{C}_0, \\ * & \text{if } f_c^i(z) = 0, \\ 1 & \text{if } f_c^i(z) \in \mathbb{C}_1. \end{cases}$$

The *kneading sequence* is the itinerary of c and is denoted as $\nu_1\nu_2\nu_3\dots$. (So $\nu_i = I_{i-1}(c, f_c)$.)

Remark 14.3.2. One can wonder if this definition depends on the choice of the external ray R_ϑ landing at c if there is more than one. Indeed, an arbitrary point $z \in J_c$ can have a different itinerary, depending on this choice, but we will mainly be interested in points in the Hubbard tree. (The Hubbard tree is a specific subset of J_c; see below.) For points in the Hubbard tree, the above definition is independent on the choice of R_ϑ. This follows from the fact that c is an endpoint of the Hubbard tree. More information can be found in [49].

Remark 14.3.3. The itinerary of z can be computed as soon as we know ϑ and the external angle, ζ say, of z. All we have to do is to iterate ζ under the angle doubling d. The points $\frac{\vartheta}{2}$ and $\frac{\vartheta+1}{2}$ divide the circle into two arcs: $(\frac{\vartheta}{2}, \frac{\vartheta+1}{2})$ and $(\frac{\vartheta+1}{2}, \frac{\vartheta}{2})$, where $\vartheta \in (\frac{\vartheta}{2}, \frac{\vartheta+1}{2})$. If $d^i(\zeta) \in (\frac{\vartheta}{2}, \frac{\vartheta+1}{2})$, then $I_i(z, f_c) = 1$, and if $d^i(\zeta) \in (\frac{\vartheta+1}{2}, \frac{tht}{2})$, then $I_i(z, f_c) = 0$. A good reference for this is [90].

Exercise 14.3.4. *Assume that $c \in \partial\mathcal{M}$. Show that the kneading sequence ν contains no $*$.* **HINT:** *Use Exercise 14.2.8.*

Exercise 14.3.5. *Assume that $c \in \partial\mathcal{M} \cap \mathbb{R}$. Show that the above definition agrees with the definitions of itinerary and kneading sequence in Definitions 5.1.3 and 6.1.3.*

In this section we use Hubbard trees to determine which $0-1$ sequences can occur as the kneading sequence of a quadratic polynomial f_c with $c \in \mathcal{M}$. Given any $0-1$ sequence, one can construct an *abstract Hubbard tree.* However, it is not the case that every abstract Hubbard tree is associated to the kneading sequence for some quadratic polynomial. The obstruction is the existence of noncyclic branch points in the Hubbard tree. Hubbard trees that are subsets of filled Julia sets can have only cyclic periodic branch points. Yet, there do exist abstract Hubbard trees with noncyclic periodic points. It is precisely the $0-1$ sequences obtained from Hubbard trees with only cyclic periodic branch points that occur as kneading sequences for quadratic polynomials f_c with $c \in \mathcal{M}$.

Definition 14.3.6. A *tree* is a finite connected graph without loops. A node p of a tree T such that $T \setminus \{p\}$ consists of at least three components is a *branch point.* The components of $T \setminus \{p\}$ are the *arms* of p.

The first two (from left to right) illustrations in Figure 14.8 are examples of trees, whereas the next two are not trees. The Julia set J_c for $c = i$ is not a tree as it is not finite.

Remark 14.3.7. Given a quadratic polynomial f_c such that orb(0) is finite, there exists a tree $T \subset K_c$, containing orb(0), such that there is no proper

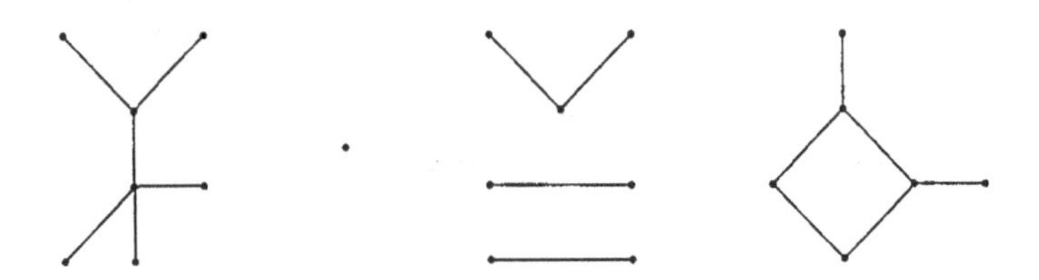

Figure 14.8: Examples of trees and nontrees

subtree T' of T that contains orb(0) with $f_c(T) \subset T$. If the filled-in Julia set has interior components, then the choice of how T passes through these components is not unique. However, every choice for T such that $f_c(T) \subset T$ satisfies our purpose.

Definition 14.3.8. Given a quadratic polynomial f_c such that orb(0) is finite, a tree from Remark 14.3.7 is called the *Hubbard tree.*

Example 14.3.9. *Consider f_c with $c = 0$. The Hubbard tree is simply a point, that is, a degenerate tree.*

The Hubbard tree serves as core for complex maps just as the interval $[c, f_c(c)]$ does for parameters $c \in \mathcal{M} \cap \mathbb{R}$. From a combinatorial viewpoint, it contains all important points, as we will see in Proposition 14.3.13.

Exercise 14.3.10. *Show that if $c \in \mathcal{M} \cap \mathbb{R}$ with orb(0) finite, then $T = [c, f_c(c)] \subset \mathbb{R}$.*

Exercise 14.3.11. $\diamondsuit$ *Suppose c is such that 0 is strictly preperiodic. Show that J_c contains two fixed points: One of them (say α_c) is multiply accessible and the other one (say β_c) has only one external ray.* **HINT:** *For β_c, find out where R_0 lands. Suppose to the contrary that β_c has more than one external ray. Let ϑ_i be the external angles of β_c. Show that f_c must permute the external rays R_{ϑ_i} cyclically and derive a contradiction.*

Remark 14.3.12. [49] Suppose c is such that 0 is strictly preperiodic and let α_c, β_c be as in Exercise 14.3.11. Then, $\alpha_c \in T$, but $\beta_c \in T$ only if $\beta_c = f_c^k(0)$ for some $k \geq 1$.

Proposition 14.3.13. *Suppose c is such that 0 is strictly preperiodic. If $z \in J_c$ is biaccessible (or n-fold accessible for some $n \geq 2$), then there is $k \geq 0$ such that $f_c^k(z) \in T$.*

In particular, this proposition shows that every branch point of a Julia set is eventually mapped into the Hubbard tree.

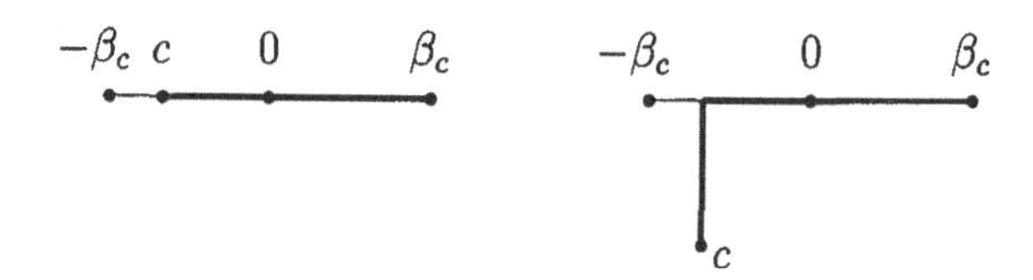

Figure 14.9: Two configurations of the spine and its image

Proof. Let $z \in J_c$ be biaccessible, say R_ϑ and $R_{\vartheta'}$ land at z. If $f^n(z) = 0$ for some $n \geq 0$, then there is nothing to prove, so assume that $0 \notin \text{orb}(z)$. Recall from Exercise 14.3.11 that β_c has only one incoming ray. Therefore, $z \neq \beta_c$ and also $f_c^n(z) \neq \beta_c$ for all $n \geq 0$. Let us show that there exists $n \geq 0$ such that $f_c^n(z) \in [-\beta_c, \beta_c]$, the spine of the Julia set.

Note that the ray R_0 lands at β_c and the preimage ray $R_{1/2}$ lands at $-\beta_c$. The set $R := R_{\frac{1}{2}} \cup [-\beta_c, \beta_c] \cup R_0$ divides $\mathbb{C}$ into two parts. If $\vartheta \in (0, \frac{1}{2})$ and $\vartheta' \in (\frac{1}{2}, 1)$, then $R_\vartheta \cup \{z\} \cup R_{\vartheta'}$ crosses R, and therefore $z \in [-\beta_c, \beta_c]$.

So let us assume that ϑ and ϑ' lie in the same component of $S^1 \setminus \{0, \frac{1}{2}\}$, say $0 < \vartheta < \vartheta' < \frac{1}{2}$. Apply d on $[\vartheta, \vartheta']$. The length of this interval doubles with every iteration until $d^n([\vartheta, \vartheta])$ has length $\geq \frac{1}{2}$ for the first time. This means that $d^n(\vartheta)$ and $d^n(\vartheta')$ lie in different components of $S^1 \setminus \{0, \frac{1}{2}\}$. Hence the common landing point $f_c^n(z)$ belongs to $[-\beta_c, \beta_c]$.

Now f_c maps the spine $[-\beta_c, \beta_c]$ in a two-to-one fashion onto its images. There are actually two possibilities, shown in Figure 14.9.

The thick lines indicate the images $f_c([-\beta_c, \beta_c])$. Note that the arc $[0, c]$ with the Julia set belongs to the Hubbard tree.

Take $w = f_c^n(z)$. We have four possibilities:

1. $w \in (0, \beta_c)$.

2. $w \in (-\beta_c, 0) \cap (-\beta_c, c)$ and $f_c(w) \in (0, \beta_c)$. Then take $f_c(w)$ instead of w, and we are in case 1.

3. $w \in (-\beta_c, 0) \cap (-\beta_c, c)$ and $f_c(w) \in (0, c)$. Then $f_c(w)$ belongs to the Hubbard tree, and we are finished.

4. $w \in (c, 0)$. Then w belongs to the Hubbard tree again, so we are done.

So only case 1, ($w \in (0, c)$), needs to be considered. We show that $f_c^m(w) \in [0, c]$ for some $m \geq 0$ and hence $f_c^m(w) \in T$. Indeed, let $\vartheta > 0$ be the smallest external angle of w. The critical value c has an external angle, say η. Then 0 itself has at least two external angles: $\eta/2$ and $(1+\eta)/2$. As w lies strictly between 0 and β_c, we have $0 < \vartheta < \eta/2$. Since ϑ doubles under every iteration of d, there must be some minimal m such that $d^m(\vartheta) \in [\eta/2, (1+\eta)/2)$. It follows that $f_c^m(w) \in [0, c]$. □

In the proof of Proposition 14.3.13 we considered T a bit as an abstract tree, that is, no longer as a subset of J_c. Hence, let us define the notion of an *abstract Hubbard tree*. An abstract Hubbard tree is a tree with a map (T, g) such that

1. $g : T \to T$ is continuous and onto.
2. g is at most two-to-one, that is, each point in T has at most two preimages.
3. Except for a single *critical* point, called 0, g is a local homeomorphism onto its image.
4. Every endpoint of T lies on the forward orbit of the critical point (in fact, $g(0)$ is already an endpoint of T).

Item (3) can be written as: If U is a subtree of T not containing the critical point, then g maps U homeomorphically onto its image.

Exercise 14.3.14. *Verify that items (1) to (4) hold for the Hubbard trees in Julia sets.* **HINT:** *For item (4), suppose that $g(0)$ is not an endpoint, so it has at least two arms. Conclude that every point in orb(0) has at least two arms. Where are the endpoints of T?*

Exercise 14.3.15. *Choose c such that orb(0) is finite. Prove that every branchpoint on T and therefore on J_c must be (pre)periodic. (This statement is also true without the assumption that orb(0) is finite; it is known as Thurston's nonwandering triangle lemma [164].)*

If we only know the Hubbard tree, and no longer the Julia set, then it is still possible to find external angles of points. The algorithm is similar to the one in Section 14.1, but we first need to find the fixed point β and its preimage $-\beta$. Since $g(T) \subset T$, there is at least one fixed point in T. Unless T is a single point, this is a point that separates 0 from $g(0)$; we call it α. The fixed point that we are looking for should lie on the other side of 0, and in general T does not contain it. Therefore we have to add arcs to T and extend g. Let us give an example; see Figure 14.10. We use the notation $0, c_1, c_2, \dots$ for the critical orbit.

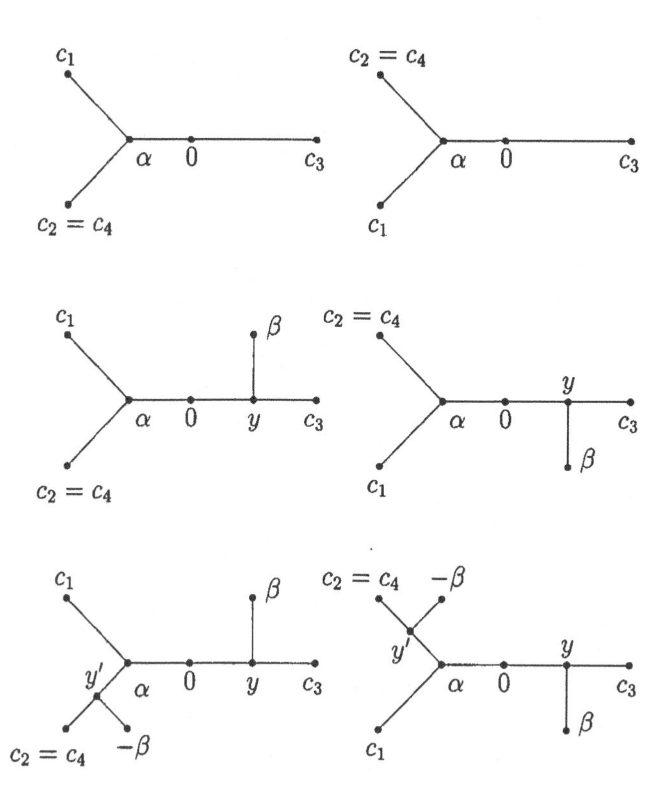

Figure 14.10: Abstract Hubbard trees for $1 \to 3 \to 5 \to 7 \to \cdots$

There is only one branch point in the pictures of the top row. Obviously, this branchpoint must be the fixed point α. In both pictures, g maps the arc $[0, c_3]$ onto $[c_1, c_4]$. Therefore we add an arc $[y, \beta]$ to the middle of $[0, c_3]$ and let g map $[y, \beta]$ onto $[\alpha, \beta]$. Therefore $y = -\alpha$ and we have found the fixed point β; see the second row of pictures. Note that in these pictures the arm $[y, \beta]$ is drawn at different sides of y. This is because we want g to preserve the cyclic order of arms at the branchpoints.

Finally we need to find $-\beta$. Since g maps $[\alpha, c_2]$ onto $[\alpha, c_3]$, we attach an arm $[y', -\beta]$ to $[\alpha, c_2]$ and map it onto $[y, \beta]$. This is done in the bottom row.

Exercise 14.3.16. *Verify that any abstract Hubbard tree can be extended to a tree T' so as to include β and $-\beta$ and such that $g(T') \subset T'$.*

Now that we have found the spine $[-\beta, \beta]$ of the tree, we can define coding to find external angles. Think of the extended tree T' as embedded in $\mathbb{C}$

and the action g as extended to a small neighborhood of T' in $\mathbb{C}$. Let R be an infinite curve in $\mathbb{C}$ dividing $\mathbb{C}$ into two parts, such that $R \cap T' = [-\beta, \beta]$. Let H_0 and H_1 be the two components of $\mathbb{C} \setminus R$. Now, for $z \in T'$, let U be a small disk around z. Take a component U_0 of $U \setminus T$. Define

$$b_i(z) = \begin{cases} 0 & \text{if } g^i(U_0) \in H_0, \\ 1 & \text{if } g^i(U_0) \in H_1. \end{cases}$$

Then $\vartheta = \sum_{i \geq 0} b_i(z) 2^{-i-1}$ is an external angle of z.

The choice of H_0 and H_1 is up to the reader. Let us choose H_0 and H_1 in both picture so that the letter α is sitting in H_0. Then we get for $b(c_1)$:

$$b(c_1) = \begin{cases} 0010101010\dots & \text{and } \vartheta = \frac{1}{2}\frac{1/4}{1-1/4} = \frac{1}{6} & \text{in the left picture,} \\ 110101010\dots & \text{and } \vartheta = \frac{1}{2} + \frac{1/4}{1-1/4} = \frac{5}{6} & \text{in the right picture.} \end{cases}$$

Exercise 14.3.17. *Compute the external angles of c_1 in both pictures if you make the other choice for H_0 and H_1.*

Topologically, the two pictures are the same; the difference is in the way of embedding T into $\mathbb{C}$. The left picture belongs to $f_i : z \mapsto z^2 + i$, while the right picture belongs to $f_{-i} : z \mapsto z^2 - i$. The Julia set J_{-i} is the mirror image of J_i, which is not difficult to prove.

Exercise 14.3.18. *Check that the type of orb(0) in the pictures indeed agrees with the critical orbit for f_i resp. f_{-i}.*

There are many other groups of parameters that give homeomorphic Hubbard trees but embed into $\mathbb{C}$ in a different way. The corresponding Julia sets are also homeomorphic to each other, but in general they are not "simple" reflections of each other. It would be nice to see for which pairs of parameters c you get homeomorphic Julia sets. This does not appear (immediately) from the external ray of c_1, but there are other means.

The exercise justifies the Definition 14.3.20.

Exercise 14.3.19. *Prove that $T \setminus \{0\}$ has at most two components. Find examples where $T \setminus \{0\}$ has only one component.* **HINT:** *Recall Exercise 14.3.14.*

Definition 14.3.20. Let T_0 and T_1 be the two components of $T \setminus \{0\}$, where T_1 is taken to contain c_1. The *itinerary* of $z \in T$ is the sequence $I(z, g) = I_0(z, g) I_1(z, g) I_2(z, g) \dots$, where

$$I_i(z, g) = \begin{cases} 0 & \text{if } g^i(z) \in T_0, \\ * & \text{if } g^i(z) = 0, \\ 1 & \text{if } g^i(z) \in T_1. \end{cases}$$

The *kneading sequence* is the itinerary of $c_1 = g(0)$; it is denoted as $\nu = \nu_1 \nu_2 \nu_3 \dots$. (So $\nu_i = I_{i-1}(c_1, g)$.)

Exercise 14.3.21. *Show that Definition 14.3.20 agrees with Definition 14.3.1 and also with the itinerary of* 0 *if* c *is real. Recall that in the latter case* $T = [g(0), g^2(0)]$.

Exercise 14.3.22. *Compute the kneading sequences of the left and right pictures in Figure 14.10.*

The point that we want to make here is that any two conjugate Hubbard trees have the same kneading sequence (and vice versa), even if the external angles are different. For a precise proof of this statement, we refer to [49].

Having defined the kneading sequence ν, we have a new way of finding internal addresses.

Definition 14.3.23. Given a $0-1$ sequence ν, let $\mathcal{P} = \mathcal{P}_\nu : \mathbb{N} \to \mathbb{N}$ be defined as in (14.2): $\mathcal{P}(n) = \min\{i > n \mid \nu_i \neq \nu_{i-n}\}$. We denote the orbit of 1 under $\mathcal{P}$ by

$$\mathrm{orb}_{\mathcal{P}}(1) = S_0 \to S_1 \to S_2 \to \dots, \text{ so } S_i = \mathcal{P}^i(1).$$

If $\mathcal{P}(S_i) = \infty$, then we say that $\mathrm{orb}_{\mathcal{P}}(1)$ is finite: $S_0 \to S_1 \to \cdots \to S_i$.

Exercise 14.3.24. *Verify that* ν *uniquely determines* $S_0 \to S_1 \to \dots$*, and vice versa.*

Exercise 14.3.25. *Show that:*

- *If* 0 *is periodic, then* $\mathrm{orb}_{\mathcal{P}}(1)$ *is finite.*
- *If* 0 *is strictly preperiodic, then* $\mathrm{orb}_{\mathcal{P}}(1)$ *is infinite, but the sequence* $(S_i - S_{i-1})$ *is (pre)periodic.*

Exercise 14.3.26. *Show that* $\mathrm{orb}_{\mathcal{P}}(1)$ *is invariant under topological conjugacies. Describe the topological conjugacy between the left and right pictures in Figure 14.10.*

Theorem 14.3.27. *The internal address of Definition 14.2.10 agrees with* $\mathrm{orb}_{\mathcal{P}}(1)$*: If* $c \in \mathcal{M}$ *is such that* 0 *is (pre)periodic, and* c *has internal address* $S_0 \to S_1 \dots$ *by Definition 14.2.10, then, for the corresponding kneading sequence* ν*, the* $\mathcal{P}$*-orbit is also* $S_0 \to S_1 \dots$*, and vice versa.*

Based on this theorem, and given an abstract Hubbard tree T with kneading sequence ν, we call $\mathrm{orb}_{\mathcal{P}}(1)$ the *internal address* of T.

Proof. The idea is the same as in Theorem 14.2.15. We omit the details; see [105]. □

We have seen how every Hubbard tree comes with a kneading sequence and an internal address. What about the other direction? Does every kneading sequence and every internal address come with a Hubbard tree? The answer is yes, but there are a few remarks to make.

- We have only dealt with finite trees, that is, with a finite number of edges and vertices. After all, we only considered maps with a finite critical orbit. Theorem 14.3.30 below holds for infinite critical orbits just as well (although the abstract Hubbard tree can become an abstract Hubbard *dendrite*), but for simplicity we only state the finite case. Therefore we have to restrict ourselves to $*$-periodic or preperiodic kneading sequences (a kneading sequence ν is called *$*$-periodic* if $\nu = (\nu_1 \dots \nu_{n-1} *)^\infty$, where $\nu_i \in \{0, 1\}$ for $1 \leq i < n$) and to finite or (pre)periodic internal addresses.

- Even though an abstract Hubbard tree T exists, it need not correspond to any Julia set, that is, there need not be a Julia set J_c such that its Hubbard tree is conjugate to the abstract Hubbard tree T. This has to do with the periodic branchpoints in the tree. If a periodic point p of f_c is a branchpoint of the Julia set, say of period n, then $f^n : U \to f^n(U)$ is a homeomorphism for any sufficiently small neighborhood U of p. This means that the arms of p must be permuted in an orientation preserving way. For example, if p has five arms, numbered 1 to 5 in counterclockwise order, and say that f^n maps arm 1 to arm 3, then f^n has to map arm 2 to arm 4, and so on; see the left picture in Figure 14.11. Let us call such branchpoint *planar*. If f^n permutes the arms of p in one cycle, then we call p *cyclic*. Figure 14.11 gives examples of planar and nonplanar and cyclic and noncyclic branchpoints. Note that "planar" depends on the embedding of the tree into the plane, whereas "cyclic" depends only on the dynamics on the tree.

Exercise 14.3.28. *Figure 14.11 shows that cyclic does not imply planar. Does planar imply cyclic?*

- There are abstract Hubbard trees with noncyclic branchpoints (see Exercise 14.3.39 and Figure 14.12). Such abstract Hubbard trees cannot belong to any quadratic polynomial!

- To every internal address there is a unique abstract Hubbard tree corresponding to it. But even if there is a unique abstract Hubbard tree T, there may be many different ways to embed T into the

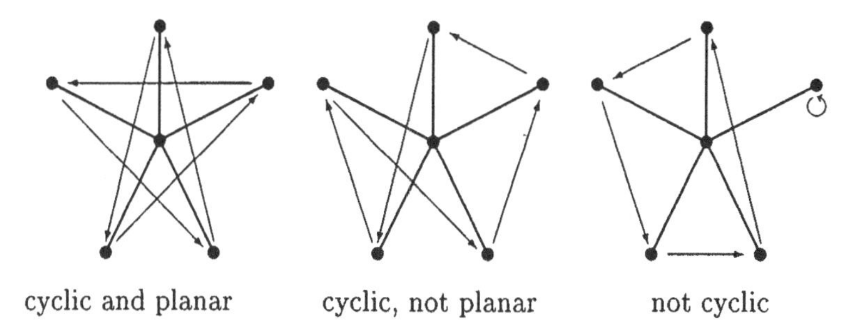

Figure 14.11: Embeddings of branchpoints in the plane

plane. Of course, we want the embedding to be such that all branchpoints are planar in the sense above. For example, if p has five arms and period 3, then there are several ways to embed p and its arms: $1 \to 2 \to 3 \to 4 \to 5 \to 1$, or $1 \to 3 \to 5 \to 2 \to 4 \to 1$ (as in Figure 14.11) and two more. Once the choice is made for p, then the arms of $f(p)$ and of $f^2(p)$ must be embedded in the same order. If $orb(p)$ and $\text{orb}(q)$ are disjoint orbits of branchpoints, then we can choose their embeddings independently of each other. The total number of plausible embeddings can be computed from the internal address; see [29].

Exercise 14.3.29. *Find a formula for the number of different planar cyclic embeddings of a branchpoint with n arms in the plane as function of n.*

Let us now state the theorem.

Theorem 14.3.30. *There are bijections between each of the following:*

a) the collection of finite abstract Hubbard trees (up to conjugacy on the critical orbits and branchpoints);

b) the collection of $*$-periodic and preperiodic kneading sequences;

c) the collection of finite or preperiodic internal addresses.

If an abstract Hubbard tree T has only cyclic periodic branchpoints, then T corresponds to a Julia set of some quadratic polynomial.

Proof. See [29]. To explain the phrase "up to conjugacy on the critical orbits and branchpoints," we say that two abstract Hubbard trees T and T' are the same if there is a homeomorphism $h : T \to T'$ such that h is a topological conjugacy between the critical orbits and branchpoints of (T, g) resp. (T', g'). □

In the remainder of this section we want to discuss branchpoints p of abstract Hubbard trees. Are they cyclic? How many arms do they have and how do you find out given the itinerary of p? In Proposition 14.3.33 below we will answer this last question, but first we need some more notation. Let $I_0(p)I_1(p)I_2(p)\dots$ be the itinerary of p. We assume that $0 \notin \mathrm{orb}(p)$, so $I_i(p) \neq *$ for all i. Define a function $\mathcal{P}_p : \mathbb{N} \cup \{0\} \to \mathbb{N}$ as follows:

$$\mathcal{P}_p(n) = \min\{i > n \mid e_i \neq \nu_{i-n}\}.$$

We say that m and n are in the same *grand orbit* of $\mathcal{P}_p$ if there exists $i, j \in \mathbb{N}$ such that $\mathcal{P}_e^i(m) = \mathcal{P}_e^j(n)$.

Exercise 14.3.31. *Fix an abstract Hubbard tree with internal address $S_0 \to S_1 \to \dots$ and kneading sequence ν.*

- *Check that the fixed point $\alpha \in T$ has S_1 arms.*
- *$\diamondsuit$ Suppose that S_{k+1} is a multiple of S_k. Show that there is a periodic branchpoint $p \in T$ with S_{k+1}/S_k arms and itinerary $I(p) = (\nu_1 \dots \nu_{S_k - 1}(1 - \nu_{S_k}))^\infty$.*

Exercise 14.3.32. ♣ *[40] Given a "real" (strictly preperiodic) kneading sequence ν, show that every periodic point of $T = [g(0), g^2(0)]$ has precisely two arms. (The only exception is when $g^2(0) = g^3(0)$; this fixed point has only one arm.)*

Proposition 14.3.33. *The number of arms at p in the abstract Hubbard tree (T, g) is the same as the number of grand orbits of $\mathcal{P}_p$.*

Proof. Call $\zeta_k = \zeta_k(p) \in T$ a *closest precritical point to p* if $g^k(\zeta_k) = 0$ and $g^k : [p, \zeta_k] \to [g^k(p), 0]$ is homeomorphic. Clearly $\zeta_0 = 0$. Take the arc $[p, \zeta_0]$ and start iterating. If $g^i|[p, \zeta_0]$ is homeomorphic but $g^{i+1}|[p, \zeta_0]$ is not, then 0 belongs to the interior of $g^i([p, \zeta_0])$. This means that $I_i(p) \neq I_i(\zeta_0) = \nu_i$, while $I_1(p) = \nu_1, I_2(p) = \nu_1, \dots, I_{i-1}(p) = \nu_{i-1}$. Hence $\mathcal{P}_p(0) = i$. At the same time, there is a point $\zeta_i \in [p, \zeta_0]$ such that $g^i(\zeta_i) = 0$. Hence $\zeta_i = \zeta_{\mathcal{P}_p(0)}$ is the first closest precritical point after ζ_0 on the arc $[p, \zeta_0]$.

Now continue iterating with $g^i([p, \zeta_i])$ until 0 belongs to the interior of the image again. This happens precisely at $g^j([p, \zeta_i])$ for $j = \mathcal{P}_p(i)$, and the point $\zeta_j \in [p, \zeta_i]$ such that $g^j(\zeta_j) = 0$ is the next closest precritical point to p. In general, the closest precritical points on $[p, \zeta_0]$ have indices $0, \mathcal{P}_p(0), \mathcal{P}_e^2(0), \dots$

Now let $[p, q]$ be another arc at p such that $[p, q] \cap [p, 0] = \{p\}$, so 0 and q belong to different arms of p. Let k be the first iterate such that $0 \in g^k([p, q])$. (If no such k exists, then the abstract Hubbard tree is not *minimal* in the sense that we could have collapsed $[p, q]$ to a point without changing essentially the dynamics on the tree.) This means that there is a closest precritical point $\zeta_k \in [p, q]$. The above argument shows that also ζ_j is a closest precritical point on $[p, q]$ for any $j \in \mathrm{orb}_{\mathcal{P}_p}(k)$. Since $[p, q] \cap [0, p] = \{p\}$ and $0 \notin \mathrm{orb}(p)$, it follows that $\mathrm{orb}_{\mathcal{P}_p}(0)$ and $\mathrm{orb}_{\mathcal{P}_p}(k)$ are disjoint. We conclude that the number of arms at p is no more than the number of grand orbits of $\mathcal{P}_p$.

Now for the converse, suppose that k is the smallest integer of a grand orbit of $\mathcal{P}_p$. Consider the arc $[0, g^k(p)]$. The above argument shows that the closest precritical points $\zeta_i(g^k(p))$ to $g^k(p)$ that lie on $[0, g^k(p)]$ must have indices $0, \mathcal{P}_p(k) - k, \mathcal{P}_p^2(k) - k, \dots$. One can verify (see Exercise 14.3.34) that $g^k(p)$ has just as many arms as p; so there must be an arc $[p, q]$ at p such that $g^k([p, q]) \subset [0, g^k(p)]$. Therefore, for $i = \mathcal{P}_p^n(e)$ sufficiently large, there is a closest precritical point $\zeta_i(p)$ to p such that $g^k(\zeta_i(p)) = \zeta_{i-k}(g^k(p)) \in [0, g^k(p)]$. This shows that every grand orbit of $\mathcal{P}_p$ corresponds to an arm at p. This concludes the proof. □

The next couple of exercises lead to the definition of the characteristic point of a periodic orbit.

Exercise 14.3.34. *Let p be a periodic point of an abstract Hubbard tree such that $0 \notin orb(p)$. Show that every point $q \in orb(p)$ has the same number of arms as p.*

Exercise 14.3.35. ♣ *[49] Let p be a periodic orbit so that $orb(p)$ is not a subset of the endpoints of T. For any $q \in orb(p)$, let T_q be T minus the component of $T \setminus \{q\}$ that contains 0. Show that either $g(T_q) = T_{g(q)}$ or that $g(T_q) \ni 0$. In the latter case, show that $g(q) \in [0, g(0)]$. (Since $orb(p)$ is not contained in the set of endpoints of T, the latter case must happen for at least one $q \in orb(p)$.) Show that the sets T_q are either nested or disjoint.*

Exercise 14.3.36. ♣ *[49] Assume that p is not the β-fixed point. Let $p_1 \in orb(p) \cap [0, g(0)]$ be such that no other point in $orb(p)$ lies on $[p_1, g(0)]$. Show that p_1 exists. Show that, for any $q \in orb(p)$, either $\#(T_q \cap orb(p)) \le \#(T_{g(q)} \cap orb(p))$ or $p_1 \in T_{g(q)}$. Conclude that $T_{p_1} \cap orb(p) = \{p_1\}$.*

Definition 14.3.37. The point p_1 defined in these exercises is called the *characteristic point* of its orbit.

Some internal addresses necessarily give rise to noncyclic branchpoints. In [29] the precise condition (in terms of internal addresses) for the existence of noncyclic branchpoints is given. Here we will only discuss an example.

Example 14.3.38. *Consider the internal address $1 \to 2 \to 4 \to 5 \to 6$, which corresponds to the kneading sequence $\nu = (10110*)^\infty$. The abstract Hubbard tree is shown in Figure 14.12.*

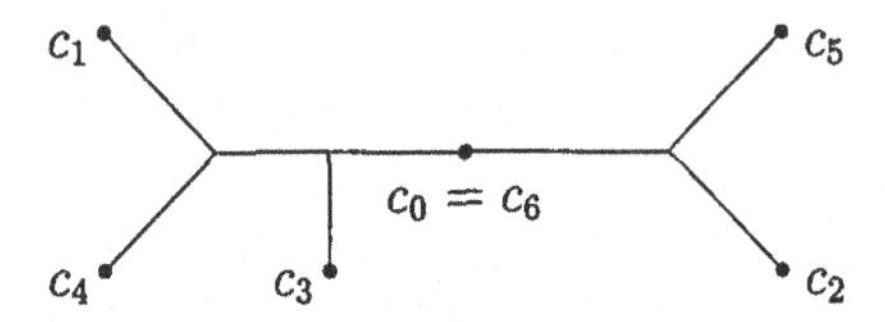

Figure 14.12: The abstract Hubbard tree for $1 \to 2 \to 4 \to 5 \to 6$

Exercise 14.3.39. *Show that for the internal address $1 \to 2 \to 4 \to 5 \to 6$ the tree (T, g) has a periodic branchpoint p of period 3 with three arms; see Figure 14.12. Show that g^3 fixes one of the arms and permutes the other two. Conclude that p is not cyclic.*

Proposition 14.3.40. *If p is a characteristic n-periodic point on (T, g), then one of the following holds:*

- *The point p has two arms.*
- *The point p has at least two arms and g^n permutes the arms of p in one cycle. The period n occurs in the internal address.*
- *The point p has at least three arms and g^n fixes the arm towards 0 and permutes the other arms in one cycle. The period n does not occur in the internal address.*

Remark 14.3.41. It is also possible that $\mathrm{orb}(p)$ consists entirely of end-points of T, for example, if p is the β-fixed point or, in the abstract Hubbard trees of Figure 14.10, $\mathrm{orb}(p) = \{c_2, c_3\}$. In these cases, no characteristic point has been defined for $\mathrm{orb}(p)$.

The proof is given in [49].

Exercise 14.3.42. ◇ *Take a sequence $\nu_1 \dots \nu_n \in \{0,1\}^n$ such that n does not belong to the internal address. Let $\nu = (\nu_1 \dots \nu_n \nu_1 \dots \nu_{n-1} *)^\infty$. Assuming the result of Proposition 14.3.40, show that the abstract Hubbard tree of ν has a noncyclic branchpoint of period n with three arms. Compare this with Example 14.3.38. Can you produce examples of noncyclic branchpoints with more than three arms?*

Bibliography

[1] R. Adler, *F-expansions revisited,* Lect. Notes in Math. **318**, Springer-Verlag, Berlin (1973), 1–5.

[2] R.L. Adler, A.G. Konheim, and M.H. McAndrew, *Topological entropy,* Trans. Amer. Math. Soc. **114** (1965), 309–319.

[3] L.V. Ahlfors, *Complex analysis. An introduction to the theory of analytic functions of one complex variable,* McGraw-Hill, New York-Toronto-London (1953).

[4] K.T. Alligood, T.D. Sauer, and J.A. Yorke, *Chaos: An introduction to dynamical systems,* Springer-Verlag, New York (1997).

[5] L. Alseda, J. Llibre, and M. Misiurewicz, *Combinatorial dynamics and entropy in dimension one,* Advanced Series in Nonlinear Dynamics, World Scientific, New Jersey (1993).

[6] P. Arnoux, V. Berthé, S. Ferenczi, S. Ito, C. Mauduit, M. Mori, J. Peyrière, A. Siegel, J.-I. Tamura, and Z.-Y. Wen, *Introduction to finite automata and substitution dynamical systems,* Preprint, http://iml.univ-mrs.fr/editions/preprint00/book/prebookdac.html (2001).

[7] P. Atela, *Bifurcations of dynamic rays in complex polynomials of degree two,* Ergod. Th. & Dynam. Sys. **12** (1992), no. 3, 401–423.

[8] G. Barat, T. Downarowicz, and P. Liardet, *Dynamiques associées à une échelle de numération,* Acta Arith. **103** (2002), no. 1, 41–78.

[9] A. Beardon, *Iteration of complex functions. Complex analytic dynamical Systems,* Graduate Texts in Mathematics **132**, Springer-Verlag, New York (1991).

[10] W.A. Beyer, R.D. Mauldin, and P.R. Stein, *Shift maximal sequences in function iteration: Existence, uniqueness, and multiplicity,* J. Math. Anal. Appl. **115** (1986), no. 2, 305–362.

[11] W.A. Beyer and P.R. Stein, *Period doubling for trapezoid function iteration: Metric theory,* Adv. in Appl. Math. **3** (1982), no. 1, 1–17.

[12] W.A. Beyer and L. Wang, *Quadratic convergence in period doubling to chaos for trapezoid maps,* J. Math. Anal. Appl. **227** (1998), no. 1, 1–24.

[13] B. Bielefeld, personal communications (1989).

[14] M. Bier and T. Bountis, *Remerging Feigenbaum trees in dynamical systems,* Phys. Lett. A **104** (1984), no. 5, 239–244.

[15] G.D. Birkhoff, *Dynamical systems,* Amer. Math. Soc, Colloquium Publications, Vol IX, Providence, R.I., revised edition (1966). [Original edition, 1927].

[16] P. Blanchard, *Complex analytic dynamics on the Riemann sphere,* Bull. Amer. Math. Soc. (N.S.) **11** (1984), no. 1, 85–141.

[17] F. Blanchard, *β-expansions and symbolic dynamics,* Theoret. Comput. Sci. **65** (1989), no. 2, 131–141.

[18] L.S. Block and W.A. Coppel, *Dynamics in one-dimension,* Lect. Notes in Math., **1513**, Springer-Verlag, Berlin (1992).

[19] L. Block and E.M. Coven, *Topological conjugacy and transitivity for a class of piecewise monotone maps of the interval,* Trans. Amer. Math. Soc. **300** (1987), no. 1, 297–306.

[20] L. Block, J. Guckenheimer, M. Misiurewicz, and L.-S. Young, *Periodic points and topological entropy of one-dimensional maps, Global theory of dynamical systems* (Proc. Internat. Conf., Northwestern Univ., Evanston, Ill., 1979), Lect. Notes in Math. **819**, Springer-Verlag, Berlin (1980), 18–34.

[21] L. Block, J. Keesling, and M. Misiurewicz, *Strange adding machines,* Preprint (2003).

[22] L. Block, J. Keesling, S. Li, and K. Peterson, *An improved algorithm for computing topological entropy,* J. Statist. Phys. **55** (1989), no. 5–6, 929–939.

[23] A.M. Blokh and M. Yu. Lyubich, *Attractors of maps of the interval, Dynamical systems and ergodic theory* (Warsaw 1986), Banach Center Publ. **23** (1989), 427–442.

[24] A.M. Blokh and M.Yu. Lyubich, *Measurable dynamics of S-unimodal maps of the interval,* Ann. Sc. E.N.S. 4 série **24** (1991), no. 5, 545–573.

[25] R. Bowen, *Entropy for group endomorphisms and homogeneous spaces,* Trans. Amer. Math. Soc. **181** (1973), 509–510.

[26] R. Bowen, *A horseshoe with positive measure,* Invent. Math. **29** (1975), no. 3, 203–204.

[27] M. Boyle and D. Lind. *Expansive subdynamics,* Trans. Amer. Math. Soc. **349** (1997), no. 1, 55–102.

[28] A. Brocot, *Calcul des rouages par approximation, nouvelle méthode,* Revue Chronométrique **6** (1860), 186–194.

[29] A.M. Bruckner and J. Smital, *The structure of ω-limit sets for continuous maps of the interval,* Math. Bohem. **117** (1992), no. 1, 42–47.

[30] K.M. Brucks and Z. Buczolich, *Trajectory of the turning point is dense for a co-σ-porous set of tent maps,* Fund. Math. **165** (2000), no. 2, 95–123.

[31] K.M. Brucks and D. Diamond, *A symbolic representation of inverse limit spaces for a class of unimodal maps,* Lect. Notes in Pure and Applied Math. **170**, Marcel Dekker, New York (1995), 207–226.

[32] K.M. Brucks, B. Diamond, M.V. Otera-Espinar, and C. Tresser, *Dense orbits of critical points for the tent map,* Cont. Math. **117** (1991), 57–61.

[33] K.M. Brucks, R. Galeeva, D.N. Rockmore, and C. Tresser, *On the star product in kneading theory,* Fund. Math. **152** (1997), no. 3, 189–209.

[34] K.M. Brucks, M.V. Otero-Espinar, and C. Tresser, *Homeomorphic restrictions of smooth endomorphisms of an interval,* Ergod. Th. & Dynam. Sys. **12** (1992), no. 3, 429–439.

[35] K.M. Brucks and M. Misiurewicz, *Trajectory of the turning point is dense for almost all tent maps,* Ergod. Th. & Dynam. Sys. **16** (1996), no. 6, 1173–1183.

[36] K.M. Brucks, M. Misiurewicz, and C. Tresser, *Monotonicity properties of the family of trapezoidal maps,* Commun. Math. Phys. **137** (1991), 1–12.

[37] K.M. Brucks, J. Ringland, and C. Tresser, *An embedding of the Farey web in the parameter space of simple families of circle maps,* Phys. D. **161** (2002), no. 3-4, 142–162.

[38] K.M. Brucks and C. Tresser, *A Farey tree organization of locking regions for simple circle maps,* Proc. Amer. Math. Soc. **124** (1995), no. 2, 637–647.

[39] H. Bruin, *Invariant measures of interval maps,* Ph.D. Thesis, Delft (1994).

[40] H. Bruin, *Combinatorics of the kneading map,* Int. J. of Bifur. & Chaos Appl. Sci. Engrg. **5** (1995), no. 5, 1339–1349.

[41] H. Bruin, *Non-monotonicity of entropy of interval maps,* Phys. Lett. A **202** (1995), no. 5-6, 359–362.

[42] H. Bruin, *Topological conditions for the existence of absorbing Cantor sets,* Trans. Amer. Math. Soc. **350** (1998), no. 6, 2229–2263.

[43] H. Bruin, *For almost every tent map, the turning point is typical,* Fund. Math. **155** (1998), no. 3, 215–235.

[44] H. Bruin, *An algorithm to compute the topological entropy of a unimodal map,* Proceedings contribution for the conference in memory of W. Szlenk, Barcelona 1996, Internat. J. Bifur. Chaos Appl. Sci. Engrg. **9** (1999), no. 9, 1881–1882.

[45] H. Bruin, *Homeomorphic restrictions of unimodal maps,* Contemp. Math. **246** (1999), 47–56.

[46] H. Bruin, G. Keller, T. Nowiscki, and S. van Strien, *Wild Cantor attractors exist,* Ann. of Math. (2) **143** (1996), no. 1, 97–130.

[47] H. Bruin, G. Keller, and M. St. Pierre, *Adding machines and wild attractors,* Preprint 18/95 (1995).

[48] H. Bruin, G. Keller, and M. St. Pierre, *Adding machines and wild attractors,* Ergod. Th. & Dynam. Sys. **17** (1997), no. 6, 1267–1287.

[49] H. Bruin and D. Schleicher, *Symbolic dynamics of quadratic polynomials,* Preprint, http://www.ml.kva.se/preprints/archive2001-2002.php (2001).

[50] S. Bullett, *One-dimensional dynamics: Complex maps,* Lecture notes for a summer dynamics course held in Hungary (1996).

[51] J. Buzzi and P. Hubert, *Piecewise monotone maps without periodic points; Rigidity, measures and complexity,* Preprint, Marseille (2001).

[52] J.S. Cánovas, Preprint (2003).

[53] L. Carleson and T. Gamelin, *Complex dynamics,* Universitext: Tracts in Mathematics, Springer-Verlag, New York (1993).

[54] R.V. Churchill, J.W. Brown, and R.F. Verhey, *Complex variables and applications,* Third Edition, McGraw-Hill, New York (1974).

[55] D.L. Cohen, *Measure theory,* Birkhäuser, Boston (1980).

[56] P. Collet and J.-P. Eckmann, *Iterated maps of the interval as dynamical systems,* Progress in Physics **1**, Birkhäuser, Boston (1980).

[57] P. Collet and J.-P. Eckmann, *Positive Liapunov exponents and absolute continuity for maps of the interval,* Ergod. Th. & Dynam. Sys. **3** (1983), no. 1, 13–46.

[58] P. Coullet and C. Tresser, *Itération d'endomorphismes et groupe de renormalisation,* C.R. Acad. Sci. Paris Sér. A-B **287** (1978), no. 7, A577–A580.

[59] E.M. Coven, I. Kan, and J.A. Yorke, *Pseudo-orbit shadowing in the family of tent maps,* Trans. Amer. Math. Soc. **308** (1988), no. 1, 227–241.

[60] E. Coven and I. Mulvey, *Transitivity and the center for maps of the circle,* Ergod. Th. & Dynam. Sys. **6** (1986), no. 1, 1–8.

[61] K. Dajani and C. Kraaikamp, *Ergodic theory of numbers,* Carus Mathematical Monographs **29**, Math. Assoc. Amer., Washington, D.C. (2002).

[62] P. Dawson, R. Galeeva, J. Milnor, and C. Tresser, *A monotonicity conjecture for real cubic maps,* Real and complex dynamical systems (Hillerød 1993), NATO Adv. Sci. Inst. Ser. C math. Phys. **464**, Klumer Acad. Publ., Dordrecht (1995).

[63] A. Denjoy, *Sur les courbes définies par les équations différentielles à la surface du tor,* J. Math. Pure et Appl. **11** série 9 (1932), 333–375.

[64] R.L. Devaney, *An introduction to chaotic dynamical systems,* Second Edition, Addison Wesley, Redwood, CA (1989).

[65] R.L. Devaney, *The Mandelbrot set, Farey tree, and the Fibonacci sequence,* Amer. Math. Monthly **106** (1999), no. 4, 289–302.

[66] E.I. Dinaburg, *The relation between topological entropy and metric entropy,* Soviet Math. **11** (1970), 13–16.

[67] A. Douady, *The topological entropy of unimodal maps: monotonicity for quadratic polynomials,* in Real and complex dynamical systems, NATO Adv. Sci. Inst. Math. Phys. Sci. **464**, Kluwer Acad. Publ. Dordrecht (1995), 65–87.

[68] M.J. Feigenbaum, *Quantitative universality for a class of nonlinear transformations,* J. Stat. Phys. **19** (1978), no. 1, 25–52.

[69] M.J. Feigenbaum, *The universal metric properties of nonlinear transformations,* J. Stat. Phys. **21** (1979), no. 6, 669–706.

[70] G.B. Folland, *Real analysis. Modern techniques and their applications,* John Wiley and Sons Inc., New York (1984).

[71] S. Friedberg, A. Insel, and L. Spence, *Linear algebra,* Prentice-Hall, New Jersey (2003).

[72] J.-M. Gambaudo and C. Tresser, *A monotonicity property in one-dimensional dynamics,* Contemp. Math. **135** (1991), 213–222.

[73] A.M. Garsia, *Topics in almost everywhere convergence,* Lect. Adv. Math. **4**, Markham Publishing Company, Chicago (1970).

[74] W.H. Gottschalk and G.A. Hedlund, *Topological dynamics,* Amer. Math. Soc. Colloquium Publ. **36** (1955).

[75] P.J. Grabner, P. Liardet, and R.F. Tichy, *Odometers and systems of enumeration,* Acta Arithmetica **70** (1995), no. 2, 103–123.

[76] J. Graczyk, D. Sands, and G. Świątek, *Metric attractors for smooth unimodal maps,* Orsay Preprint (2001).

[77] J. Guckenheimer, *Sensitive dependence on initial conditions for one dimensional maps,* Commun. Math. Phys. **70** (1979), no. 2, 133–160.

[78] J. Guckenheimer, *Limit sets of S-unimodal maps with zero entropy,* Commun. Math. Phys. **110** (1987), no. 4, 655–659.

[79] J. Guckenheimer and P. Holmes, *Nonlinear oscillations, dynamical systems, and bifurcations of vector fields,* Applied Math. Sciences Ser. **42**, Springer-Verlag, New York (1983).

[80] P. Halmos, *Ergodic theory,* Chelsea Publishing, New York (1956).

[81] G.H. Hardy and E.M. Wright, *An introduction to the theory of numbers,* Claredon Press, Oxford (1979).

[82] E. Hewitt and K. Stromberg, *Real and abstract analysis,* Springer-Verlag, Berlin (1965).

[83] F. Hofbauer, *On intrinsic ergodicity of piecewise monotonic transformations with positive entropy,* Israel J. Math. **34** (1980), no. 3, 213–237.

[84] F. Hofbauer, *The topological entropy of the transformation* $x \mapsto ax(1-x)$, Monatsh. Math. **90** (1980), no. 2, 117–141.

[85] F. Hofbauer, *Piecewise invertible dynamical systems,* Probab. Theory Relat. Fields **72** (1986), no. 3, 359–386.

[86] F. Hofbauer and G. Keller, *Zeta-functions and transfer-operators for piecewise linear transformations,* J. Reine Angew. Math. **352** (1984), 100–113.

[87] F. Hofbauer and G. Keller, *Quadratic maps without asymptotic measure,* Commun. Math. Phys. **127** (1990), no. 2, 319–337.

[88] F. Hofbauer and G. Keller, *Some remarks on recent results about S-unimodal maps. Hyperbolic behavior of dynamical systems (Paris, 1990),* Ann. Inst. Henri Poincaré, Phys. Théor. **53** (1990), no. 4, 413–425.

[89] R.A. Horn and C.R. Johnson, *Matrix analysis,* Cambridge University Press, New York (1985).

[90] J. Hubbard and D. Schleicher, *The spider algorithm,* Complex dynamical systems (Cincinnati, OH, 1994), Proc. Sympos. Appl. Math. **49**, Amer. Math. Soc., Providence RI (1994), 155–180.

[91] A. Katok and B. Hasselblatt, *Introduction to the modern theory of dynamical systems,* Encyclopedia of Mathematics and its Applications **54**, Cambridge University Press, Cambridge (1995).

[92] Y. Katznelson, *Sigma-finite invariant measures for smooth mappings of the circle,* J. Analyse Math. **31** (1977), 1–18.

[93] M.S. Keane, *Ergodic theory and subshifts of finite type,* in *Ergodic theory, symbolic dynamics and hyperbolic spaces*, Ed. T. Bedford, M.S. Keane, and C. Series, Oxford Science Publ., The Clarendon Press, Oxford University Press, New York (1991).

[94] G. Keller, *Exponents, attractors, and Hopf decompositions for interval maps,* Ergod. Th. & Dynam. Sys. **10** (1990), no. 4, 717–744.

[95] K. Keller, *Invariant factors, Julia equivalences and the (abstract) Mandelbrot set,* Lect. Notes in Math. **1732**, Springer-Verlag, Berlin (2000).

[96] B. Kitchens, *Symbolic dynamics. One-sided, two-sided, and countable state Markov shifts,* Universitext, Springer-Verlag, Berlin (1998).

[97] B. Kitchens and K. Schmidt, *Markov subgroups of* $(\mathbb{Z}/2\mathbb{Z})^{\mathbb{Z}^2}$, in *Symbolic dynamics and its applications*, Contemp. Math **135**, Amer. Math. Soc., Providence, RI (1992), 265–283.

[98] A. N. Kolmogorov, *New metric invariants of transitive dynamical systems and automorphisms of Lebesgue spaces,* Dokl. Akad. Nauk SSSR **119** (1958), 861–864.

[99] S. Kolyada, *One-parameter families represented by integrals with negative Schwarzian derivative violating monotone bifurcations,* Ukr. Mat. Z. **41** (1989), 258–261 [translated from Russian].

[100] S. Kolyada, L. Snoha, and S. Trofimchuk, *Noninvertible minimal maps,* Fund. Math. **168** (2001), no. 2, 141–163.

[101] O. Kozlovski, *Getting rid of the negative Schwarzian derivative condition,* Ann. of Math. (2) **152** (2000), no. 3, 743–762.

[102] S.G. Krantz, *Complex Analysis: The geometric viewpoint,* Carus Mathematical Monographs **23**, Math. Assoc. Amer., Washington, D.C. (1990).

[103] U. Krengel, *Ergodic theorems*, de Gruyter Studies in Math **6**, Walter de Gruyter & Co., Berlin (1985).

[104] J.C. Lagarias and C. Tresser, *A walk along the branches of the extended Farey tree,* IBM J. Res. Dev. **39** (1995), 283–294.

[105] E. Lau and D. Schleicher, *Internal addresses in the Mandelbrot set and irreducibility of polynomials,* SUNY Stony Brook Preprint **19** (1994).

[106] F. Ledrappier. *Un champ markovien peut être d'entropie nulle et mélangeant,* C. R. Acad. Sc. Paris, Ser. A-B **287** (1978), no. 7, A561–A562.

[107] T. Li and J.A. Yorke, *Period three implies chaos,* Amer. Math. Monthly **82** (1975), no. 10, 985–992.

[108] D. Lind and B. Marcus, *An introduction to symbolic dynamics and coding,* Cambridge University Press, Cambridge (1995).

[109] J.H. van Lint and R.M. Wilson, *A course in combinatorics,* Cambridge University Press, Cambridge (1992).

[110] S. Luzzatto and L. Wang, *Topological invariance of generic non-uniformly expanding multimodal maps,* Preprint (2003).

[111] M.Yu. Lyubich, *Combinatorics, geometry and attractors of quasi-quadratic maps,* Ann. of Math. (2) **140** (1994), no. 2, 347–404.

[112] R. Mañé, *Ergodic theory and differentiable dynamics,* Springer-Verlag, Berlin (1987).

[113] M. Martens, *Interval dynamics,* Ph.D. Thesis, Delft (1990).

[114] M. Martens, *Distortion results and invariant Cantor sets of unimodal mappings,* Ergod. Th. & Dyn. Sys. **14** (1994), no. 2, 331–349.

[115] W. de Melo and S. van Strien, *One-dimensional dynamics,* Springer-Verlag, New York (1993).

[116] M. Metropolis, M.L. Stein, and P.R. Stein, *On finite limit sets for transformations of the interval,* J. Combin. Theory A **15** (1973), 25–44.

[117] J. Milnor, *On the concept of attractor,* Commun. Math. Phys. **99** (1985), 177–195.

[118] J. Milnor, *Complex dynamics,* SUNY Stonybrook Preprint **5** (1990).

[119] J. Milnor, *Dynamics in one complex variable: Introductory lectures,* Friedr. Vieweg & Sohn, Braunschweig (1999).

[120] J. Milnor, *Periodic orbits, externals rays and the Mandelbrot set: An expository account,* Géometrie complexe et systèmes dynamiques, Orsay (1995).

[121] J. Milnor, *Periodic orbits, external rays, and the Mandelbrot set: An expository account,* Astérisque **261** (2000), 277–333.

[122] J. Milnor and W. Thurston, *On iterated maps of the interval: I,II,* Preprint Princeton (1977). Published in *Dynamical Systems: Proc. Univ of Maryland* (1986-87).

[123] J. Milnor and W. Thurston, *On iterated maps of the interval,* Lect. Notes in Math. **1342**, Springer-Verlag, Berlin, New York (1988), 465–563.

[124] J. Milnor and C. Tresser, *On entropy and monotonicity for real cubic maps* (with an appendix by Adrien Douady and Pierrette Sentenac), Commun. Math. Phys. **209** (2000), no. 1, 123–178.

[125] M. Misiurewicz and W. Szlenk, *Entropy of piecewise monotone mappings,* Studia Math. **67** (1980), no. 1, 45–63.

[126] J.R. Munkres, *Topology, a first course,* Prentice-Hall, New Jersey (1975).

[127] P.J. Myrberg, *Iteration der reellenPolynome zweiten Grades,* Ann. Acad. Sci. Fenn. **256 A** (1959), 1–10.

[128] *Iteration der reellenPolynome zweiten Grades III,* Ann. Acad. Sci. Fenn. **336 A** (1963), 1–18.

[129] Z. Nitecki, *Topological dynamics on the interval,* Ergodic theory and dynamical systems, **II** (College Park, Md., 1979/1980), 1–73, Progr. Math.**21**, Birkhaüser, Boston (1982).

[130] T. Nowicki and F. Przytycki, *Topological invariance of the Collet-Eckmann property for S-unimodal maps,* Fund. Math. **155** (1998), no. 1, 33–43.

[131] H.E. Nusse and J.A. Yorke, *Period halving for* $x_{n+1} = MF(x_n)$*, where F has negative Schwarzian derivative,* Phys. Lett. A **127** (1988), no. 6-7, 328–334.

[132] D. Ornstein, *Bernoulli shifts with the same entropy are isomorphic,* Adv. in Math. **4** (1970), 337–352.

[133] W. Parry, *On the* β*-expansion of real numbers,* Acta Math. Acad. Sci. Hungar **11** (1960), 401–416.

[134] W. Parry, *Intrinsic Markov chains,* Trans. Amer. Math. Soc. **112** (1964), 55–66.

[135] W. Parry, *Symbolic dynamics and transformations of the unit interval,* Trans. Amer. Math.Soc. **122** (1966), 368–378.

[136] C. Penrose, *On quotients of shifts associated with dendrite Julia sets of quadratic polynomials,* Thesis, University of Coventry (1994).

[137] K. Petersen, *Ergodic Theory,* Cambridge Studies in Advanced Mathematics **2**, Cambridge University Press, Cambridge (1983).

[138] H. Poincaré, *Sur les courbes définies par des équations différentielles,* J. Math. Pures et Appl. 4ème série **1** (1885), 167–244. Also in *Oeuvres Complètes, t.1,* Gauthier-Villars, Paris (1951).

[139] M. Pollicott and M. Yuri, *Dynamical systems and Ergodic theory,* London Mathematical Society Student Text **40**, Cambridge University Press, Cambridge (1998).

[140] F. Przytycki, J. Rivera-Letelier, and S. Smirnov, *Equivalence and topological invariance of conditions for non-uniform hyperbolicity in the iteration of rational maps,* Invent. Math. **151** (2003), no. 1, 29–63.

[141] F. Przytycki, S. Rohde, *Porosity of Collet-Eckmann Julia sets,* Fund. Math. **155** (1998), no. 2, 189–199.

[142] A.N. Quas and P.B. Trow, *Subshifts of multi-dimensional shifts of finite type,* Ergod. Th. & Dynam. Sys. **20** (2000), no. 3, 859–874.

[143] P. Raith, *Hausdorff dimension for piecewise monotonic maps,* Studia Math. **94** (1989), no. 1, 17–33.

[144] A. Renyi, *Representations for real numbers and their ergodic properties,* Acta Math. Acad. Sci. Hungar **8** (1957), 472–493.

[145] I. Richards, *Continued fractions without tears,* Math. Mag. **54** (1981), no. 4, 163–171.

[146] J. Ringland and M. Schell, *The Farey tree embodied — in bimodal maps of interval,* Phys. Lett. A, **136** (1989), no. 7-8, 379–386.

[147] J. Ringland and C. Tresser, *A geneology for the finite kneading sequences of bimodal maps on the interval,* Trans. Amer. Math. Soc. **347** (1995), no. 12, 4599–4624.

[148] J. Ringland, N. Issa, and M. Schell, *From U sequence to Farey sequence: a unification of one-parameter scenarios,* Phys. Rev. A (3) **41** (1990), no. 8, 4223–4235.

[149] J. Rothschild, *Computation of topological entropy,* Ph.D. Dissertation, City Univ. New York. (1971).

[150] W. Rudin, *Real and complex analysis,* Third Edition, McGraw-Hill Book Co., New York (1987).

[151] G. Ryd, *Iterations of one-parameter families of complex functions,* Ph.D. thesis, Stockholm (1997).

[152] D. Sands, *Topological conditions for positive Lyapunov exponent in unimodal maps,* Ph.D thesis, Cambridge (1994).

[153] J. Schmeling, *Symbolic dynamics for β-shifts and self-normal numbers,* Ergod. Th. & Dyn. Sys. **17** (1997), no. 3, 675–694.

[154] E. Seneta, *Non-negative matrices. An introduction to the theory and applications,* Halsted Press [A division of John Wiley & Sons], New York (1973).

[155] A. Sharkovsky, *Coexistence of cycles of a continuous map of the line into itself,* (Russian) Ukrain. Math. Zh. **16** (1964), 61–71.

[156] A.N. Sharkovsky, S.F. Kolyada, A.G. Sivak, and V.V Fedorenko, *Dynamics of one-dimensional maps,* Math. and its Appl. **407**, Kluwer Acad. Pub. Group, Dordrecht (1997).

[157] R.M. Siegel, C. Tresser, and G. Zettler, *A decoding problem in dynamics and in number theory,* Chaos **2** (1992), no. 4, 473–493.

[158] D. Singer, *Stable orbits and bifurcation of maps of the interval,* SIAM J. Appl. Math. **35** (1978), no. 2, 260–267.

[159] M. Smorodinsky, *β-automorphisms are Bernoulli shifts,* Acta. Math. Acad. Sci. Hungar **24** (1973), 273–278.

[160] P. Stefan, *A theorem of Sharkovskii on the existence of periodic orbits of continuous endomorphisms on the real line,* Commun. Math. Phys. **54** (1977), no. 3, 237–248.

[161] M.A. Stern, *Ueber eine zahlentheoretische Funktion,* J. für die reine und angewandte Mathematik **55** (1858), 193–220.

[162] S. van Strien, *Smooth dynamics on the interval,* New Directions in Dynamical Systems, London Math. Soc. Lecture Note Ser. **127**, Cambridge University Press, Cambridge (1988), 57–119.

[163] D. Sullivan and W. Thurston, *Extending holomorphic motions,* Acta Math. **157** (1986), no. 3-4, 243–257.

[164] W. Thurston, *On the geometry of iterated rational maps,* Preprint, Princeton University (1985).

[165] M. Tsujii, *A note on Milnor and Thurston's monotonicity theorem,* Geometry and analysis in dynamical systems (Kyoto 1993), Adv. Ser. Dynam. Systems **14**, World Sci. Publishing, New Jersey (1994), 60–62.

[166] M. Tsujii, *A simple proof for monotonicity of entropy in the quadratic family,* Ergod. Th. & Dynam. Sys. **20** (2000), no. 3, 925–933.

[167] A.M. Vershik and N.A. Sidorov, *Arithmetic expansions associated with a rotation of the circle and with continued fractions,* St. Petersburg Math. J. **5** (1994), no. 6, 1211–1136.

[168] P. Walters, *An introduction to ergodic theory,* Graduate Texts in Mathematics **79**, Springer-Verlag, New York-Berlin (1982).

[169] L. Wang, *Quadratic convergence in period doubling to chaos for trapezoidal maps* (with appendices by W.A. Beyer.), J. Math. Anal. Appl. **227** (1998), no. 1, 1–24.

[170] S. Willard, *General topology,* Addison-Wesley, Reading, Mass.-London-Don Mills, Ont. (1970).

[171] A. Zdunik, *Entropy of transformations of the unit interval,* Fund. Math. **124** (1984), no. 3, 235–241.

Index

Printed in the United States
by Baker & Taylor Publisher Services